INTRODUCTION TO
CHEMICAL ENGINEERING
COMPUTING

INTRODUCTION TO CHEMICAL ENGINEERING COMPUTING

Second Edition

BRUCE A. FINLAYSON
Rehnberg Professor Emeritus of Chemical Engineering
Department of Chemical Engineering
University of Washington
Seattle, WA

Using the latest user interface with Aspen Plus 8.0

Published by John Wiley & Sons, Inc., Hoboken, New Jersey
Published simultaneously in Canada

For general information on our other products and services or for technical support, please contact our
Customer Care Department within the United States at (800) 762-2974, outside the United States at
(317) 572-3993 or fax (317) 572-4002.

Wiley also publishes its books in a variety of electronic formats. Some content that appears in print may
not be available in electronic formats. For more information about Wiley products, visit our web site at
www.wiley.com

Library of Congress Cataloging-in-Publication Data:

Finlayson, Bruce A.
   Introduction to chemical engineering computing / Bruce A. Finlayson. – 2nd ed.
      p. cm.
   Includes index.
   ISBN 978-1-118-88831-5 (pbk.)
1. Chemical engineering–Data processing.   I. Title.
   TP184.F56 2012
   660.0285–dc23

                                                            2011045242

Printed in the United States of America

V10006725_122618

CONTENTS

PREFACE

Chemical engineering students and chemical engineers are being asked to solve problems that are increasingly complex, whether the applications are in refineries, fuel cells, microreactors, or pharmaceutical plants. Many years ago, students wrote their own programs, first in the FORTRAN programming language, then in languages such as MATLAB®. However, with the growth in personal computers, software has been written that solves many problems for students, provided they use the programs correctly. Thus, the emphasis shifted from a small group of people who were interested in writing their own programs to a large group of students who will use the programs, but do not write them. In my 42 years of teaching at the University of Washington, I taught those small groups of students how to use numerical analysis to solve complicated problems. Now, I teach *all* the students how to use the computer wisely. Only a few of the students I teach are interested in the numerical analysis (to my sorrow!), but all the students know they must be able to solve difficult problems, and they need to use the computer to do that.

The goals of this book are to illustrate (a) the problems chemical engineers have to solve, (b) the type of computer programs used to solve them, and (c) how engineers check to be sure they have solved the problems correctly. This is done in the context of how contemporary students learn—minimal reading, just-in-time learning, with lots of computer usage. The programs demonstrated here are Excel®, MATLAB®, Aspen Plus®, and Comsol Multiphysics®.

When writing this book, I assumed that readers are not absolute beginners. Junior and senior chemical engineering students have had experience with spreadsheet programs such as Excel, and they can easily learn on the computer when provided a direction and key ideas or phrases. In fact, many students are more computer-savvy than their instructors. However, a beginner chemical engineering student may not know the application very well and may not have gained a solid understanding of the physical phenomenon behind an engineering problem. Furthermore, they may not have solved very difficult problems. Thus, it is important to give some explanation of why students need to solve certain problems and how to overcome the obstacles when the problems tax the numerical methods. I have drawn on my experience to give insights into the problems in this book.

My teaching philosophy is that the problems engineers are solving today are usually intractable with analytical methods, but they can be solved with the sophisticated software available today. Thus, every engineer will be solving a problem that no one knows the answer to, and it is the engineer's job to ensure that the problem is posed correctly on paper and in the computer, and it is correctly solved. Engineering students must know how to determine if the computer solved the problem correctly by validating the work done by the computer. If they can do this, they can convince their instructor—or their future boss—that they have a solution that is every bit as reliable as an analytical solution, although without the analytical form and for a problem that cannot be solved analytically. In fact, 98% of the problems in this book are nonlinear and only a few of them have analytical solutions.

HOW TO USE THIS BOOK IN TEACHING

This book grew out of a course I developed at the University of Washington, beginning in 2003. The course is part lecture and part hands-on computer work in a computer laboratory. I usually provide simple problems for the students to solve in the laboratory, when help is available, to help them get over the barrier of using an unfamiliar program. But then, students have to solve a problem that expands their knowledge of chemical engineering and demonstrates they have used the program correctly (and described the checks they made). Since the applications cover much of the chemical engineering field, I joke with the students, saying, "I'm teaching you the entire field in 20 hours." Although I retired from teaching the course in 2009, the same objectives still apply.

This book can also be used in other courses since each chapter is keyed to a course in the curriculum. Once chemical reaction equilibrium has been discussed in the Thermodynamics class, for example, instructors can hold a laboratory session that teaches computer applications, using the chapter on chemical reaction equilibrium. The material on choice of thermodynamic model (and comparisons) also adds a bit of realism to the Thermodynamics course. Other chapters could be used in other courses. In this way, the students would use the book during their entire education, in course after course: Mass and Energy Balances, Transport of Heat, Mass, and Momentum, Reactor Design, and courses concentrating on projects such as biomedical engineering. The hope is, of course, that students would then be able to concentrate more on the chemical engineering principles and use the computer as a tool.

There are four programs that are featured in this book. It is possible that your school does not use all four. Although the screen images may be different, the ideas and procedures are the same. Certainly the problems can be solved using other programs. In a working environment, engineers use what their company provides. Thus, engineers may use a less powerful program because it is available. The more powerful program may cost more, too. Thus, in several chapters, the same problem is solved using different programs, which lets students see first-hand that the more general purpose programs require significantly more programming to solve complicated problems. In my experience, when given a suite of programs, students will use the one that allows them to solve their problem fastest. The program Comsol Multiphysics comes in many modules. Nearly all the problems in this book can be solved by the basic module, although there are cases where the Chemical Reaction Engineering Module is useful. Connections with MATLAB are made with LiveLink™

for MATLAB, another module. There are only a couple of problems involving turbulent flow, and those require the CFD module. A complete list of what you get with various combinations is available from Comsol and my list is on the book website (see Appendix D).

Each chapter begins with a list of instructional objectives. In addition, the book website has a list of principles learned from each problem, both from a chemical engineering viewpoint and a computer/computer technique viewpoint. Professors that use the book are encouraged to discuss possible use in other chemical engineering courses so that more advanced problems can be solved in them, too. The indices are available on the book website, too, since students prefer using the Internet rather than turning to the back of the book; more importantly, they can be downloaded and searched for a phrase.

WHAT IS NEW?

One big change from the first edition is the fact that all four programs now have different interfaces than they did in 2005. More importantly, they have greatly enhanced capabilities. I have cut back on some explanations and refer the user to the help menus that come with the programs, since those have improved, too, and they give more information than the book can. But, I provide hints where to look.

The number of problems has approximately doubled. More importantly, the added problems are concentrated in the field of energy: integrated gas-combined cycle, including low temperature air separation, making ethanol from switchgrass, and pressure swing adsorption to make hydrogen to fuel cars. In each case a discussion of the field precedes the definition of the problem so that students can see the applicability. Microfluidics has expanded since 2005, and there are added problems in the field of biomedical applications. This has lead to many more examples and problems involving fluid flow and diffusion in two and three dimensions. An important addition was made in Aspen Plus 7.3: now you have direct access within the program to experimental data on pressure–volume–temperature of pure components and binary vapor–liquid equilibria as summarized by the National Institute of Standards and Technology. This is very important for chemical engineers, since the choice of thermodynamic model must usually be accompanied by a comparison with experimental data, and that is now made very easy—so easy that it would be unprofessional not to do the comparison. Thus, the thermodynamic sections of the book include industrial guidelines, some molecular considerations, and experimental data for comparison. Aspen Plus also has the capability to easily summarize the greenhouse impact of a process. There are talks made by professors about how they used AspenTech products in their courses; contact: University.Program@aspentech.com. One thing that is pointed out by Professor Luyben is that material and energy balances are primarily flow-based, whereas safety problems must be pressure based (and dynamic). The dynamic options are not treated here in detail, but are often covered in a control course. Aspen Plus runs under Microsoft Windows, but the author ran it under Windows by using Parallels Desktop for Mac on an Apple computer. The second edition also has examples running Aspen Plus with a simple user-defined FORTRAN program. The book uses Aspen Plus 8.0.

Some professors like to have more numerical programming in their courses, so a number of problems like that have been added to the end of many chapters. They make a good contrast—solve them using the numerical programming and then solve them using one of the four programs emphasized here to compare the ease of use of each method. Appendix E provides more detail about the numerical methods. While the programs make the numerical analysis easy to use, it is also important to recognize that most problems involve an approximation from continuous to discrete variables. A few problems in the book ask the students to do the actual numerical analysis (and compare with other programs). Instructors may say, "If you don't program the method, you haven't really understood the problem." I reply by pointing out that when a doctor prescribes an MRI, you do not say you would not do it until he/she explains how the magnetic field works in the machine, discusses hydrogen molecules flipping orientation, and describes how the imaging takes place. The doctor and technician know how to interpret the results and how to detect if the machine is not operating correctly; engineering students can do that, too.

The number of problems has been doubled, and they are organized into easy problems (subscript 1), harder problems (subscript 2), and problems that are suitable as projects, either for one student or for teams. Finally, more techniques that are in Comsol Multiphysics are explained.

The code used to solve the examples in the book is not provided on the book website, because the author believes that learning takes place when you try to duplicate the steps in the book. However, some material needed to start problems is on the book website, such as geometries for three dimensional flow/diffusion problems. Depending upon the memory of your computer, and what can be allocated to Comsol Multiphysics, some of the three dimensional problems may not be soluble on your computer. The book website is http://www.ChemEComp.com.

ACKNOWLEDGMENTS

In writing this book, I owe a great thanks to the students in my classes. Sometimes students taught me how to use an advanced technique, and their questions brought out the best of the programs. I had over 100 undergraduate research students work with me over the past decade, and the results of their work are available on my websites: http://faculty.washington.edu/finlayso/ and http://courses.washington.edu/microflo. The Department provided a challenge grant to write textbooks, funded by a gift in the memory of alumnus Maurice Richford, BS 1926. Without that challenge grant, the first edition would not have been written so it could not have been revised. My daughter, Christine Finlayson, improved my writing greatly by serving as a copy editor of the first edition, and the clarity is due to her work; any confusion left is my responsibility. I thank especially Professor Stanley Sandler, University of Delaware, for reviewing the revised treatment of thermodynamics—I learned a lot, too! The folks at Comsol and Aspen Tech have been very helpful since both Comsol Multiphysics and Aspen Plus have been improved over the past few years. Most of all, I thank my wife, Pat, for putting up with the long hours of work that such a project requires. She has always supported me and made sacrifices that enabled me to finish. And I was smart enough to take a few weeks off from this rewrite to celebrate our 50th wedding anniversary!

BRUCE FINLAYSON

Seattle, October, 2011

1

INTRODUCTION

Computers have revolutionized the way chemical engineers design and analyze processes, whether designing large units to make polyethylene or small microreactors used to detect biological agents. In fact, the engineering problems that many of you will study as undergraduates are similar in complexity to problems PhD students solved 30 or 40 years ago. Computer programs can now solve difficult problems in a fraction of the time it used to take. Nowadays, you no longer have to write your own software programs to use computers effectively. Computer programs can do the numerical calculations for you, but you will still need to understand how to apply these programs to specific engineering challenges.

The goal of this book is to help you practice better chemical engineering. Computers are valuable tools that enable progressive, far-reaching chemical engineering. Unfortunately, computers are not as basic as DVD players, where you insert a DVD, push a button, and get the same result every time. Sometimes computer programs do not work properly for the parameters you have given them. Thus, you must be careful to use them wisely.

This book will also

1. Illustrate the problems that you as chemical engineers may need to solve.
2. Compare the types of computer programs you can use and illustrate which ones are best for certain applications.
3. Describe how to check your work to ensure you have solved the problems correctly.

This book demonstrates four computer programs: Excel®, MATLAB®, Aspen Plus®, and Comsol Multiphysics®. You may have access to other programs created by other companies. While the exact details will not be the same, the steps you take will be similar.

Computer skills are invaluable, but as an engineer, you also need to understand the physical phenomena. Each chemical engineering application chapter starts with a

Introduction to Chemical Engineering Computing, Updated Second Edition. Bruce A. Finlayson.
© 2014 John Wiley & Sons, Inc. Published 2014 by John Wiley & Sons, Inc.

description of the physical problem in general terms. Then those general terms are put into a mathematical context so the computer can represent them. Next, the chapter gives several examples in which such problems are solved, providing step-by-step instructions so you can follow along on your own computer. Sometimes, the same problem is solved using different programs so you can see the advantages of each program. Finally, the chapters give more complicated problems your instructor may use as homework problems.

Examples throughout this book demonstrate how to check your work and how to learn from the answers the computer gives you. When using computers, it is always important to know if the computer obtained the correct answer. If you follow this strategy you will have no trouble convincing your instructor—or your boss—that you have a solution every bit as reliable as an analytical solution for a problem that cannot be solved analytically:

1. Solve the problem
2. Validate your work
3. Understand how you reached that answer

ORGANIZATION

The book is organized into eleven chapters followed by five appendices as listed in Table 1.1. Each chapter treats a type of chemical engineering phenomenon, such as process simulation

TABLE 1.1   Computer Programs Used in Different Chapters

	Chapters	Excel	MATLAB	Aspen Plus	Comsol Multiphysics	Numerical Methods
1	Introduction					
2	Equations of state	✓	✓	✓		✓
3	Vapor–liquid equilibria	✓	✓	✓		✓
4	Chemical reaction equilibria	✓	✓	✓		✓
5	Mass balances with recycle streams	✓		✓		
6	Simulation of mass transfer equipment			✓		
7	Process simulation			✓		
8	Chemical reactors and initial value problems	✓	✓	✓	✓	✓
9	Transport processes in 1D and boundary value problems	✓	✓	✓	✓	✓
10	Navier–Stokes equation in 2D and 3D				✓	
11	Convective diffusion equation in 2D and 3D and elliptic partial differential equations				✓	✓
A	Hints when using Excel	✓				
B	Hints when using MATLAB		✓			
C	Hints when using Aspen Plus			✓		
D	Hints when using Comsol Multiphysics				✓	
E	Mathematical methods	✓	✓	✓	✓	✓

or convective diffusion. Four of the appendices give additional details about each computer program. The fifth appendix provides the nitty-gritty details of many of the numerical methods. An appendix on parameter estimation that was in the 1st edition is available on the book website.

As a modern chemical engineering student, many of you are computer-savvy. This book assumes that you are not a complete beginner, but have some experience with spreadsheet programs such as Excel. The chapters provide examples and step-by-step instructions for using the computer programs to solve chemical engineering problems. If needed, you can find more detailed information about the individual programs in the appendices.

Algebraic Equations

Chapters 2–5 deal with chemical engineering problems that are expressed as algebraic equations—usually sets of nonlinear equations, perhaps thousands of them to be solved together. In Chapter 2, you can study equations of state that are more complicated than the perfect gas law. This is especially important because the equation of state provides the thermodynamic basis for not only volume but also fugacity (phase equilibrium) and enthalpy (departure from ideal gas enthalpy). Chapter 3 covers vapor–liquid equilibrium, and Chapter 4 covers chemical reaction equilibrium. All these topics are combined in simple process simulation in Chapter 5. This means that you must solve many equations together. These four chapters make extensive use of programming languages in Excel and MATLAB as well as Aspen Plus.

Process Simulation

Chapter 6 provides an extensive discussion of the possible (and reasonable) choices of thermodynamic models, and how you check your choice. It then introduces mass transfer problems such as distillation and absorption and single units. Chapter 7 gives a more detailed look at process simulation, where the power of process simulators like Aspen Plus really is evident. These chapters make use of commercial codes that are run by inserting data into their custom-designed interface.

Differential Equations

Chapters 8–11 treat problems that are governed by differential equations. Chapter 8 gives methods to model chemical reactors. These are usually initial value problems, which are illustrated in Eq. (1.1):

$$u\frac{dc}{dz} = -kc^2, \quad c(z = 0) = c_0 \tag{1.1}$$

Note that the dependent variable, c, is a function of only one independent variable, z, and that the initial value is specified. For reactors, you start at the inlet and integrate down the reactor using either Excel, MATLAB, Aspen Plus, or Comsol Multiphysics.

Chapter 9 then solves transport problems in one space dimension (1D) using Comsol Multiphysics. If you consider heat transfer through a slab, one side of the slab is kept at one temperature, T_0, and the other side of the slab is maintained at another temperature, T_L. The governing equation is

$$k\frac{d^2 T}{dx^2} = 0 \tag{1.2}$$

with boundary conditions

$$T(0) = T_0, \quad T(L) = T_L \tag{1.3}$$

The differential equation, (1.2), is an ordinary differential equation because there is only one independent variable, x. In this case, equations in one space dimension are boundary value problems, because the conditions are provided at two different locations. While it is also possible to solve this problem using Excel, it is much simpler to use Comsol Multiphysics or MATLAB since the numerical analysis will have been done for you. Transient heat transfer in one space dimension is governed by

$$\rho C_p \frac{\partial T}{\partial t} = k\frac{\partial^2 T}{\partial x^2} \tag{1.4}$$

and this problem can be solved using Comsol Multiphysics or MATLAB, too.

Chapters 10 and 11 use Comsol Multiphysics to solve fluid flow, heat transfer, and mass transfer problems in 2D and 3D. Here, again the power of the software program shows through. You get to solve real problems that go beyond the simple 1D cases in your textbook. Those 1D problems are good for learning the subject, but in real-life situations, complications often arise that can only be handled numerically. These problems are partial differential equations, because there are two or more independent variables (say x and y). For example, the Navier–Stokes equations in Cartesian geometry and two dimensions are

$$\rho\left(\frac{\partial u}{\partial t} + u\frac{\partial u}{\partial x} + v\frac{\partial u}{\partial y}\right) = -\frac{\partial p}{\partial x} + \mu\left(\frac{\partial^2 u}{\partial x^2} + \frac{\partial^2 u}{\partial y^2}\right)$$
$$\rho\left(\frac{\partial v}{\partial t} + u\frac{\partial v}{\partial x} + v\frac{\partial v}{\partial y}\right) = -\frac{\partial p}{\partial y} + \mu\left(\frac{\partial^2 v}{\partial x^2} + \frac{\partial^2 v}{\partial y^2}\right) \tag{1.5}$$
$$\frac{\partial u}{\partial x} + \frac{\partial v}{\partial y} = 0$$

Appendices

If you need more background information while solving the problems in the book, consult the appendices. Appendices A–D discuss hints, examples, and step-by-step instructions for the four computer programs demonstrated in this book. You are encouraged to consult the appendices while looking at examples or solving problems using those programs—many of the details are summarized in the appendices. Appendix E illustrates the mathematical methods built into each computer program. While you will not usually need to program the methods, you may be curious about the mathematical analysis behind the programs. An appendix on parameter estimation using Excel or MATLAB is available on the book website.

Whether you tackle one chemical engineering problem or work chapter by chapter through the book, try to enjoy yourself. You and a classmate can sit down and work together—possibly on adjacent computers—to share insights and answer each others' questions. Remember, too: go back and forth from the application chapters to the computer program appendices; build up your knowledge bit by bit. Your reward is to be a better-trained engineer, be able to use your inherent creativity, and be able to compete in a fast-paced global environment. As you take other chemical engineering courses you can use the programs to solve more advanced problems that are not soluble using analytical methods. Ninety-eight percent of the problems in this book are nonlinear and few of them have analytical solutions.

Version 8.0 of Aspen Plus is different from 7.3, but mainly in the top menu items. Most of the windows are the same, once you get to them, and the nomenclature is somewhat different. Thus the book has been revised to use the new nomenclature and be consistent with Aspen Plus 8.0. The windows look different, but the information is the same, so that those haven't been changed except where necessary. Note that Aspen Plus 8.0 is used in the Aspen Suite 8.2.

2

EQUATIONS OF STATE

Solving equations of state (EOS) allows us to find the specific volume of a gaseous mixture of chemicals at a specified temperature and pressure. Without using equations of state, it would be virtually impossible to design a chemical plant. By knowing this specific volume, you can determine the size—and thus cost—of the plant, including the diameter of pipes, the horsepower of compressors and pumps, and the diameter of distillation towers and chemical reactors. Imagine how challenging it would be to design a plant without knowing this important information!

Determining the specific volume is the first step in calculating the enthalpy and vapor–liquid properties of mixtures. Calculating this enthalpy is especially important when making energy balances to reduce energy use and help the environment. In this chapter, we work only with the vapor phase, and the liquid phase is introduced in the next chapter.

To solve equations of state, you must solve algebraic equations as described in this chapter. The later chapters cover other topics governed by algebraic equations, such as phase equilibrium, chemical reaction equilibrium, and processes with recycle streams. This chapter introduces the ideal gas EOS, then describes how computer programs such as Excel®, MATLAB®, and Aspen Plus® use modified EOS to easily and accurately solve problems involving gaseous mixtures.

Step-by-step instructions will guide you in using each of these computer programs to determine the specific volume of gaseous mixtures. Practice problems are given at the end of the chapter. The lessons learned from this chapter are carried forward to other applications involving algebraic equations in Chapters 3–6 and 8. After completing this chapter, not only will you be able to solve algebraic equations but you will also be able to determine the size of the equipment in a chemical plant, certainly the size of those pieces of equipment containing gases.

Introduction to Chemical Engineering Computing, Updated Second Edition. Bruce A. Finlayson.
© 2014 John Wiley & Sons, Inc. Published 2014 by John Wiley & Sons, Inc.

Instructional Objectives: After working through this chapter, you should have

1. Updated your skills using Excel.
2. Learned to use MATLAB for simple problems.
3. Learned to use Aspen Plus to perform thermodynamic calculations.
4. Learned to check your numerical work.
5. Reviewed and expanded your chemistry and chemical engineering knowledge of EOS.

EQUATIONS OF STATE—MATHEMATICAL FORMULATION

The ideal gas EOS, which relates the pressure, temperature, and specific volume, is a familiar equation:

$$ pV = nRT \quad \text{or} \quad p\hat{v} = RT \text{ where } \hat{v} = \frac{V}{n} \tag{2.1}$$

The term p is the absolute pressure, V is the volume, n is the number of moles, R is the gas constant, and T is the absolute temperature. The units of R have to be appropriate for the units chosen for the other variables. This equation is quite adequate when the pressure is low (such as 1 atm). However, many chemical processes take place at very high pressure. For example, ammonia is made at pressures of 220 atm or more. Under these conditions, the ideal gas EOS may not be a valid representation of reality. In particular, the ideal gas, while it includes rotational and vibrational degrees of freedom, ignores intramolecular potential energy, which is important when the molecules are closer together at high pressure. The rule-of-thumb is that the ideal gas is a good approximation for pressures up to 10 atm, although this can change depending on the temperature.

Other equations of states have been developed, usually in conjunction with process simulators, to address chemical processes at high pressure. There are two key criteria: (1) the equation is able to represent the real p–V–T behavior and (2) the parameters must be easily found, including for mixtures. This last criterion is no small requirement. There are more than 25 million chemicals, leading to an infinite number of different mixtures. Obviously, you cannot look up the properties of all those mixtures on the Web.

The first generalization of the ideal gas law was the van der Waals EOS:

$$ p = \frac{RT}{\hat{v} - b} - \frac{a}{\hat{v}^2} \tag{2.2}$$

In this equation, the "b" accounts for the excluded volume (a second molecule cannot use the same space already used by the first molecule), and the "a" accounts for the force of interaction between two molecules. This extension is just a first step, however, because it will not be a good approximation at extremely high pressures. The constants a and b are given in Table 2.1.

The Redlich–Kwong EOS (1949) is a modification of the van der Waals EOS:

$$ p = \frac{RT}{\hat{v} - b} - \frac{a}{\hat{v}(\hat{v} + b)} \tag{2.3}$$

TABLE 2.1 Equations of State for Pure Components

Model	EOS	a	b	α	ω	Z_c
Ideal gas	$pV = nRT$					
van der Waal	$p = \dfrac{RT}{\hat{v} - b} - \dfrac{a}{\hat{v}^2}$	$a = 0.42188\left(\dfrac{R^2 T_c^2}{p_c}\right)$	$b = 0.125\left(\dfrac{RT_c}{p_c}\right)$			0.375
Redlich–Kwong	$p = \dfrac{RT}{\hat{v} - b} - \dfrac{a}{\hat{v}(\hat{v} + b)}$	$a = 0.42748\left(\dfrac{R^2 T_c^2}{p_c}\right)\alpha$	$b = 0.08664\left(\dfrac{RT_c}{p_c}\right)$	$\alpha = \dfrac{1}{T_r^{0.5}}$		0.333
RK–Soave	$p = \dfrac{RT}{\hat{v} - b} - \dfrac{a}{\hat{v}(\hat{v} + b)}$	$a = 0.42748\left(\dfrac{R^2 T_c^2}{p_c}\right)\alpha$	$b = 0.08664\left(\dfrac{RT_c}{p_c}\right)$	$\alpha = [1 + m(1 - T_r^{0.5})]^2$	$m = 0.480 + 1.574\omega - 0.176\omega^2$	0.333
Peng–Robinson	$p = \dfrac{RT}{\hat{v} - b} - \dfrac{a}{\hat{v}(\hat{v} + b) + b(\hat{v} - b)}$	$a = 0.45724\left(\dfrac{R^2 T_c^2}{p_c}\right)\alpha$	$b = 0.07780\left(\dfrac{RT_c}{p_c}\right)$	$\alpha = [1 + m(1 - T_r^{0.5})]^2$	$m = 0.37464 + 1.54226\omega - 0.26992\omega^2$	0.307

where equations for a and b are given in Table 2.1 in terms of the critical temperature and pressure and the reduced temperature, $T_r = T/T_c$. In these equations, T_c is the critical temperature (in absolute terms), p_c is the critical pressure, T_r is called the reduced temperature (the absolute temperature divided by the critical temperature). α is particular to the Redlich–Kwong EOS.

The Redlich–Kwong EOS was modified further by Soave to give the Redlich–Kwong–Soave EOS (Soave, 1972) (called RKS in Aspen Plus, see Table 2.1)[1], which is a common one in process simulators. The purpose of the modification was to account for situations in which the molecular structure was asymmetric by introducing the Pitzer acentric factor, ω, which is a tabulated quantity for many substances. Thus, the value of α can be computed for each chemical and reduced temperature.

The Peng–Robinson EOS (1976) is another variation appropriate for molecules that are asymmetric, and it is an important one that is used in later chapters to improve the simulation of vapor–liquid equilibria:

$$p = \frac{RT}{\hat{v} - b} - \frac{a}{\hat{v}(\hat{v} + b) + b(\hat{v} - b)} \tag{2.4}$$

One criterion that can be applied to the EOS is the value of the compressibility factor at the critical point. Experimental values range from 0.23 to 0.31 (Sandler, 2006). As Table 2.2 shows, the Peng–Robinson EOS is closest to satisfying this condition.

All these equations can be rearranged into a cubic function of specific volume. The form of the Redlich–Kwong and Redlich–Kwong–Soave EOS is

$$\hat{v}^3(p) - \hat{v}^2(RT) + \hat{v}(a - pb^2 - RTb) - ab = 0 \tag{2.5}$$

When the temperature and pressure of a gaseous mixture, and the parameters a and b are given, then to find the specific volume you would have to solve the cubic EOS for specific volume, $\hat{v}$. This represents one algebraic equation in one unknown, the specific volume. The Peng–Robinson EOS results in

$$\hat{v}^3(p) + \hat{v}^2(bp - RT) + \hat{v}(a - 3pb^2 - 2RTb) + (pb^3 + RTb^2 - ab) = 0 \tag{2.6}$$

For a pure component, the parameters a and b are determined from the critical temperature and critical pressure, and possibly the acentric factor. These are all tabulated quantities, and there are even correlations for them in terms of vapor pressure and normal boiling point, for example. For mixtures, it is necessary to combine the values of a and b for each component according to the composition of the gaseous mixture. Since the parameters a and b come about because of intramolecular potential energy, it can be justified that when species 1 is in a mixture with species 2, the molecule 1 will interact differently with molecule 2 than it does with another molecule 1. Furthermore, the interaction of molecule 1 with another molecule 1 will be the same as in a pure species (usually), and will be proportional to the mole fraction squared (i.e., the relative amount of both of them). Thus, an expected form of the mixing rule for a binary is (Koretsky, 2004)

$$a = y_1^2 a_1 + y_1 y_2 a_{12} + y_1 y_2 a_{21} + y_2^2 a_2 = y_1^2 a_1 + 2 y_1 y_2 a_{12} + y_2^2 a_2 \tag{2.7}$$

---

[1] Aspen Plus also has a Soave–RK equation, which differs slightly. In this text, they are used interchangeably.

TABLE 2.2 Equations of State for Mixtures

Model	a_i	b_i	α	a	b
Redlich–Kwong	$a_i = 0.42748 \left(\dfrac{R^2 T_{ci}^2}{p_{ci}} \right) \alpha_i$	$b_i = 0.08664 \left(\dfrac{R T_{ci}}{p_{ci}} \right)$	$\alpha_i = \dfrac{1}{T_{ri}^{0.5}}$	$a = \left(\sum\limits_{i=1}^{NCOMP} y_i a_i^{0.5} \right)^2$	$b = \sum\limits_{i=1}^{NCOMP} y_i b_i$
RK–Soave	$a_i = 0.42748 \left(\dfrac{R^2 T_{ci}^2}{p_{ci}} \right) \alpha_i$	$b_i = 0.08664 \left(\dfrac{R T_{ci}}{p_{ci}} \right)$	$\alpha_i = [1 + m_i (1 - T_{ri}^{0.5})]^2$	$a = \left(\sum\limits_{i=1}^{NCOMP} y_i a_i^{0.5} \right)^2$	$b = \sum\limits_{i=1}^{NCOMP} y_i b_i$
Peng–Robinson	$a_{ii} = 0.45724 \left(\dfrac{R^2 T_{ci}^2}{p_{ci}} \right) \alpha_i$ $a_{ij} = a_{ji} = \sqrt{a_{ii} a_{jj}} \,(1 - k_{ij})$	$b_i = 0.07780 \left(\dfrac{R T_{ci}}{p_{ci}} \right)$	$\alpha_i = [1 + m_i (1 - T_{ri}^{0.5})]^2$	$a = \sum\limits_{i,j=1}^{NCOMP} y_i y_j a_{ij}$	$b = \sum\limits_{i=1}^{NCOMP} y_i b_i$

because the interaction of molecule 1 with molecule 2 is the same as the inverse. The mixing rule for the b parameter representing the excluded volume is expected to be a linear equation based on the relative amount of each species:

$$b = y_1 b_1 + y_2 b_2 \tag{2.8}$$

Common mixing rules for multicomponent mixtures are shown in Table 2.2 in terms of the pure component properties $\{a_i, b_i, \alpha_i\}$ for the major equations of state. The $\{k_{ij}\}$ for the Peng–Robinson equation are called binary interaction parameters and are tabulated for many binary mixtures (Sandler, 2006). Thus, the only difference between the mathematical problem for a pure component and for a mixture is in the evaluation of the parameters a and b.

Here is a mathematical problem to be solved. Given a set of chemicals, temperature and pressure, find the specific volume of the mixture. To do this, you must find the critical temperature and pressure of each chemical, plus possibly the acentric factors. Once you have the parameters, you must solve the cubic equation, Eq. (2.5 or 2.6), which is a nonlinear equation in one variable. Because it is a cubic equation, it is possible to find the solution in a series of analytical steps (Perry and Green, 2008, pp. 3–10), but this is not usually done because it is quicker to find the solution numerically, albeit iteratively.

Programs such as Excel and MATLAB allow us easily to solve the specific volumes. However, one advantage of process simulators like Aspen Plus is that the physical properties of many components are saved in a database that users can access. In fact, users do not need to look up the numbers because Aspen Plus will do that when it needs them. In addition, some of the equations of state involve much calculation, all of which is subject to error. However, there are situations when you will need to program the equations yourself. The next section illustrates how to use each of these programs to solve equations of state.

SOLVING EQUATIONS OF STATE USING EXCEL[2] (SINGLE EQUATION IN ONE UNKNOWN)

There are at least two methods to solve algebraic equations using Excel. The first one uses "Goal Seek" while the other uses "Solver," and both are illustrated using a simple example: find the x that makes $f(x)$ zero:

$$f(x) = x^2 - 2x - 8 \tag{2.9}$$

Solution Using "Goal Seek"

Step 1 Open a spreadsheet and put the following statement in cell B1.

```
=A1*A1 - 2*A1 - 8
```

Cell B1 is the equation that should be zero, and cell A1 contains the variable that is adjusted to make this happen.

---

[2]Excel is a registered trademark of Microsoft Corporation, Inc.

Step 2 Under "Tools" choose "Goal Seek." When a small screen appears, fill in the spaces to show the following:

Set cell	$B$1
To value	0
By changing cell	A1

Step 3 Click OK. The answer appears in the spreadsheet.

```
-2.000007 4.1137E-06
```

Thus, the solution found is −2, with a tiny error—a small fraction of a percent. The test of whether the calculation is correct is shown in cell B1, which is 4.1×10^{-6}. This is not zero, but it is small enough for most purposes.

Step 4 If you want to decrease the tolerance to make the solution more accurate, under "Tools" and "Options" choose "Calculation." Then, in "Maximum Change" add a few zeros in the middle (changing it from 0.001 to 0.000001), add a zero to the maximum number of iterations, choose OK, and repeat the "Goal Seek." This time the answer is

```
-2    -1.376E-8
```

Step 5 To get the other root, put the value 3 in cell A1 and choose "Goal Seek."

Solution Using "Solver"

You can solve the same problem using the "Solver" option in Excel.

Step 1 Under the Tools menu, click on "Solver." Note: If the choice "Solver" does not appear, choose Add-Ins and load Solver from the Analysis ToolPak or the original Excel program disk (or see your system administrator for help). For the Macintosh, install the Solver provided by Frontline Systems.

Step 2 When the window opens, choose the option to make a cell equal to a value (or a maximum or minimum) by changing another cell. If you insert the appropriate cell locations, you will obtain the same answer as with "Goal Seek." This time, however, it is much more accurate.

```
-2    5.3291E-14
```

Example of a Chemical Engineering Problem Solved Using "Goal Seek"

Find the specific volume of *n*-butane at 393.3 K and 16.6 atm using the Redlich–Kwong EOS. The molecular structure of *n*-butane is symmetric, so we expect the Redlich–Kwong EOS to give values close to experimental ones, and using Redlich–Kwong–Soave or Peng–Robinson may not be necessary. Since the pressure is above 10 atm, we expect some differences between the predictions by any of these equations and the results for an ideal gas.

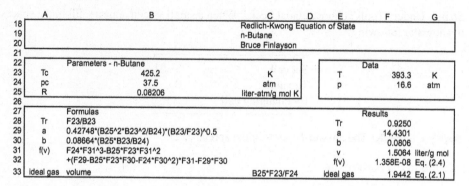

	A	B	C	D	E	F	G
18			Redlich-Kwong Equation of State				
19			n-Butane				
20			Bruce Finlayson				
21							
22		Parameters - n-Butane				Data	
23	Tc	425.2	K		T	393.3	K
24	pc	37.5	atm		p	16.6	atm
25	R	0.08206	liter-atm/g mol K				
26							
27		Formulas				Results	
28	Tr	F23/B23			Tr	0.9250	
29	a	0.42748*(B25^2*B23^2/B24)*(B23/F23)^0.5			a	14.4301	
30	b	0.08664*(B25*B23/B24)			b	0.0806	
31	f(v)	F24*F31^3-B25*F23*F31^2			v	1.5064	liter/g mol
32		+(F29-B25*F23*F30-F24*F30^2)*F31-F29*F30			f(v)	1.358E-08	Eq. (2.4)
33	ideal gas	volume	B25*F23/F24		ideal gas	1.9442	Eq. (2.1)

FIGURE 2.1 Excel spreadsheet to find the volume of a nonideal gas.

Step 1 You must first find the critical temperature and pressure; Perry and Green (2008) gives $T_c = 425.2$ K and $p_c = 37.5$ atm.

Step 2 Calculate values of "a" and "b" using the formulae in Table 2.1. The value of gas constant in these units is 0.08206 L-atm/g mol K.

Step 3 Prepare the spreadsheet shown in Figure 2.1. The title, name, and data will be useful when you come back to the problem at a future date.

Step 4 You enter the parameters in the parameter box. The cells containing the critical parameters and the temperature and pressure can be named, *Tc, pc, T*, and *p*, respectively. That way the equation for $f(v)$ would be easier to understand.

Step 5 The lower box gives the equations actually used as well as the results. Use the "Goal Seek" command to make $f(v)$ (cell F32) equal to zero by changing cell *v* (F31).

Step 6 For reference, the result for an ideal gas is also shown, and indeed *n*-butane is not well represented by an ideal gas under these conditions.

Step 7 How can you check this result? First, you have to be sure you have put the correct formulas into the spreadsheet, and that the units are consistent. That can only be determined by reference to the original equations and critical properties. It is easy to tell that $f(v) = 0$, but the solution is correct only if the equation for $f(v)$ is correct. In fact, the most challenging part of checking this calculation is the paper and pencil work before you developed the spreadsheet—to test the equations in the spreadsheet. Calculate the value of a on your calculator—is it the same as in the spreadsheet? You can also find a published problem, insert those values of a and b and see that you get the published result. (That works here, too, when you go to solve the problems at the end of the chapter.) Be careful, though: it is possible that for a certain pressure and temperature some of the terms in the cubic are too small to reveal an error when the only check is the final result. Later in the chapter, it is shown that Aspen Plus gives an answer of 1.49 L/g mol, using slightly different critical properties. Thus, we expect the calculation in Table 2.1 is correct. We note that the ideal gas volume is about 29% higher than the result obtained using the Redlich–Kwong equation, as expected at pressures above 10 atm. If we search the literature for an experimental

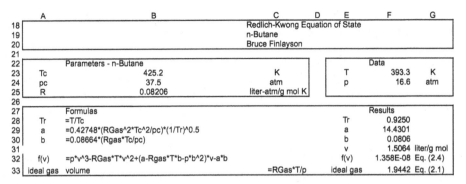

	A	B	C	D	E	F	G
18			Redlich-Kwong Equation of State				
19			n-Butane				
20			Bruce Finlayson				
21							
22		Parameters - n-Butane				Data	
23	Tc	425.2	K		T	393.3	K
24	pc	37.5	atm		p	16.6	atm
25	R	0.08206	liter-atm/g mol K				
26							
27		Formulas				Results	
28	Tr	=T/Tc			Tr	0.9250	
29	a	=0.42748*(RGas^2*Tc^2/pc)*(1/Tr)^0.5			a	14.4301	
30	b	=0.08664*(Rgas*Tc/pc)			b	0.0806	
31						v	1.5064	liter/g mol
32	f(v)	=p*v^3-RGas*T*v^2+(a-Rgas*T*b-p*b^2)*v-a*b			f(v)	1.358E-08	Eq. (2.4)
33	ideal gas	volume		=RGas*T/p	ideal gas	1.9442	Eq. (2.1)

FIGURE 2.2   Excel spreadsheet using named cells.

measurement (Sage et al., 1937), we find, for example, that the volume is 1.47 L/g mol at 16.6 atm and 393.3 K; this provides a check on the calculations as well as the data used to make them. The reason for the irregular choice of temperature and pressure was to compare with this data. PVT data can be found either by searching on the Web (chemical name PVT experimental data) or using the citations given by Aspen (see below).

The techniques used to create this spreadsheet are shown in more detail in Appendix A, including (a) inserting an equation for calculation, (b) inserting a text version of the equation for display, (c) creating a border around a group of cells, and (d) using "Goal Seek."

While this works, using equations involving the cell numbers, like B25 and F29, is inconvenient and can lead to errors. Instead, we can name the cells using physical terms that relate to the EOS. To do this, put the cursor in one of the cells that contains a number, such as B23. Then choose insert/name/define and type Tc. Then in the formulas, everywhere we want to use the critical temperature we replace B23 with Tc. The formulas now look like ours, and that makes it easier to find and correct mistakes. The result is Figure 2.2. This is the preferred method used in the rest of the book.

SOLVING EQUATIONS OF STATE USING MATLAB[3] (SINGLE EQUATION IN ONE UNKNOWN)

Nonlinear algebraic equations can be solved using MATLAB, too. First, you have to define the problem to solve by writing a file called an m-file; then, you check it; finally, you issue a command to solve it. These steps are analogous to the steps used in Excel. You can use MATLAB most effectively if you learn to use the Command Window and learn to create m-files and save them properly; see Appendix B for additional details.

Step 1 Define the function. It is created as an m-file, called f.m here.

```
function y = f(x)
y = x*x - 2*x - 8;
```

[3]MATLAB is a registered trademark of The MathWorks, Inc.

Change the current directory in MATLAB (at the top of the command window) to a directory where you wish to save your work and save it as f.m.

Step 2 Check the function. Issue the command: ≫feval(@f,2) to get the result: ans = −8. You can easily calculate Eq. (2.9) to see that for $x = 2$, the function value is −8. Now you know that f.m is correct. Note that we used a value for x that meant that every term in the function was important. If we had used $x = 1e−5$, then we would have obtained something close to −8, too, but the $x*x$ term would not be used; hence, we would not have checked the entire function. The value $x = 1.0$ is not a good choice either, since an incorrect function f.m = $x−2*x−8$ would return the same value as $x*x−2*x−8$, hence the error would not be discovered. This is a trivial example, and it is more important for more complicated problems.

Step 3 To find the value of x that makes $f(x) = 0$ in MATLAB, use the "fzero" function. In the command window, issue the following command.

```
>>fzero(@f, 0)
ans = -2
```

This command solves the following problem for x: $f(x) = 0$ starting from an initial guess of 0. Sometimes the function will have more than one solution, and that can be determined only by using the command with a different initial guess. You can test the result by : ≫feval(@f, ans).

To summarize the steps, Step 1 defined the problem you wished to solve, Step 2 checked your programming, and Step 3 instructed MATLAB to solve the problem. It is tempting to skip the second step—checking your programming—but remember: if the programming is wrong, you will solve the wrong problem.

When examining the command "fzero(@f, x0)" in MATLAB, the f defines which problem to solve, the $x0$ is your best guess of the solution, and fzero tells MATLAB to vary x, starting from $x0$ until the f is zero. In Excel's "Goal Seek," the analogous steps were to make a cell zero by varying the value of another cell. "Goal Seek" becomes fzero, a cell with an equation becomes f, and another cell becomes $x0$.

In all the commands and m-files in the previous text, the "f" can be replaced by other things, say "prob1." Just be sure you change it in three places: the filename of the m-file, the first line of the m-file (not absolutely necessary), and in the command. Additional forms of the command are as follows.

```
>>fzero(@function, x0, options)
>>z = fzero('f',x0)
```

In the last example, the result is put into the variable z. The options vector allows you to set certain quantities, like the tolerance; see how by saying: ≫help foptions. For the example used earlier, you can find the other root by running the program with $x0 = 3$. Multiple roots can be found only if you search for them starting with different guesses.

Example of a Chemical Engineering Problem Solved Using MATLAB

Find the specific volume of *n*-butane at 393.9°K and 16.6 atm using the Redlich–Kwong EOS.

Step 1 First, you need to prepare an m-file that will calculate the $f(x)$, or here $f(v)$, given the temperature, pressure, and thermodynamic properties. The file is shown in the following text.

```
% calculate Eq. (2.4), Chapter 2
function y = specvol(v)
% in K atm l/gmol
% parameters for n-butane
Tc = 425.2
pc = 37.5
T = 393.3
p = 16.6
R = 0.08206
aRK = 0.42748*(R*Tc)^2/pc
aRK = aRK*(Tc/T)^0.5
bRK = 0.08664*(R*Tc/pc)
y = p*v^3 - R*T*v^2 + (aRK - p*bRK^2 - R*T*bRK)*v -aRK*bRK;
```

This function, called "specvol," defines the problem you wish to solve.

Step 2 To test the function "specvol," you issue either of the following commands.

```
feval('specvol',2) or ans = specvol(2)
```

The feval function causes MATLAB to compute the value of y (the output defined in specvol) using the m-file named specvol when $v = 2$. The output you get is

```
Tc = 425.2000
pc = 37.5000
T = 393.3
p = 16.6
R = 0.08206
aRK = 13.8782
aRK = 14.4301
bRK = 0.0806
y = 25.98
```

You should check these results line by line, especially the calculation of aRK, bRK, and y. Alternatively, you can use the spreadsheet you developed, put in $v = 1.506$ and see what $f(v)$ is; it should be the same as in MATLAB since the cubic function and parameters are the same.

Step 3 When you use "fzero," the function "specvol" will be evaluated for a variety of v. Thus, it is inconvenient to have the constants printed out on the screen every iteration. To avoid this, you change the function "specvol" by adding a semicolon (;) at the end of each line. This suppresses the output. Do this and save the m-file, "specvol".

Step 4 Next, you issue the command

```
v = fzero(@specvol,2)
v = 1.5064
```

In feval, the 2 was the v to be used in the calculation, whereas with fzero, the 2 is an initial guess of the answer. To check, you might evaluate the function to find how close to zero $f(v)$ is.

```
>> ans = specvol(v)
ans = 1.7764e-15
```

Of course, you expect this to be zero (or very close to zero) because you expect MATLAB to work properly. If MATLAB cannot find a solution, it will tell you. You can also use the command "fsolve" in the same way. To find out more about "fzero" and "fsolve," enter the command "help fzero" or "help fsolve." If you use an initial guess of 0.2, you might get the specific volume of the liquid rather than the gas.

Another Example of a Chemical Engineering Problem Solved Using MATLAB

Next rearrange the MATLAB code to compute the compressibility factor for a number of pressure values. The compressibility factor is defined in Eq. (2.10):

$$Z = \frac{pv}{RT} \tag{2.10}$$

For low pressures, where the gas is ideal, the compressibility factor will be close to 1.0. As the pressure increases, it will change. Thus, the results will indicate the pressure region where the ideal gas is no longer a good assumption. There are two new features illustrated: the use of global and plotting.

Step 1 The code called "run_volplot" computes the specific volume for pressures from 1 to 31 atm at a temperature of 500 K and then calculates the corresponding compressibility factor. The first statement in "run_volplot" is a global command. The variables are identified, T, p, Tc, and so on, and then assigned values. In other programs, such as "specvol," the same global command is used, and these variables can then be accessed; see Appendix B for more information.

The next part of the program is a loop with 31 steps. The pressure is changed from 1 to 2, 3, 4, ..., 31. For each pressure, the constants a and b are calculated, as aRK and bRK, respectively. Then "specvol" is called to find the specific volume for those conditions; the answer is stored in the variable vol as an array, or vector. The compressibility factor is calculated and stored in the vector Z. Finally, a plot is made with pres along the x-axis and Z along the y-axis.

```
% run_volplot
global T p Tc pc R aRK bRK
% in K atm l/gmol
% parameters for n-butane
Tc = 425.2
pc = 37.5
T = 500
R = 0.08206
for i=1:31
```

```
pres(i) = i;
p = i;
aRK = 0.42748*(R*Tc)^2/pc;
aRK = aRK*(Tc/T)^0.5;
bRK = 0.08664*(R*Tc/pc);
vol(i) = fzero('specvol',0.2);
Z(i)=pres(i)*vol(i)/(R*T);
end
plot(pres,Z)
xlabel('pressure (atm)')
ylabel('Z')
```

Step 2 The m-file "specvol" is changed to use the global command, as shown in the following text. Now the parameters are available to "specvol" because they are defined as global variables and have been set in the program "run_volplot."

```
% calculate Eq. (2.4), Chapter 2
function y = specvol(v)
global T p Tc pc R aRK bRK
y = p*v^3 - R*T*v^2 + (aRK - p*bRK^2 - R*T*bRK)*v -aRK*bRK;
```

Step 3 Because you have already checked the program "specvol," you do not need to check it again. You will want to put semicolons (;) at the end of the lines in "specvol" because you do not need intermediate results. You can check the values of *aRK* and *bRK* for one of the pressures, to ensure they are correct in the program "run_volplot," but here these statements were copied from "specvol" directly, so that they need not be checked again. The use of vol(*i*), *Z*(*i*), pres(*i*) is easy to check from the graph, Figure 2.3. (Only the Redlich–Kwong EOS will be plotted by the program given in the previous text.) Also shown

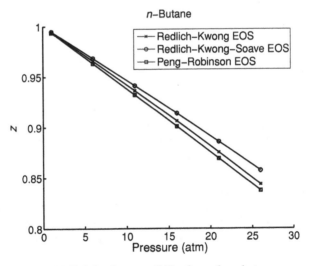

FIGURE 2.3 Compressibility factor for *n*-butane.

in Figure 2.3 are the predictions made using the Redlich–Kwong–Soave EOS ($\omega = 0.193$) and the Peng–Robinson EOS ($\omega = 0.193$). They differ by only a few percentage points. The effect of pressure is significant, although an ideal gas is within 10% for pressures less than 10 atm.

Step 4 An alternative way to calculate the compressibility factor would be to use the following command after the loop: Z = pres.*vol/(R*T). This command computes the same thing as

Z(i) = pres(i)*vol(i)/(R*T) inside the loop, but does it element by element outside the loop. Note the symbol.*; the period causes the element by element calculation, pres(2) times vol(2) and so forth.

EQUATIONS OF STATE WITH ASPEN PLUS[4]

You can also find the specific volume using Aspen Properties (or the Properties Analysis in Aspen Plus). Given here are commands that will enable you to find the specific volume of *n*-butane at the stated conditions. You may need to review Appendix C, too, which has more detail about Aspen Plus.

Example Using Aspen Plus

Find the specific volume of *n*-butane at 393.3 K and 16.6 atm using the SRK EOS option in Aspen Properties (see the following text for the reason).

Step 1

1. Start Aspen Plus and choose New.
2. When the window appears, choose Blank Simulation, then Create.
3. In the lower left-hand corner choose Properties if it isn't already chosen.

Step 2 On the left side of the list, choose Component/Specifications and enter the names or formulas of the chemicals, as shown in Figure 2.4. If there is no list on the left, click on the eyeglasses or choose the menu Data/Components. If Aspen Plus does not recognize your chemical, a window appears that allows you to search again, and it will suggest a number of possibilities. When the components are completely specified, it is important that there is an entry for every chemical in the column labeled "Component name." The first column is what *you* are naming the chemicals, but the third column is what Aspen Plus uses when it gets physical properties. If that column is blank, the program will not work.

Step 3 On the left side of the list, choose Methods | Specifications. Before selecting the property method, lets see what Aspen recommends. Under Home, choose Method Assistant. Choose, component type, hydrocarbon system, no petroleum assays, or pseudocomponents, respectively. The three methods suggested are Peng–Robinson, Soave–Redlich–Kwong,

---

[4] Aspen Plus is a registered trademark of Aspen Technology, Inc.

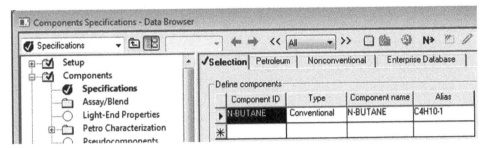

FIGURE 2.4 Aspen Properties window for component names.

and Lee–Kesler–Plocker methods. For this application, choose SRK in the Properties window. Now all the icons on the left should be blue, meaning the menus are completely filled out. If any menus are still red, go back and click on them to accept Aspen's parameters. If you want to know what the parameters are for *n*-butane, choose Home | Retrieve Parameters. In this case the dipole moment is zero, which suggests that nonideal gas behavior would only occur at high pressures or low temperatures.

Step 4 In the menus at the top choose: Home | Analysis then Pure. Fill out the window as shown in Figure 2.5. Click on Run Analysis.

```
Property type: Thermodynamic
Property: V (need to move the cursor down to see it)
Check vapor, uncheck liquid
Units: choose ml/mol
Components: select n-butane
Temperature, choose Units: K and • List, put in 393.3 and 400
Pressure: choose 16.6 atm
Property: scroll down to find SRK
Click Go.
```

Step 5 A graph appears with the result plotted (see Figure 2.6). You can read results from the graph or use the table behind the graph, giving the exact answer. The table is accessed from the Results tab under the Analysis node in the Navigation Pane. The result is 1490 mL/mol, or 1.49 cm^3/gmol, which compares favorably with the result of 1.51 when using Excel or MATLAB. The critical properties are slightly different in the three cases. The axis of Figure 2.6 has been changed from the default value in Aspen. Double click on each axis and change the range and interval.

Specific Volume of a Mixture

If you wish to get the specific volume of a mixture, it is necessary to use another approach, since the procedure in the previous text works only for pure substances. This time, use one of the units and specify the stream into the unit. The calculation will tell you the specific volume.

Find the specific volume of a mixture consisting of 630 kmol/h of carbon monoxide, 1130 kmol/h of water, 189 kmol/h of carbon dioxide, and 63 kmol/h of hydrogen at 1

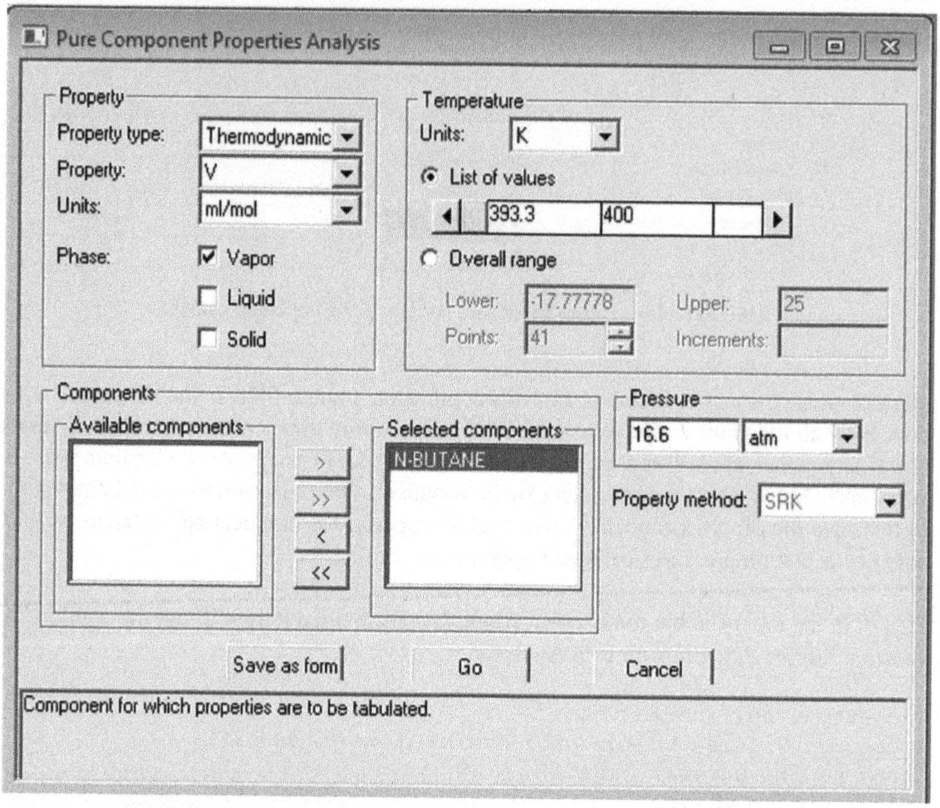

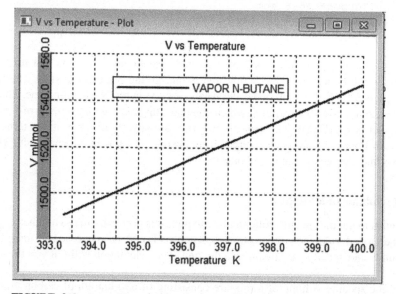

FIGURE 2.5   Aspen Plus window for pure component property analysis.

FIGURE 2.6   Aspen Plus window showing pure component property analysis.

atm and 500 K. The specific volume is the solution to the Redlich–Kwong–Soave EOS, Eq. (2.5). Since the pressure is low, an ideal gas is probably sufficient, but we will also use the Redlich–Kwong–Soave EOS for comparison. Aspen recommends (oil and gas processing): Peng–Robinson, PR–BM (Boston–Mathias), RKS–BM, SRK, under these conditions RK–Soave would be similar.

Step 1

1. Start Aspen Plus and choose New.
2. When a window appears, choose Blank Simulation, then Create.

Step 2 The first screen is for Properties (lower left). This will bring a menu to the left of the screen called the Navigation Pane. The red boxes indicate that you still need to supply information. Start at the top and work down, turning the red boxes into blue boxes by filling in the forms. In the list at the left, choose Component/Specifications (see Figure 2.4). Type in the names or formulas of each chemical.

If Aspen Plus does not recognize a chemical (like *n*-pentane), you can select that chemical and choose Find. A window appears that allows you to search again, and the program will suggest a number of possibilities.

Step 3 In the list at the left, choose Methods | Specifications. In the method screen, scroll down to get RK–Soave. To eliminate the red "pair parameters," open the window and click on the designated (recommended) data set.

Step 4 Switch to Simulation (lower left).

1. In the tabs at the bottom, choose Pressure Changes.
2. Click on the Compressor in the Model Palette.
3. Click on the flowsheet, and a compressor appears.

Step 5 To add the input and output streams, click on Material Streams (lower left-hand corner of the Model Palette), click on the flowsheet and drag a stream to the red arrow that is input to the compressor unit and click. Next, click the red arrow coming out and drag the stream away and click, giving Figure 2.7. Note: To make changes in the location of the streams or units, you can click on the arrow just above the Material Stream button. You can toggle back and forth between the arrow and Material Stream in the lower left corner as you improve the presentation of your flowsheet. When the flowsheet is showing, click on Material Streams. If any red arrows show in the flowsheet, it means that the unit is not properly connected; an input or output stream is necessary. Fix that before proceeding—you may have placed your cursor incorrectly when you drew the streams.

Step 6 In the list at the left, choose Streams by double clicking on it. Inside that folder are two or more folders, one for each stream. Choose the input stream, click on it, and insert the temperature, pressure, and flow rate in units you choose, as shown in Figure 2.8. You can specify the units for input numbers, thus avoiding having to do unit conversions yourself.

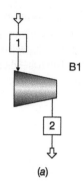

(a)

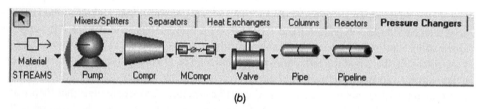

(b)

FIGURE 2.7   Flowsheet for single compressor: (*a*) flowsheet; (*b*) Model library (F10 key).

Step 7 In the list at the left, choose Blocks, then B1 (or whatever you have named your compressor). Choose Specifications, and choose Type: isentropic; insert the discharge pressure, as shown in Figure 2.9. (For this problem, you will use the inlet stream; thus you can put any discharge pressure you want, as long as it is above the inlet pressure.)

Step 8 Choose Run arrow at the top or the F5 key. If the input is incomplete, a window will appear to notify you and direct you to the missing data. If the input is complete, a

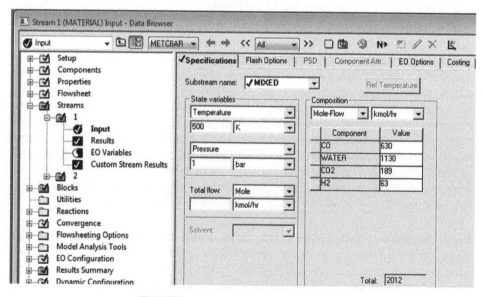

FIGURE 2.8   Setting stream information.

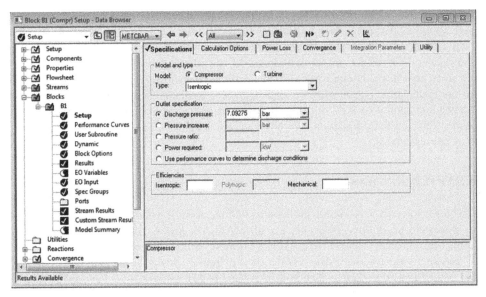

FIGURE 2.9   Setting compressor information.

window will appear to notify you of that, too. Click on the button to perform the calculation. Once the calculations finish (read the error messages, if any, in the Control Panel), click the Results box (lower one) and look at Results Summary | Streams. The stream data will appear in tabular form as shown in Table 2.3.

Step 9 You can obtain the specific volume by dividing the volumetric flow rate by the molar flow rate, $83,537/2012 = 40.98$ m^3/kmol. To get these units, choose Setup | Unit Sets in the Navigation Pane and make the choices. Then click Setup | Specifications and choose the unit set you've defined. To obtain Table 2.3 in spreadsheet format, place your cursor in the upper left cell of the Stream table, Right-click Copy it and it will be copied to the clipboard. Then paste into Excel. You can also have the Stream Table copied to the flowsheet, and it can be copied from there as a picture.

TABLE 2.3   Stream Information for Compressor Problem

Temperature (°C)	226.8	622.7
Pressure (bar)	1	7.093
Vapor fraction	1	1
Mole flow (kmol/h)	2012	2012
Mass flow (kg/h)	46448.67	46448.67
Volume flow (cum/h)	83537.33	21141.45
Enthalpy (Gcal/h)	−96.444	−89.534
Mole flow (kmol/h)		
CO	630	630
Water	1130	1130
CO_2	189	189
H_2	63	63

Your report of the results should explain what problem you have solved and how you solved it (focus on the chemical engineering information rather than the detailed step-by-step process on each screen), describe how you checked your results, and give the results. In this case you took a stream of specified composition, pressure and temperature, sent it to a compressor, and obtained the data shown in the Table 2.3. The type of thermodynamic option must be specified (here Redlich–Kwong–Soave) and justified (depending upon your level of expertise in chemical engineering). You might assume that Aspen Plus did the calculation correctly, but you should also review the results to see if they are reasonable.

CHAPTER SUMMARY

You have solved a very simple problem to find the specific volume of a pure component or a mixture using three programs: Excel, MATLAB, and Aspen Plus. Excel is readily available, and easy to use. MATLAB is a bit more difficult for beginners to use because it uses files, which require data transfer. It is extremely powerful, though, and is needed for other types of problems. With both Excel and MATLAB, you must look up the critical temperature, critical pressure, and perhaps the acentric factor of each chemical. You then must carefully and laboriously check your equations, one-by-one. When you use Aspen Plus, the parameters are stored in a database, and the calculations are preprogrammed. Your main concern is to use the graphical user interface (GUI) correctly. Aspen Plus is extremely powerful and is needed for other classes of problems.

When using any of the programs, you want to compare at least some of the calculations to experimental data, to verify that the type of thermodynamics you have chosen is appropriate to the physical case you are solving.

PROBLEMS

2.1₁ Find the molar volume of ammonia gas at 56 atm and 450 K using the Redlich–Kwong equation of state, $T_c = 405.5$ K, $p_c = 111.3$ atm, $a = 4.2527$, $b = 0.02590$; units of a and b correspond to v in L/g mol. (a) Use Excel; (b) use MATLAB.

2.2₁ For the problem in 2.1, use Aspen Plus with the RKS and Peng–Robinson thermodynamic options. Give reasons that the RKS and Peng–Robinson options might be necessary.

2.3₂ Find the compressibility factor of ammonia gas at conditions from 50 to 250 atm and 400 K using the Redlich–Kwong–Soave equation of state in Excel. (Hint: before beginning your spreadsheet, think about how you can organize it so that you can copy formulas from cell to cell easily.) Experimental data is available that gives a volume of 0.459 l/g mol at 66.8 atm and 446 K (Groenier and Thodos, 1960).

2.4₂ Consider the following mixture going into a water-gas shift reactor to make hydrogen for the hydrogen economy. CO, 630; H_2O, 1130; CO_2, 189; H_2, 63 kmol/h. The gas is at 1 atm and 500 K. Use Excel (or MATLAB) to compute the specific volume using

(a) ideal gas law

(b) Redlich–Kwong equation of state

(c) Redlich–Kwong–Soave equation of state.

The acentric factors for the RK–Soave method are: CO, 0.049; water, 0.344; CO_2, 0.225; Hydrogen, -0.22.

Where did you get the other data you needed? How do the three answers compare? Is the gas ideal or not? Comment. Then redo the calculations for a pressure of 200 atm and comment on the results.

2.5₁ Consider a mixture of 25% ammonia, the rest nitrogen and hydrogen in a 1:3 ratio. The gas is at 270 atm and 550 K. Use Excel (or MATLAB) to compute the specific volume using

(a) ideal gas law

(b) Redlich–Kwong equation of state

(c) Redlich–Kwong–Soave equation of state.

Where did you get the data you needed? How do the three answers compare? Is the gas ideal or not? Comment on the reasons the RKS equation might be expected to be better.

2.6₁ Find the molar volume of methanol gas at 100 atm and 300°C using the Redlich–Kwong–Soave EOS, $T_c = 512.6$ K, $p_c = 79.9$ atm, $\omega = 0.559$, $a = 8.96$, $b = 0.04561$; units of a and b correspond to v in L/g mol. Give reasons why the ideal gas or Redlich–Kwong equations of state are not expected to be valid choices. At 9.174 MPa and 300.0°C, the volume is 0.373 l/g mol (Straty et al., 1986).

2.7₁ Consider the following mixture that is coming out of a methanol reactor. CO, 100; H_2, 200; methanol, 100 kmol/h. The gas is at 100 atm and 300°C. Compute the specific volume using

(a) ideal gas law

(b) Redlich–Kwong equation of state

(c) Redlich–Kwong–Soave equation of state.

Give reasons why the ideal gas or Redlich–Kwong equations of state are not expected to be valid choices. The acentric factors for the RK–Soave method are: CO, 0.049; H_2, -0.22; methanol, 0.559. Where did you get the other data you needed? How do the three answers compare? Is the gas ideal or not? Comment.

2.8₁ Barron (1985) lists the specific volume of nitrogen at 5 MPa and 150 K as 0.116 m^3/kmol. The gas constant in these units is 8317 Pa m^3/kmol K.

(a) Use the Redlich–Kwong equation of state to calculate the specific volume and compare with this value. The critical constants for nitrogen are 126.2 K and 3.4 MPa.

(b) Do the same thing with the Redlich–Kwong–Soave equation of state. The acentric factor for nitrogen is 0.040. Is this molecule symmetric or asymmetric? Is the Soave version necessary?

(c) Barron (1985) gives the Dieterici equation of state for high pressures.

$$p(v - b)\exp(a/vRT) = RT \qquad (2.11)$$

For nitrogen, $a = 1.754 \bullet 105$ Pa m^6/kmol^2; $b = 0.04182$ m^3/kmol. Compute the specific volume from this equation of state and compare with parts (a) and (b).

2.9$_1$ Aspen recommends one of the following equations of state for n-butane: Peng–Robinson, Soave–Redlich–Kwong, Lee–Kesler–Plöcker. Compare the predictions of specific volume at 393.3 K and 16.6 atm.

Numerical Problems

2.10$_1$ Solve the problem in Figure 2.2 using the Newton–Raphson method.

2.11$_1$ Solve Problem 2.1 using the Newton–Raphson method.

3

VAPOR–LIQUID EQUILIBRIA

Have you ever driven by a refinery and wondered what happens in those tall towers? Some of them are distillation towers that are used to separate a mixture of chemicals into two or more streams, each a relatively pure stream of one of the chemicals. The physical process governing that separation is vapor–liquid equilibria. It has been estimated that 10% of the energy used commercially in the United States is used in distillation processes. Thus, it is important to make this process as efficient as possible.

Take a mixture of two or more chemicals in a temperature regime where both have a significant vapor pressure. The composition of the mixture in the vapor is different from that of the liquid. By harnessing this difference, you can separate two chemicals, which is the basis of distillation. To calculate this phenomenon, though, you need to predict the thermodynamic quantities such as fugacity, and then perform mass and energy balances over the system. This chapter explains how to predict the thermodynamic properties and then how to solve equations for a phase separation. While phase separation is only one part of the distillation process, it is the basis for the entire process. In this chapter you will learn to solve vapor–liquid equilibrium problems. The principles introduced in this chapter are employed in calculations for distillation towers that are discussed in Chapters 6 and 7. Vapor–liquid equilibria problems are expressed as algebraic equations, and the methods used are the same ones introduced in Chapter 2. Chapter 2 was limited to the gas phase, whereas vapor–liquid equilibria obviously includes liquids, too. Thus, the thermodynamic models need to be able to predict thermodynamic properties for liquids, and can be very complicated.

Instructional Objectives: After working through this chapter, you will have

1. Reaffirmed your skills using Excel® and MATLAB® for simple problems.
2. Learned to use Aspen Plus to perform vapor–liquid equilibria calculations.

Introduction to Chemical Engineering Computing, Updated Second Edition. Bruce A. Finlayson.
© 2014 John Wiley & Sons, Inc. Published 2014 by John Wiley & Sons, Inc.

3. Learned to use the NIST Thermo Data Engine to provide data for comparison.

4. Reviewed and expanded your chemistry and chemical engineering knowledge of activity coefficients.

FLASH AND PHASE SEPARATION

Suppose you put some water in an open pan on the stove, initially at room temperature. The partial pressure of water in the air (at equilibrium) will be equal to the vapor pressure of water at that temperature. Now heat the pan; the vapor pressure increases as the temperature rises. If the vapor pressure of water at the pan temperature exceeds the partial pressure of water in the room (usually set by humidity), the water will evaporate.

Next, imagine doing the same thing with a mixture of two chemicals in a closed vessel. The closed vessel is one with a piston that can move so that the pressure inside remains constant. The two chemicals have different boiling points, and different vapor pressures at a given temperature. As you increase the temperature of the vessel, the relative amount of each chemical in the vapor changes, because one is more volatile than the other. At temperatures below the bubble point, T_{bubble}, the mixture is entirely a liquid. At temperatures above the dew point, T_{dew}, the mixture is entirely a vapor. At temperatures in between, both liquid and vapor coexist. However, the composition of the liquid and vapor are not the same. Thus, as you gradually increase the temperature from a low value, some vapor forms, and this vapor is richer in the more volatile component. As the temperature increases further, more and more vapor forms. Finally, as the last drop of liquid evaporates, all the material is in the vapor phase, which has the same composition as the original liquid. However, between the bubble point and dew point, the composition of the liquid and vapor are changing as the temperature increases, and it is this change that you need to calculate.

This chapter looks first at equations governing an isothermal flash, and then shows how you can predict the thermodynamic quantities you need to solve the isothermal flash problem. The problems are all sets of algebraic equations, and you can solve these problems using Excel and MATLAB. The chapter then addresses more complicated vapor–liquid separations, but now using Aspen Plus because of its large database.

ISOTHERMAL FLASH—DEVELOPMENT OF EQUATIONS

Consider the flowsheet shown in Figure 3.1. Suppose you know the temperature, pressure, and overall composition of the inlet stream. The mole fractions of the chemicals in the inlet are called $\{z_i\}$. In the phase separator, however, the liquid and vapor are separated. The mole fractions of the chemicals in the vapor phase are called $\{y_i\}$ and those in the liquid phase are called $\{x_i\}$. When the vapor and liquid are in equilibrium, you can relate the mole fractions of each chemical in the vapor and liquid by the equation

$$y_i = K_i x_i \tag{3.1}$$

The set of $\{K_i\}$ are called K-values, and they can be predicted from thermodynamics as shown in the following text. For now, though, assume that you know their values.

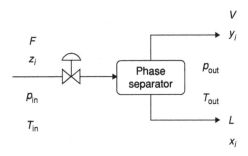

FIGURE 3.1 Flash phase separator.

To derive the equation governing the phenomenon, you first add the mole fractions of vapor and liquid of all components:

$$\sum_{i=1}^{NCOMP} y_i = 1, \quad \sum_{i=1}^{NCOMP} x_i = 1 \tag{3.2}$$

Then, subtract those two equations:

$$\sum_{i=1}^{NCOMP} y_i - \sum_{i=1}^{NCOMP} x_i = 0 \tag{3.3}$$

Next, substitute the equilibrium expression and rearrange to get

$$\sum_{i=1}^{NCOMP} K_i x_i - \sum_{i=1}^{NCOMP} x_i = 0 \quad \text{or} \quad \sum_{i=1}^{NCOMP} (K_i - 1)x_i = 0 \tag{3.4}$$

Next, make a mass balance for each component over the phase separator. F is the total molar flow rate, V is the molar flow rate of vapor, and L is the molar flow rate of liquid. The mole balance is then

$$z_i F = y_i V + x_i L \quad \text{and} \quad F = V + L \tag{3.5}$$

Divide by F and define v' as the fraction of feed that is vapor:

$$z_i = y_i v' + x_i (1 - v'), \quad v' = \frac{V}{F} \tag{3.6}$$

By using the equilibrium expression again, you can write this as

$$z_i = K_i x_i v' + x_i (1 - v') = (K_i - 1)x_i v' + x_i \tag{3.7}$$

Solve for the mole fractions in the liquid:

$$x_i = \frac{z_i}{1 + (K_i - 1)v'} \tag{3.8}$$

Put that expression into Eq. (3.3) to obtain the final equation:

$$\sum_{i=1}^{NCOMP} \frac{(K_i - 1)z_i}{1 + (K_i - 1)v'} = 0 \tag{3.9}$$

This is called the Rachford–Rice equation. Notice that if the K-values and inlet compositions $\{z_i\}$ are known, this is a nonlinear equation to solve for v'. Thus, you can apply here the
same methods used with Excel and MATLAB in Chapter 2. Once the value of v' is known,
you can calculate the value of the liquid compositions, $\{x_i\}$, and vapor compositions, $\{y_i\}$,
using Eqs. (3.1) and (3.8). The mole balance is then complete.

Example Using Excel

Suppose you have a mixture of hydrocarbons in the inlet stream. You want to find the
fraction of the stream that is vapor and find the mole fraction of each chemical in the vapor
and liquid streams. Table 3.1 shows the flow rates and K-values at 180°F and 70 psia, using
a basis of one mole per unit of time. To solve this problem, prepare the spreadsheet as
shown.

Step 1 "Term 1" is the numerator in Eq. (3.9) and "Term 2" is the denominator, and these
are calculated in columns D and E.

Step 2 The ratios of "Term 1" over "Term 2" are calculated and put into column F; this
column is summed in cell F9, which represents Eq. (3.9).

Step 3 You can then use "Goal Seek" to make the Rachford–Rice equation (cell F9) zero
by changing the fraction of the feed that is vapor (cell E1), giving the result shown. Once
you find the fraction vapor, the mole fractions in the two phases are easy to calculate using
Eq. (3.8) and (3.1), and these are included in columns G and H.

TABLE 3.1   Spreadsheet for Flash Example

	A	B	C	D	E	F	G	H
1					$v = 0.425837$			
2			Phase					
3			Equil.	Term1 =	Term2 =	Ratio =		
4		z_i	K-Value	$(K_i-1)z_i$	$(K_i-1)v+1$	Term1/Term2	x_i	y_i
5	Propane	0.1	6.8	0.58	3.4699	0.1672	0.0288	0.1960
6	n-Butane	0.3	2.2	0.36	1.5110	0.2383	0.1985	0.4368
7	n-Pentane	0.4	0.8	−0.08	0.9148	−0.0874	0.4372	0.3498
8	n-Octane	0.2	0.052	−0.1896	0.5963	−0.3180	0.3354	0.0174
		1				$f(v) = $ 8.2719E-07	0.9999996	1.0000005

How might you check these results?

Step 1 Of course, the most important aspect is to have the correct K-values, which can only be determined by comparing them with experimental data. The entire calculation depends upon those K-values.

Step 2 You need the Rachford–Rice equation to sum to zero, which is evident in the spreadsheet.

Step 3 You can check that the sum of vapor mole fractions equals one, and the sum of liquid mole fractions equals one.

Step 4 You can also check one term in the Rachford–Rice equation by detailed calculations, and then copy the equation down. This ensures that the formula is correct for all components.

THERMODYNAMIC PARAMETERS

Where did the vapor–liquid equilibrium K-value come from? It is defined as

$$K_i = \frac{\gamma_i f_i^o}{\varphi_i p} \qquad (3.10)$$

where γ_i is the activity coefficient, f_i^o is the fugacity of the pure chemical, φ_i is the fugacity coefficient in the vapor phase, and p is the total pressure. Most of these quantities can be calculated using thermodynamics. For example, your thermodynamic textbook derives the following equation for fugacity of a pure substance:

$$\ln\left(\frac{f}{p}\right) = -RT \int_0^p \left(\frac{RT}{p} - \hat{v}\right) dp \qquad (3.11)$$

This means that if you know the p–V–T behavior of a chemical, then you can find the fugacity by calculating the integral. In this text you do not need to do that, because Aspen Plus will do that for you. Instead you can either use K-values from the literature or do the calculations with a process simulator.

The other important parameter in Eq. (3.10) is the activity coefficient. It depends upon the chemicals involved, their symmetry, forces of attraction like van der Waal forces and electric dipoles, whether they chemicals are polar or not (have a dipole moment) and various other considerations. An important part of modeling any chemical system is to choose activity coefficient models that are appropriate to the situation being modeled. Chapter 6 discusses this topic in more detail and illustrates options available in Aspen Plus. In this chapter it suffices to say that if the chemicals have dipole moments, good choices are NRTL, UNIQUAC (and UNIFAC), and Wilson. Both NRTL and UNIQUAC can model azeotropes and two liquid phases, while the Wilson equation can model azeotropes where there is only one liquid phase.

For most of these options, the activity coefficient depends upon the liquid mole fractions, and that means the K-values depend upon the liquid mole fractions, too. Thus, there is not just one K-value that you can use in the calculations. Instead, one option is to choose the liquid mole fractions, determine the activity coefficients, then the K-values and see if the liquid mole fractions changed significantly. If so, you might use the new values and repeat the process. If this procedure did not work, you would have to use more sophisticated methods to solve the Rachford–Rice equation using process simulators such as Aspen Plus or other iterative algorithms.

There is one special case that is easy, however. If the liquid is an ideal liquid mixture, the activity coefficients are 1.0. If the pressure is low enough (say less than 10 atm.) the vapor phase is essentially an ideal gas and the fugacity coefficient of the vapor is 1.0, too. The fugacity of the liquid in the standard state is the vapor pressure, too, at low pressures. With those assumptions, the K-value becomes

$$K_i = \frac{vp_i}{p} \tag{3.12}$$

where vp_i is the vapor pressure at a given temperature. In this special case, you can find the K-values knowing the vapor pressure as a function of temperature and the total pressure of the system. Clearly, light components (lower boiling points) have a higher vapor pressure than heavy components, and thus their K-value will be larger. If two phases coexist at equilibrium, at least one of the components needs a K-value greater than 1 and at least one of the other components needs a K-value less than one. Even then two phases are not guaranteed, and the result depends upon the composition as well as the K-values.

Example Using MATLAB

You can solve the same problem (Table 3.1) using MATLAB. First, you define a function that represents the problem you wish to solve [Eq. (3.9)]. Then, you check the function to make sure it is correct. Finally, you use the "fzero" function to find the solution. The function defining the problem is

```
% vapor-liquid equilibrium
function y = vpequil(v)
z = [0.1 0.3 0.4 0.2]
K = [6.8 2.2 0.8 0.052]                                        (3.13)
sum1 = 0.;
for i=1:4                              or
num = (K(i)-1)*z(i)                    num = (K-1).*z
denom = 1 + (K(i)-1)*v                 denom = 1 + (K-1)*v
sum1 = sum1 + num/denom                sum1 = sum(num./denom)
end
y = sum1
```

Step 1 Program the function and save it in your working directory as vpequil.m.

Step 2 Issue the command "vpequil(0.2)" in the command window; the output should be

```
z = 0.1000    0.3000    0.4000    0.2000
K = 6.8000    2.2000    0.8000    0.0520
num = 0.5800                                                   (3.14)
```

```
denom = 2.1600
sum1 = 0.2685
....
ans = 0.2415
```

Calculation using the Excel spreadsheet (verified in Table 3.1) gives the same answers, so you have verified that the program is correct.

Step 3 Add semicolons at the end of each line of the m-file and save it. Solve the problem by issuing the fzero command in the command window.

```
>> fzero(@vpequil,0.2)
ans = 0.4258
```
                                                                    (3.15)

As expected, the result is the same as obtained using Excel.

Example Using Aspen Plus

Aspen Plus allows you easily to solve this same problem using the Flash2 unit operation.

Step 1 Open Aspen Plus and choose New, Blank Simulation, and Create. In the lower left-hand corner choose Properties Environment if it isn't already selected.

Step 2 In the Navigation Pane on the left, the boxes that are red indicate that you still need to supply information. Start at the top and work down, turning the red boxes into blue boxes by filling in the forms.

Step 3 In the list at the left, choose Component/Specifications and identify the components as you did in Chapter 2 (see Figure 2.4). Type the name or formula of the chemicals. If Aspen Plus does not recognize your chemical (such as *n*-pentane), select the *n*-pentane box and then Find. A window appears that allows you to search again, and it will suggest a number of possibilities. For *n*-pentane, several chemicals are listed, one of which is *n*-pentane; choose it ad close the window. When the components are completely specified, you should have an entry for every chemical in the column labeled "Component name." The first column is what you are naming the chemicals, but the third column is what Aspen Plus uses when it gets physical properties. If that column is blank, the program will not work.

Step 4 In the list at the left, choose Methods | Specifications and choose the thermodynamic model as you did in Chapter 2. In GLOBAL, scroll down to get RK–Soave. You also need to tell the computer which database to use for the binary interaction parameters in RK–Soave. We use the default database. Select Methods | Parameters | Binary Interactions and click all the red boxes to accept them.

Step 5 If the bottom of the screen does not show the units, use the View | Model Palette pull-down menu or press the F10 key. In the units at the bottom, choose Separators, and then click on the Flash2.

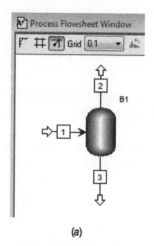

(a)

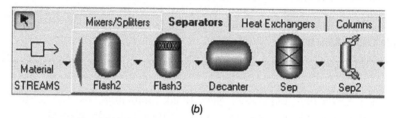

(b)

FIGURE 3.2 Flowsheet for Flash2 unit: (*a*) flash unit; (*b*) Model Library.

Step 6 Click on the flowsheet, and a flash (phase separation) unit appears. To add the input and output streams, click on Material Streams (lower left-hand corner), click on the flowsheet and drag a stream to the red arrow that is input to the flash unit. Click on the red arrows coming out and drag the stream away. Note: To make changes in the location of the streams or units, you can click on the arrow just above the Material Stream button. You can toggle back and forth between the arrow and Material Stream in the lower left corner as you improve the presentation of your flowsheet. When the flowsheet is showing, click on Material Streams. If any red arrows show in the flowsheet, it means that the unit is not properly connected; an input or output stream is necessary. Fix that before proceeding—you may have placed your cursor incorrectly when you drew the streams. The result should look something like Figure 3.2.

Step 7 In the list at the left, choose Streams by clicking on it or the + in front of it. Inside that folder are two or more folders, one for each stream. Choose the input stream, click on it, and insert the temperature, pressure, and flow rates of every chemical as shown in Figure 3.3. Note that you can specify the units you would like to use as you put in the numbers. That way you can use the numbers as given to you, without doing the unit conversion yourself. To set the units for the report, go to Setup | Unit Sets and define a new set with your choices. Then use Setup | Specifications and choose your set as the Global unit settings.

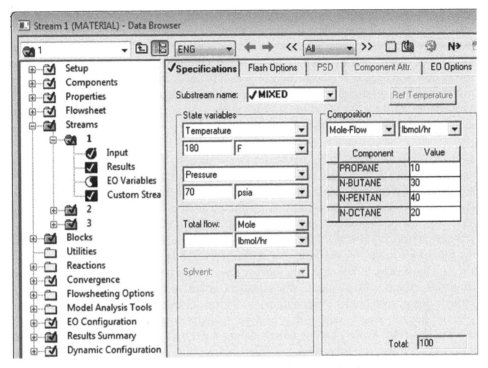

FIGURE 3.3   Stream information for Flash2 unit.

Step 8 In the list at the left, choose Blocks, then B1 (or whatever you have named your Flash2 unit), then Input. Choose Specifications, and insert the temperature and pressure of the unit as shown in Figure 3.4. (For this problem, you can use the same temperature and pressure as the inlet stream, and regard the unit simply as a phase separator.)

Step 9 Choose the Run button (filled triangle) at the top. If the input is incomplete, a window will appear to notify you and direct you to the missing data. If you had not specified a title in Specifications, a window may appear to insert it. If the input is complete, a window

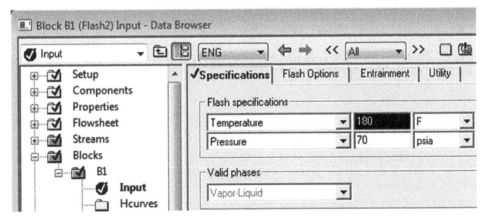

FIGURE 3.4   Block information for Flash2 unit.

will appear to notify you of that, too. Click on the filled triangle to make it perform the calculation. This will cause the calculation of the process (here one unit) to proceed. Once the calculations finish (read the error messages, if any, in the Control Panel), click the Results box (lower one) to return to the regular menu. Then look at Results/Streams. The stream data will appear in tabular form as shown in Table 3.2. If you click Stream Table, it will also be reproduced on your flowsheet. The default report does not list the mole fractions; to see them, go to Setup | Report Options and click on Streams; then check the mole fraction basis and repeat the calculations.

Step 10 For your report, select the sketch of the process by putting your cursor on the upper left corner and dragging it to the lower right corner. Choose Copy, switch to a word processing program, and choose Paste. Do the same thing with the stream table: click on the upper left box, right-click Copy, and paste into Excel. Your written report should include more than the process sketch and stream table. You still need to explain what problem you have solved and how you solved it, focusing on the chemical engineering information, not the detailed steps on the computer screen. Describe how you checked your results, and give the results. In this case, you took a stream of specified composition, pressure and temperature that was in two phases, and separated the gas from the liquid, and calculated the composition of each phase. Depending on your level of expertise in chemical engineering, you also need to specify and justify the type of thermodynamic model you chose. You might assume that Aspen Plus did the calculation correctly, but you also should look at the results to see if they are reasonable.

From the Table 3.2, you can see that the vapor fraction is 0.419 (compared with 0.426 in Table 3.1). Since Aspen Plus computed the K-values, it may have used different ones than you did. You can find out what Aspen Plus used by dividing y_i by x_i for each component. The results are: propane, 5.2; n-butane, 2.1; n-pentane, 0.84; n-octane, 0.067. While these numbers differ slightly from the ones used in Table 3.1, the mole fractions of each chemical

TABLE 3.2   Flash2 Results

	1	2	3
Temperature (°F)	180	180	180
Pressure (psia)	70	70	70
Vapor fraction	0.419	1	0
Mole flow (lb mol/h)	100	41.898	58.102
Mass flow (lb/h)	7355.297	2587.239	4768.058
Volume flow (cuft/h)	3865.909	3735.26	130.649
Enthalpy (MM Btu/h)	−6.768	−2.263	−4.505
Mole flow (lb mol/h)			
Propane	10	7.881	2.119
n-Butane	30	18.016	11.984
n-Pentane	40	15.096	24.904
n-Octane	20	0.905	19.095
Mole fraction			
Propane	0.1	0.188	0.036
n-Butane	0.3	0.43	0.206
n-Pentane	0.4	0.36	0.429
n-Octane	0.2	0.022	0.329

in the vapor and liquid streams are close. The K-values are also listed in the report made from Aspen (see Appendix C).

NONIDEAL LIQUIDS—TEST OF THERMODYNAMIC MODEL

In the examples given earlier, you were either given the K-values or you were told what thermodynamic model to choose. Usually, that decision is yours to make, and your results are only as good as your choice. Consequently, it is important to test your choice, preferably against experimental data. One easy way to do that is to model the T–xy or p–xy diagram for vapor–liquid of binary pairs. You can do this easily using the Property Analysis option in Aspen Plus. Given here are commands that will enable you to create such diagrams. You may also need to review Appendix C, which has more detail about Aspen Plus.

Step 1 Start Aspen Properties and choose New, Blank Simulation, and Create. In the lower left-hand corner choose Properties Environment if it isn't already selected.

Step 2 In the list at the left, choose Component I Specifications and enter the names or formulas of the chemicals; see Step 3 mentioned earlier. Here, we use water and ethanol.

Step 3 In the list at the left, choose Methods I Specifications. For the first calculation, choose the ideal property method (activity coefficients are all 1.0). For the second calculation, choose WILS-2. This option uses an ideal gas, Henry's Law, and the Wilson binary parameters, which is applicable to alcohols.

Step 4 (Skip this calculation for an ideal liquid.)
   You also need to tell the computer which database to use for the binary interaction parameter with the WILS-2 choice. Choose the default WILS-2 binary parameters. Now all the icons on the left should be blue, meaning the menus are completely filled out.

Step 5 In the menus at the top choose: Home I Analysis I Binary. (The binary option will be grayed out until you do Steps 2–4.) A window appears; choose the following.

```
Components: Water and Ethanol should show
Pressure: set the pressure for a T-xy diagram
or
Temperature: set the temperature for a p-xy diagram
Click Run Analysis.
```

Step 6 A graph will appear showing the result. You can read from this graph, or you can look at the table by clicking Results. To get the y–x plot, look at Home I Plot. Click the down arrow and choose the type of plot you want. The yx-diagram for ideal thermodynamics is shown in Figure 3.5. This is not a good fit. You know that ethanol and water form an

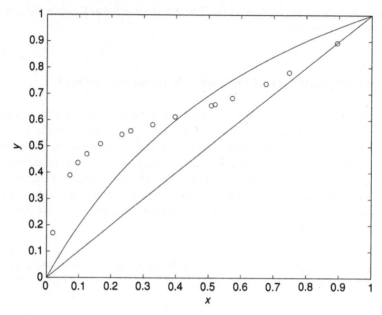

FIGURE 3.5   Ethanol–water vapor–liquid diagram for ideal thermodynamics calculations for an ideal liquid (line) with data (circles); plot made in MATLAB; y and x are mole fractions of ethanol.

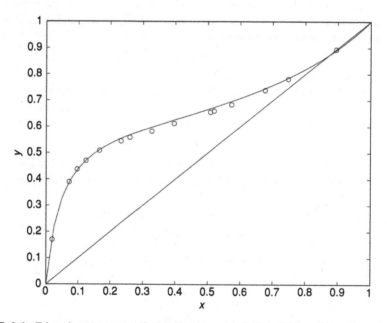

FIGURE 3.6   Ethanol–water vapor–liquid diagram with Wilson binary interaction parameters (line); data (circles); plot made in MATLAB; y and x are mole fractions of ethanol.

azeotrope, and the predictions do not give an azeotrope (where $y = x$). You should conclude that an ideal solution is not a good thermodynamic model for this system.

Go back to Step 3 and repeat using the Wilson thermodynamics; the results are shown in Figure 3.6, and the calculations agree with the data. The table from Analysis | BINARY | Results also lists K-values and activity coefficients. When you have good comparison with data, whether in the form of a T–xy diagram, K-values, or activity coefficients, you have validated a thermodynamics model for this binary pair. In a process, with many chemicals, you may have to do this for all binary pairs before deciding on the best thermodynamic model.

NIST THERMO DATA ENGINE IN ASPEN PLUS

Using the NIST Thermo Data Engine (Home | Data Source | NIST) we can compare experimental data with results of calculations made with our chosen equation of state. Consider a simulation that already has ethanol and water in it. Click the TDE icon and choose ethanol as the first component and water as the second one for a binary mixture. Then click "retrieve data". A window will open that contains experimental data for this system from a variety of sources. In this example, scroll down to Binary VLE/Isobaric/VLE045. Click on that to obtain Figure 3.7. The data are from Jones et al. (1943). Click on the "Plot T-xy" box to plot the data, Figure 3.8. Click on "Consistency Test" and the data will be evaluated for internal consistency (satisfying the Gibbs–Duhem equation). Two windows show the results, Figure 3.9. For additional information about

	Liquid Mole fraction ETHANOL	Temperature (K)	Gas Mole fraction ETHANOL	Total pressure(N/sqm)
1	0	373.15	0	101000
2	0.018	368.65	0.179	101000
3	0.054	363.75	0.3375	101000
4	0.124	358.55	0.47	101000
5	0.176	356.85	0.514	101000
6	0.23	355.9	0.542	101000
7	0.288	355.15	0.57	101000
8	0.385	354.15	0.612	101000
9	0.44	353.65	0.633	101000
10	0.514	352.95	0.657	101000
11	0.683	352.05	0.735	101000
12	0.84	351.41	0.85	101000
13	1	351.47	1	101000

Jones, C. A.; Schoenborn, E. M.; Colburn, A. P. Ind. Eng. Chem., 1943, 35, 666-72 Equilibrium still for miscible liquids data on ethylene dichloride - toluene and ethanol - water

FIGURE 3.7 Experimental data for ethanol–water system at 101,000 Pa.

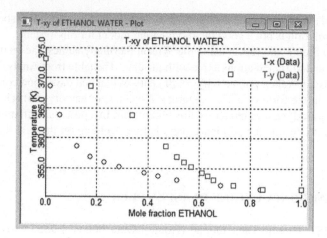

FIGURE 3.8 *T–xy* of ethanol–water.

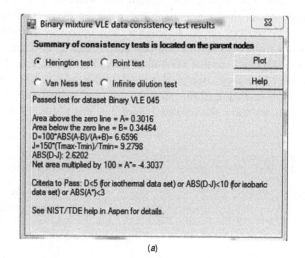

(a)

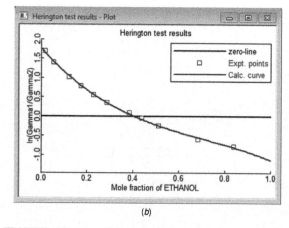

(b)

FIGURE 3.9 Consistency results for ethanol–water system.

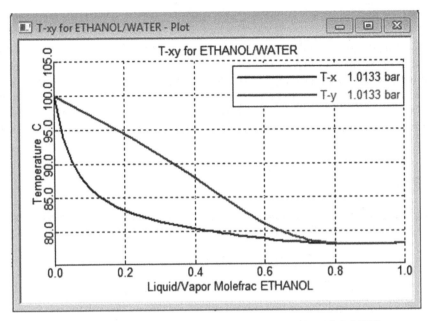

FIGURE 3.10   *T–xy* curve for ethanol–water using the WILS-2 EOS.

the test, use the Help button in Figure 3.9a. Once we have made choices of the thermodynamic model (here WILS-2), we can plot the same *T–xy* curve, Figure 3.10. In both cases it is possible to capture the data and copy it to Excel. Then the calculations can be compared point by point with experimental results. Figure 3.11 does this using MATLAB.

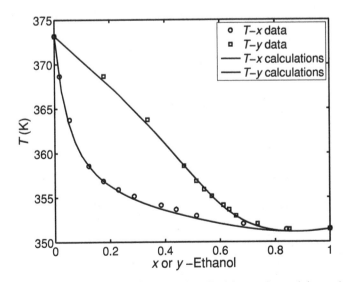

FIGURE 3.11   Comparison of calculated VLE with experimental data at 1 atm.

CHAPTER SUMMARY

In this chapter, you have derived the equations governing phase equilibrium and seen how the key parameters can be estimated using thermodynamics. You have solved the resulting problems using Excel, MATLAB, and Aspen Plus. You also learned to prepare a $T–xy$ diagram as a way of testing the thermodynamic model chosen to represent the phenomenon. You have been introduced to a variety of models of the liquid, and you have found that some of them model reality and some do not. Thus, the material in later chapters will be important when modeling a process with both vapor and liquids.

PROBLEMS

3.1₁ The stream in Table 3.3 is at 100 psia and 178°F. Calculate the fraction that is vapor by solving the Rachford–Rice equation (a) using Excel; (b) using MATLAB.

TABLE 3.3 Input Stream for Problem 3.1

Chemical	lb mol/h	K-value
Propane	20	3.7
n-Butane	30	1.4
n-Pentane	50	0.6

3.2₁ Consider a mixture of 78% nitrogen, 21% oxygen, and 1% argon at 2 atmospheres and 86 K. Barron (1985) gives the K-values as 1.23, 0.37, and 0.47, respectively. Determine the composition of liquid and vapor that are in equilibrium at these conditions.

3.3₁ A reservoir in Northern Louisiana contains a volatile oil. The reservoir conditions at discovery were 246°F, 4800 psia. The composition of the stream is in Table 3.4. You will have to choose how to model the butanes and pentanes: normal? Iso-? You will also have to choose which chemical to use to model the heptanes plus: heptane? octane? Aspen suggests using either Peng–Robinson, LK–PLOCK, or SRK. Use Peng–Robinson.

1. The gas–liquid separator at the surface is at 500 psia, 65°F. Find the composition of the gas and liquid streams and the vapor fraction.

TABLE 3.4 Input Stream for Problem 3.3

Chemical	Mole Fraction
Nitrogen	0.0167
Methane	0.6051
Carbon dioxide	0.0218
Ethane	0.0752
Propane	0.0474
Butane	0.0412
Pentane	0.0297
Hexane	0.0138
Heptane plus	0.1491

2. The liquid is taken to the stock tank, which is at 14.7 psia, 70°F. Find the vapor fraction at these conditions.

3.4₁ Using the Rachford–Rice equation, Eq. (3.9), prove that for two phases to coexist, at least one component needs a K-value greater than one and another component needs a K-value less than one.

3.5₁ The stream in Table 3.5 comes out of a distillation tower (described in detail in Chapter 6). It is at 138 psia and 197.5°F. If the pressure is reduced (adiabatically) to 51 psia, what will be the vapor fraction and temperature? (Hint: in Aspen Plus, put a valve before the Flash2 unit, and reduce the pressure with the valve block.) For hydrocarbons at these conditions Aspen suggests using Peng–Robinson, LK–PLOCK, or SRK. Use SRK.

TABLE 3.5   Input Stream for Problem 3.5

Chemical	lb mol/h
Propane	1.00
Isobutane	297.00
n-Butane	499.79
i-Pentane	400.00
n-Pentane	500.00

3.6₁ The stream in Table 3.6 comes out of a distillation tower (described in detail in Chapter 6). It is at 36 psia and 141.5°F. If the pressure is reduced (adiabatically) to 20 psia, what will be the vapor fraction and temperature? Solve the same way as in Problem 3.5.

TABLE 3.6   Input Stream for Problem 3.6

Chemical	lb mol/h
n-Butane	4.94
i-Pentane	396.00
n-Pentane	499.65

3.7₁ The stream in Table 3.7 comes of an ammonia reactor at 700°F and 3000 psia. If the temperature is reduced to 60°F in a flash unit, keeping the pressure the same, determine the flow rate of each species in the vapor and the liquid stream. Aspen suggests that for chemicals at high pressure, one can use an equation of state with advanced mixing rules, such as Wong–Sandler. Use RKSWS, and compare with the data listed in the NIST TDE system.

TABLE 3.7   Input Stream for Problem 3.7

Chemical	lb mol/h
Nitrogen	143.3
Hydrogen	450.4
Ammonia	221.6
Carbon dioxide	1.8

3.8₂ The flow rates out of a methanol reactor are CO: 100; hydrogen: 200; methanol: 100 lb mol/h. This stream is at 100 atm and 300°C. Cool it to 30°C and determine the flow rates of each species in the vapor and liquid stream.

  1. What are the K-values for the three species at these conditions? A flash separation (Flash2 in Aspen) is adequate to model the condenser. Using the mole fractions

y_i and x_i obtained by Aspen Plus for the vapor and liquid respectively; calculate the K-values, $K_i = y_i/x_i$.

2. Do the simulation for an ideal solution and also for another choice of thermodynamic model (your choice). What considerations did you use to choose?

3. Your report should have a flowsheet and give the mass balances. Discuss the difference between the two cases and why you think one might be better. How can you find out which thermodynamic model is more realistic? Note: The dipole moments of the three species are as follows: CO: 0.1; hydrogen: 0; methanol, 1.7 debyes (Reid et al., 1977).

3.9 The flow rates out of an equilibrium reactor to make hydrogen were CO: 5.25, hydrogen: 1095, carbon dioxide: 699, and water: 364.2 (use these as lb mol/h). Ideally, you could cool the stream and remove all the water. In actuality, you might not be able to condense all the water, and the water may contain trace amounts of the other components. This problem challenges you to find out how much. The outlet from the reactor is at 450 K, 1 atm, and is cooled to 80°F, still at 1 atm.

1. What are the K-values for the three species at these conditions? A flash separation (Flash2 in Aspen) is adequate to model the condenser. Using the mole fractions y_i and x_i obtained by Aspen Plus for the vapor and liquid respectively, calculate the K-values, $K_i = y_i/x_i$.

2. Do the simulation for an ideal solution and also for another choice of thermodynamic model. Aspen suggests that for chemicals use NRTL, Wilson, UNIQUAC, or UNIFAC. Use NRTL here.

3. Your report should have a flowsheet and give the mass balances. Discuss the difference between the two cases and why you think one might be better. How can you find out which is more realistic?

3.10$_1$ Model the vapor–liquid equilibria data (Table 3.8) of ethyl acetate(1) and water at 760 mm Hg using the UNIQUAC and Wilson–RK models in Aspen Plus. Also compare with experimental data obtained using the NIST TDE system (Ellis and Garbett, 1960).

TABLE 3.8 Vapor-Liquid Equilibrium Data for Problem 3.10

x_1	y_1	T (°C)
0.0006	0.0405	98.85
0.0011	0.1256	96.20
0.0049	0.5910	86.50
0.0086	0.6680	76.20
0.0459	0.7140	70.55
0.1440	0.7025	70.55
0.2690	0.7060	70.50
0.3540	0.7090	70.45
0.4080	0.7090	70.55
0.5140	0.7100	70.50
0.6080	0.7025	70.50
0.6900	0.7070	70.45
0.7750	0.6990	70.50
0.8737	0.7650	71.45
0.9444	0.8650	73.35

3.11₁ Model the vapor–liquid equilibria data (Table 3.9) of ethyl alcohol(1) and ethyl acetate at 1 atm using the UNIQUAC model in Aspen Plus. Also compare with experimental data obtained using the NIST TDE system (Murti and Van Winkle, 1959).

TABLE 3.9   Vapor-Liquid Equilibrium Data for Problem 3.11

x_1	y_1	T (°C)
0.1260	0.2146	73.82
0.1343	0.2146	73.78
0.2271	0.2960	73.04
0.3128	0.3634	72.50
0.3358	0.3643	72.28
0.5052	0.4803	72.18
0.5441	0.5074	72.35
0.6442	0.5618	72.70
0.6828	0.6092	72.90
0.7850	0.6819	74.14
0.8774	0.7908	75.50
0.9482	0.8924	76.70

3.12₁ Model the following vapor–liquid equilibria data (Table 3.10) of acetaldehyde (1) and ethyl alcohol at 1 atm using the WilsonRK model in Aspen Plus. Also compare with experimental data obtained using the NIST TDE system (De Leeuw, 1911).

TABLE 3.10   Vapor-Liquid Equilibrium Data for Problem 3.12

x_1	y_1	T (°C)
0.0690	0.2410	30.10
0.2030	0.6530	23.30
0.3160	0.7950	15.90

3.13₂ The air separation process operates at high pressure and low temperature, and the choice of thermodynamic model is important. Using Aspen Plus, prepare a T–xy diagram for nitrogen–oxygen using an ideal system, UNIFAC, NRTL–RK, and RK–SOAVE. Barron gives the following results at 20 atm and nitrogen mole fraction = 0.6: vapor temperature = 125.4 K and liquid temperature = 122.6. Which thermodynamics did a better job? Will it matter? In this process the temperature difference from the reboiler to the condenser is very small, so that a small error in the vapor–liquid properties might have a big effect on the design of the column.

3.14₂ Table 2.9 of Perry and Green (2008) gives the vapor pressure of inorganic liquids. Use data for nitrogen, oxygen, and argon and fit the data to a curve of the following form (p is in Pa, T in K):

$$\ln p = C_1 + C_2/T + C_3 \ln T + C_4 T^{0.5} \tag{3.16}$$

Then check your results in Table 2.8 of Perry and Green (2008) to see if you get the same formula cited by the Handbook. You probably know how to do the curve fit, but a detailed explanation is on the book website.

3.15$_2$ The formula you derived in Problem 3.14 is based on pressures below 1 atm. The formula given by the Handbook does not specify the range of applicability. Since the air separation unit operates at high pressure, compare the three predictions at pressures from $-180°C$ to $-140°C$: your formula, the formula from the Handbook, Aspen Plus. (Note: the critical temperures are: nitrogen, 126.2 K, oxygen, 154.6 K, and argon, 150.8 K.)

Numerical Problems

3.16$_1$ Solve the problem in Table 3.1 using the Newton–Raphson method.

3.17$_1$ Solve Problem 3.2 using the Newton–Raphson method.

4

CHEMICAL REACTION EQUILIBRIA

Hydrogen can be made for fuel cells by using the water–gas shift reaction. In fact, hydrogen is made this way every day in oil refineries. The reaction is very fast, and the effluent from the reactor is close to being in chemical equilibrium. Ammonia is made by reacting hydrogen and nitrogen, and the effluent from an ammonia reactor is usually in chemical equilibrium. The ammonia is then used to make fertilizer, which helps increase food production for the world. Thus, chemical reaction equilibrium is important for both energy and food production.

This chapter shows how to solve problems involving chemical reaction equilibrium. The chemical reaction equilibrium gives the upper limit for the conversion, so knowing the equilibrium conversion is the first step to analyzing a process. The second question, what is the rate of reaction, can then be answered to decide the volume of the reactor. This second question, using kinetics, is covered in Chapter 8. Chemical reaction equilibrium leads to one or more nonlinear algebraic equations that must be solved simultaneously, and such problems are described in this chapter.

Instructional Objectives: After working through this chapter, you will have

1. Additional practice using Excel and MATLAB®.
2. Used the RGibbs block in Aspen Plus to perform reaction equilibria calculations.
3. Continued learning to check your numerical work.
4. Reviewed and expanded your chemistry and chemical engineering knowledge of chemical reaction equilibrium.

Introduction to Chemical Engineering Computing, Updated Second Edition. Bruce A. Finlayson.
© 2014 John Wiley & Sons, Inc. Published 2014 by John Wiley & Sons, Inc.

CHEMICAL EQUILIBRIUM EXPRESSION

When you take some chemicals that can react and mix them in a vessel, the reaction can be represented as

$$A + B \Rightarrow C + D \tag{4.1}$$

If the reaction is reversible, the reverse reaction can also take place:

$$C + D \Rightarrow A + B \tag{4.2}$$

You can write this as

$$A + B \Leftrightarrow C + D \tag{4.3}$$

If the reaction rate is very fast, then the forward and reverse reactions proceed quickly, and the ultimate mixture contains all four chemicals, in specific amounts. If you change the temperature, these amounts may change. Sometimes in a gaseous reaction, changing the pressure has the same effect. In this chapter, your goal is to predict the composition of the final mixture, determined by chemical reaction equilibrium.

Of course, the reaction rate may not be infinitely high, and you may use a catalyst to speed up the reaction. However, even if the reaction rate is increased, you can never go beyond the composition determined by chemical reaction equilibrium. Here is how to set up this problem mathematically.

In thermodynamics courses, you have learned that chemical reaction equilibrium is determined by the equilibrium constant, which is defined in terms of the change of Gibbs free energy:

$$\Delta G^o = -RT \ln K \tag{4.4}$$

The Gibbs free energy is tabulated at 298 K for pure components, and it is possible to extend the Gibbs free energy for a reaction to any temperature using the van't Hoff equation:

$$\frac{d \ln K}{dT} = \frac{\Delta H_R^o}{RT^2} \tag{4.5}$$

Here, ΔH_R^o is the heat of reaction. When the reaction is

$$aA + bB \Leftrightarrow cC + dD \tag{4.6}$$

the equilibrium constant is defined in terms of the activities of the species:

$$K = \frac{a_C^c a_D^d}{a_A^a a_B^b} \tag{4.7}$$

In the gas phase, the activity is the fugacity, since the activity is the fugacity divided by the fugacity of the standard state, which is 1 atm. In turn, you can write the fugacity as the product of the fugacity coefficient (providing a correction from ideal gas behavior) times the total pressure times the mole fraction in the vapor phase:

$$f_i = \varphi_i \, p_i = \varphi_i \, p y_i, \text{ where } \varphi_i \equiv \frac{f_i}{p_i} \tag{4.8}$$

The fugacity coefficient can be calculated using the equation of state (Denbigh, 1971, p. 126):

$$\ln \varphi_i = \int_0^p \left(\frac{\hat{v}_i}{RT} - \frac{1}{p} \right) dp \tag{4.9}$$

When the pressure is less than 10 atm, the gas is usually ideal. From the formula, it is clear that for an ideal gas the integral vanishes. Then $\ln \varphi = 0$ or $\varphi = 1$. You can use that assumption and combine all the terms in Eq. (4.7) into Eq. (4.10):

$$K_p = \frac{y_C^c y_D^d}{y_A^a y_B^b} p^{c+d-a-b} \tag{4.10}$$

Example of Hydrogen for Fuel Cells

As a specific example, consider the water–gas shift reaction that can be used in a chemical process to make hydrogen for fuel cell applications:

$$CO + H_2O \Leftrightarrow CO_2 + H_2 \tag{4.11}$$

At equilibrium, Eq. (4.10) holds with $c + d - a - b = 0$, and the number of moles remains constant when the reaction takes place. Thermodynamic data give the value of $\ln K = 5$ (or $K = 148.4$) at 500 K. If you start with a stoichiometric mixture of carbon monoxide and water, what will the equilibrium composition be? In this case, and you are left with,

$$148.4 = \frac{y_{CO_2} y_{H_2}}{y_{CO} y_{H_2O}} \tag{4.12}$$

Step 1 To begin, make a mole balance table (see Table 4.1), with a basis of 1 mole each of carbon monoxide and water, which then react to equilibrium. Use x to represent the number of moles reacting, thus giving the values in the table.

TABLE 4.1 Solution for Equilibrium of Water–Gas Shift Reaction

Species	Start	End	y_i
CO	1	$1 - x$	$(1 - x)/2$
H_2O	1	$1 - x$	$(1 - x)/2$
CO_2		x	$x/2$
H_2		x	$x/2$
Total	2	2	1

TABLE 4.2   Spreadsheet for Equilibrium of Water–Gas Shift Reaction

1	G	H	I	J		K
2	Species	Start	End			Mole Fraction
3	CO	1	0.075861	= H3-conv		0.03793
4	H_2O	1	0.075861	= H4-conv		0.03793
5	CO_2	0	0.924139	= H5+conv		0.46207
6	H_2	0	0.924139	= H6+conv		0.46207
7	Total	2	2	= SUM (I3:I6)		1
8		conv	0.924139			
9		Eq.	7.35E-09	= 148.4-I5*I6/(I3*I4)		

Step 2   Put the mole fractions into Eq. (4.12) and simplify:

$$148.4 = \frac{(x/2)(x/2)}{[(1-x)/2][(1-x)/2]} = \frac{x^2}{(1-x)^2} \tag{4.13}$$

You can easily solve this equation by taking the square root, giving $x = 0.924$. In more complicated cases, it would not be possible to solve it so easily. Instead, you would have to solve a nonlinear equation numerically to find the value of x. When there are multiple reactions, which also may be in equilibrium, there will be several equations to solve simultaneously. Thus, being able to solve multiple nonlinear equations is useful when dealing with reactions at equilibrium.

Solution Using Excel

To solve for the equilibrium in Eq. (4.13) using Excel, you put the equation in one cell, put a guess of x in another cell, and use "Goal Seek" or "Solver" to make the first cell zero by changing x. Since the total number of moles does not change, you can change Eq. (4.12) to one involving moles:

$$148.4 = \frac{y_{CO_2}y_{H_2}}{y_{CO}y_{H_2O}} = \frac{(n_{CO_2}/n_t)(n_{H_2}/n_t)}{(n_{CO}/n_t)(n_{H_2O}/n_t)} = \frac{n_{CO_2}n_{H_2}}{n_{CO}n_{H_2O}} \tag{4.14}$$

The calculations are conveniently done in a spreadsheet, and the final results are shown in Table 4.2. Let us see how to construct this spreadsheet.

Step 1   Column H is the initial moles of each species.

Step 2   Column I is computed according to the equation displayed in Column J.

Step 3   Then, the equilibrium equation (4.14) is calculated in cell I9, according to the formula displayed in cell J9.

Step 4   Finally, you use Goal Seek (if it converges) or Solver (more likely to converge) to make cell I9 zero by varying cell I8. If you have difficulty getting convergence, try changing the value of the number of moles reacting (conv) to a value closer to where you think the

TABLE 4.3   Equilibrium of Water–Gas Shift Reaction with Nonstoichiometric Input

	G	H	I	J	K
1					
2	Species	Start	End		Mole Fraction
3	CO	1	0.01164	= H3-conv3	0.00364
4	H_2O	1.8	0.81164	= H4-conv3	0.25364
5	CO_2	0.3	1.28836	= H5+conv3	0.40261
6	H_2	0.1	1.08836	= H6+conv3	0.34011
7	Total	3.2	3.20000	= SUM (I3:I6)	1.00000
8		conv 3	0.98836		
9		Eq.	-9.168E-07	= 148.4-I5*I6/(I3*I4)	

answer is. If the iteration goes off to oblivion (10^{24}!), you need to reset the spreadsheet by not accepting the result (with Goal Seek) or "return to initial values" (with Solver). (The formulas listed in column J are displayed by entering them in a cell, preceded by the apostrophe symbol. In order to prepare the spreadsheet to compute the mole fractions, the first cell (K3) is set to = I3/$I$7. Then, the formula can easily be copied down to get I4/$I$7, etc.; see Appendix A for more information.)

Once you have prepared the spreadsheet, it is easy to change the conditions, either the equilibrium constant or the starting moles of various species. For another case, with some CO_2 and H_2 initially, you get the results shown in Table 4.3. Note that you can get more hydrogen by adding water to the initial composition. This is one way to help the hydrogen economy.

Solution Using MATLAB

To solve for the equilibrium in Eq. (4.13) using MATLAB, you create an m-file that will calculate

$$f(x) = 148.4 - \frac{x^2}{(1 - x)^2} \qquad (4.15)$$

for any x. Then, you use "fsolve" or "fzero" or "fminsearch" to find the x that makes $f(x) = 0$.

Step 1 Construct an m-file that evaluates the function, given x. The name is equil_eq.m and it is listed in the following. For testing purposes, you leave off the ";" of the commands that involve computation. Save this function in your desired workspace. (Note that the values for the "in" parameters have been set arbitrarily for the test.)

```
% equil_eq
function y=equil_eq(x)
COin = 1.1;
H2Oin = 1.2;
CO2in = 0.1;
H2in = 0.2;
Kequil = 148.4;
CO = COin - x
H2O = H2Oin - x
```

```
CO2 = CO2in + x
H2 = H2in + x
y = Kequil - CO2*H2/(CO*H2O)
```

This m-file acts like any function: give it an x and it comes back with the value of y.

Step 2 To test this function, you evaluate it for a specific value of x. Issue the following command in the command window.

```
equil_eq(0.9)
```

The result is

```
CO = 0.2000
H2O = 0.3000
CO2 = 1
H2 = 1.1000
y = 130.0667
ans = 130.0667
```

This is the correct answer.

Step 3 Once the check is made, you can add ";" to the end of every line. Note that to test the function you had to do three things correctly: (1) tell MATLAB the name of the function (equil_eq), (2) be sure the function was in the working directory (or someplace MATLAB would look), and (3) provide a value of x.

Step 4 To run the problem shown in Table 4.1, you change the function to the following.

```
% equil_eq
function y=equil_eq(x)
COin = 1.;
H2Oin = 1.;
CO2in = 0.;
H2in = 0.;
Kequil = 148.4;
CO = COin - x;
H2O = H2Oin - x;
CO2 = CO2in + x;
H2 = H2in + x;
y = Kequil - CO2*H2/(CO*H2O);
```

Step 5 You then issue the next two commands to find the answer.

```
format long
fzero('equil_eq',0.5)
ans = 0.92413871189774
```

As expected, this is the same answer as you found using Excel. The initial concentrations can be changed, too. Because the check was for various values (not 0s and 1s), you have checked the code for any set of parameters. In finding the solution, MATLAB will access the function many times. The strategy for finding the solution is built into MATLAB, but you need to provide the function name and an initial guess. MATLAB does the rest (the hard part). To solve the problem shown in Table 4.3, with different stoichiometry, you can use the same program. Here, the program is changed slightly to illustrate the use of the global command.

```
% equil_eq_global
function y=equil_eq_global(x)
global COin H2Oin CO2in H2in Kequil
CO = COin - x;
H2O = H2Oin - x;
CO2 = CO2in + x;
H2 = H2in + x;
y = Kequil - CO2*H2/(CO*H2O);

% run equil_eq_global
global COin H2Oin CO2in H2in Kequil
COin = 1
H2Oin = 1
CO2in = 0
H2in = 0
Kequil = 148.4
x=fzero('equil_eq_global',0.5)
```

Step 1 Here, you use the global command in run_equil_eq_global, set the variables, and repeat the global command in the function equil_eq_global. Then those variables are accessible to the function (see Appendix B for additional information).

Step 2 Leave off the ";" in the running program, so the data are repeated in the output.

Step 3 When using the global command, you must be sure to use the global command in the command window (or a run_program that runs in the command window), set the variables after the global statement, and then repeat the global statement in the function. If you do not follow these three rules, MATLAB will inform you that it was asked for an undefined variable or function, and give you the m-file and line number where it appears.

Step 4 When you run this program, the results are the same as before, which checks your modification to use the global option. To solve the problem in Table 4.3, you change the input value in run_equil_eq_global to be 1, 1.8, 0.3, and 0.1, respectively. The answer is

```
x = 0.98835845820682
```

In this case, MATLAB could not find a solution when starting from a guess of 0.5. After some experimentation, using a guess of 0.99 worked and gave the answer provided above. In these examples, you provided the data either by inserting them into the function itself, which was called by a driver program, or by using the Global option. Another option is to transfer those parameters in the calling arguments of the function. The function is changed to

```
% equil_eq_global2
function y=equil_eq_global2(x,param)
CO = param(1) - x;
H2O = param(2) - x;
CO2 = param(3) + x;
H2 = param(4) + x;
Kequil = param(5);
y = Kequil - CO2*H2/(CO*H2O);
```

and the calling program is changed to

```
% run equil_eq_global2
COin = 1
H2Oin = 1
CO2in = 0
H2in = 0
Kequil = 148.4
param = [COin H2Oin CO2in H2in Kequil]
x=fzero(@(x) equil_eq_global2(x,param), 0.5)
```

Note how the parameters are transferred—as a vector. Then in the function, the individual terms are obtained from the vector so that it is easier to check the equation. Naturally, the result is the same.

CHEMICAL REACTION EQUILIBRIA WITH TWO OR MORE EQUATIONS

In most real cases, multiple reactions can take place, and several of them may be in equilibrium at the same time. Then you will have two equations like Eq. (4.14) that depend upon two variables; you must choose the variables to satisfy both equations. MATLAB is ideally suited to solving such problems.

Multiple Equations, Few Unknowns Using MATLAB

Suppose you want to solve the following two equations:

$$10x + 3y^2 = 3$$
$$x^2 - \exp(y) = 2 \qquad (4.16)$$

There are two main ways to solve multiple equations in MATLAB. The first way is to use the "fsolve" command. The second way is to use the optimization routines and the "fminsearch" command. Both are illustrated here.

Method 1 Using the "fsolve" Command The steps are create an m-file, check it, call the appropriate MATLAB function, and check the results.

Step 1   Create an m-file and call it prob2.m.

```
%filename prob2.m
function y2 = prob2(p)
% vector components of p are transferred to x and y
     % for convenience
% in remembering the equation
x = p(1)
y = p(2)
% the components of the two equations are calculated
y2(1) =10*x + 3*y*y - 3
y2(2) = x*x - exp(y) - 2
```

Step 2   Test the file prob2.m with $p(1) = 1.5$ and $p(2) = 2.5$. If you tested with values of 1.0 and 1.0 there are several chances for error. For example, if a line is supposed to be x^*x but only x is written, then using a value of 1.0 for x gives the same value for x and x^*x and you will not detect the error. Using the same value for x and y [or $p(1)$ and $p(2)$] has the same pitfall: if you inadvertently typed x in place of y in the m-file, then you would not detect the error with these values.

```
p =  [1.5 2.5]
>>feval('prob2',p)
y2 = 30.75    -11.9325
```

These values agree with the hand calculations, so the functions are computed correctly. Now you put ";" after every line in the code in order to suppress the displayed output.

Step 3   From the command mode, you call the "fsolve" program with an initial guess, $p0$.

```
>>p0 = [0 0]
>>z = fsolve('prob2',p0)
>>z = -1.4456    -2.4122
```

Step 4   Because you do not know how accurate the results are, you should check them by evaluating the function.

```
>>ans = feval('prob2',z)
ans =    -1.0120e-04    5.1047e-06
```

Method 2 Using the "fminsearch" Function   In MATLAB the "fminsearch" function is used to find the minimum of a function of several variables. You first create an m-file that calculates the function, and then invoke the "fminsearch" function to minimize it.

Step 1   To solve the same problem, create a slightly different m-file and call it prob3.m. You can do this most easily by opening prob2.m, changing it, and saving it with the new name: prob3.m.

```
%filename prob3.m
function y2 = prob3(p)
% vector components of p are transferred to x and y
```

```
% for convenience in remembering the equation
x = p(1)
y = p(2)
% the components of the two equations are calculated
f1 =10*x + 3*y*y - 3
f2 = x*x - exp(y) - 2
y2 = sqrt(f1*f1 + f2*f2)
```

Step 2   Check the function as you did before. In fact, because you already checked the m-file "prob2.m" you only have to check that $y2$ is calculated correctly from $f1$ and $f2$. The command is invoked by first setting the initial guess (this also sets the length of the vector of independent variables, that is, it tells MATLAB how many parameters it has to adjust).

```
>>p0 =  [1.5 2.5]
>>feval('prob3',p0)
f1 = 30.75
f2 = -11.9325
ans = 32.984
```

These values agree with the hand calculations, so the functions are computed correctly.

Step 3   Next, replace the ";" in the m-file and run the problem from an initial guess of [1 1].

```
>>p0 = [1 1];
>>xvec = fminsearch('prob3',p0)
```

The results are

```
xvec = -1.4456   -2.4121
```

Step 4   You still do not know how accurate these results are. Check by evaluating the function after you have removed the ";" from the lines calculating $f1$ and $f2$ so that they will be displayed when the function is called.

```
>>ans = feval('prob3',xvec)
f1 =      -1.0120e-04
f2 = 5.1047e-06
```

Variations in MATLAB

Step 5   If these values are not small enough, you should reduce the tolerance using the options variable. The standard tolerance is 10^{-4}. To see the effect, though, you must get more significant digits by using the "format long" command.

```
>>format long
>>options=optimset('TolFun',1e-12)
>>xvec = fminsearch('prob2',p0,options)
xvec =   -1.44555236880465   -2.41215834803936
```

Step 6   To see how well the equations are solved, you take out the ";" from the file prob3.m and save it.

```
>>feval('prob3',xvec)
f1 =      1.4531e-12
f2 =     -9.0994e-13
ans =      1.7145e-12
```

Thus, you have made the functions zero to at least 10 digits. (The numbers you get will not be identical to these, because the numbers are so small they are corrupted by round-off error.)

Step 7   To find information about other options, type

```
>>help fminsearch
```

which tells you to use the options: Display, TolX, TolFun, MaxFunEvals, and MaxIter. Then

```
>>help optimset
```

tells you how to set those parameters.

Chemical Reaction Equilibria Using Aspen Plus

You can also use the process simulator Aspen Plus to solve chemical reaction equilibrium problems. It has a huge advantage over Excel and MATLAB: Aspen Plus contains the Gibbs free energies of many chemicals, and it can calculate them as a function of temperature. Thus, the data gathering aspect of the problem is handled for you. Your job is to compare the results and the predicted K values with experimental information.

The reactor module with Gibbs free energy is RGibbs. You can set up the problem in the same fashion as shown in Chapter 3 (p. 35), except that you choose Reactors, and then choose the RGibbs reactor. In the example shown in Table 4.4, the NRTL model is chosen for property evaluation. This is one of the equations of state recommended by Aspen: under Home, select Property Method Selection Assistant, for chemicals, not high pressure.

By using the mole fractions, you can find the equilibrium constant used, which is 137.5. This is slightly different from the one used in Eq. (4.13), which makes a slight change in the answer. For the problem illustrated in Table 4.1, Aspen Plus obtained a CO molar flow rate of 0.076 versus 0.079 using an equilibrium constant of 148.4. For the problem in Table 4.3, Aspen Plus obtained a CO molar flow rate of 0.013 versus 0.012 and a hydrogen flow rate of 1.087 versus 1.088. The chief difference between the calculations is due to the K value. Aspen Plus predicts that value from the Gibbs free energy, while the value used in Eq. (4.13) was obtained by reading off a graph.

CHAPTER SUMMARY

In this chapter, you have derived the equations governing chemical reaction equilibrium and seen how the key parameters can be estimated using thermodynamics. You have solved

TABLE 4.4 Equilibrium of Water–Gas Shift Reaction at 500 K and 1 atm (a) Feed with Reactants Only, and (b) Feed with Reactants and Products

	(a)			(b)	
	S1	S2		S1	S2
Temperature (°C)	226.8	226.8	Temperature (°C)	226.8	226.8
Pressure (bar)	1.013	1.013	Pressure (bar)	1.013	1.013
Vapor fraction	1	1	Vapor fraction	1	1
Mass flow (kg/h)	46.026	46.026	Mass flow kg/hr	73.842	73.842
Volume flow (cum/h)	81.964	82.059	Volume flow (cum/h)	131.091	131.223
Enthalpy (Gcal/h)	−0.081	−0.09	Enthalpy (Gcal/h)	−0.153	−0.163
Density (kg/cum)	0.562	0.561	Density (kg/cum)	0.563	0.563
Mole flow (kmol/h)			Mole flow (kmol/h)		
CO	1	0.079	CO	1	0.013
H_2O	1	0.079	H_2O	1.8	0.813
CO_2	0	0.921	CO_2	0.3	1.287
H_2	0	0.921	H_2	0.1	1.087
Mole fraction			Mole fraction		
CO	0.5	0.039	CO	0.313	0.004
H_2O	0.5	0.039	H_2O	0.563	0.254
CO_2	0	0.461	CO_2	0.094	0.402
H_2	0	0.461	H_2	0.031	0.34
Vapor phase			Vapor phase		
Volume flow (cum/h)	81.964	82.059	Volume flow (cum/h)	131.091	131.223
Compressibility	0.999	1	Compressibility	0.998	0.999
HeatCapRatio	1.346	1.294	HeatCapRatio	1.324	1.291

the resulting problems using Excel, MATLAB, and Aspen Plus. You have also learned to solve multiple equations using MATLAB when there are several reactions in equilibrium.

PROBLEMS

4.1₁ Methanol is formed by reacting carbon monoxide with hydrogen. (This might be an on-board generation of hydrogen for the hydrogen economy.) With the reaction

$$CO + 2H_2 \leftrightarrow CH_3OH \tag{4.17}$$

the condition at equilibrium is

$$K_p = \frac{p_{CH_3OH}}{p_{CO}p_{H_2}^2} = \frac{y_{CH_3OH}}{y_{CO}y_{H_2}^2}\frac{1}{p^2} \tag{4.18}$$

When the carbon monoxide and hydrogen enter the reactor in a 1:2 ratio, find the equilibrium conversion when the pressure is 50 atm and $K_p = 0.0016$. First, (a) use Excel; then, (b) use MATLAB to solve the problem.

4.2₁ Find the molar flow rates of all species out of an equilibrium reactor when the inlet values of nitrogen, hydrogen, and ammonia are 1.1, 3, and 0.2. The equilibrium constant is 0.05 at 589 K:

$$K_p = \frac{p_{NH_3}}{p_{N_2}^{1/2} p_{H_2}^{3/2}} = \frac{y_{NH_3}}{y_{N_2}^{1/2} y_{H_2}^{3/2}} \frac{1}{p} \qquad (4.19)$$

for the reaction written as $\frac{1}{2}N_2 + \frac{3}{2}H_2 \leftrightarrow NH_3$. Use 220 atm for the pressure. First, (a) use Excel; then, (b) use MATLAB.

4.3₁ Redo Problem 4.1 using Aspen at 256.9 °C. Since the pressure is high for this chemical system, use the RK–Soave equation of state, which is one of the options suggested by Aspen. How do the results compare with those of Problem 4.1?

4.4₁ Redo Problem 4.2 using Aspen. Since the pressure is high for this chemical system, use the RK–Soave equation of state, which is one of the options suggested by Aspen. How do the results compare with those of Problem 4.2?

4.5₂ The example of a water–gas shift reactor was done at a fixed temperature. Usually, this reaction is carried out adiabatically, and the temperature rises to what it needs to be for an adiabatic operation. The result of that is that the equilibrium conversion is less. Thus, a common method of performing this reaction is to use two reactors (Stephenson, 1966, p. 126). Let the reaction proceed adiabatically in the first one, then cool the gases, and then go into a second adiabatic reactor. Do this for the same problem shown in Table 4.4, using Aspen Plus, but have the first reaction be an adiabatic one rather than isothermal one. Then use a heat exchanger to cool to 342.9°C and go into another adiabatic reactor. How does the overall conversion compare with the example in Table 4.4?

4.6₁ A steam reformer takes methane and reacts it with water to form carbon dioxide and hydrogen by the reaction:

$$CH_4 + H_2O \Leftrightarrow CO + 3H_2 \qquad (4.20)$$

In addition, the water–gas shift reaction occurs:

$$CO + H_2O \Leftrightarrow CO_2 + H_2 \qquad (4.21)$$

The overall reaction is endothermic, though, so that the partial oxidation reaction is used to increase the temperature:

$$CH_4 + \left(\frac{3}{2}\right)O_2 \Rightarrow CO + 2H_2O \qquad (4.22)$$

Model the reformer with these reactions using the RGibbs block in an adiabatic mode. The oxidation reaction should proceed to completion and the other two reactions will be at equilibrium at the exit temperature. Take the input as (kmol/h) methane, 10,000; water, 5000; oxygen stream 5800 (99% oxygen, 1% nitrogen), 530 C, 20 bar. One must be careful when operating such an autothermal reformer because you may have an explosive mixture in a contained space. You do have an advantage, though, because when you start up the reactor you can use more oxygen to cause the reactor to reach its operating temperature faster.

4.7$_2$ A process to make hydrogen uses gas from a reformer; the gas is fed to a reactor where the water–gas shift reaction takes place [Eq. (4.11)]. Under usual operating conditions, the stream leaving the reactor is in equilibrium at the outlet temperature (or close to it). The feed rate to this part of the process is (lb mol/h) CO: 1260, H$_2$: 932, CO$_2$: 140, steam: 1972 at 177°C and 1.36 atm. The stream goes to a reactor, and the output of the reactor is in chemical equilibrium at 377°C. The equilibrium constant at 377°C is 15.6. Find the equilibrium composition leaving the reactor using (a) Excel, then (b) MATLAB, and then (c) using Aspen Plus with the RGibbs block. Since the pressure is low, an ideal gas equation of state is appropriate.

4.8$_1$ Ethyl chloride is made by reacting ethylene with hydrogen chloride:

$$C_2H_4 + HCl \Leftrightarrow C_2H_5Cl \tag{4.23}$$

Start with an equimolar amount of ethylene and hydrogen chloride and determine the equilibrium composition at 350°F and 250 psig using Aspen Plus with the RK–Soave equation of state.

4.9$_2$ Find the equilibrium conversion for a mixture of sulfur dioxide, oxygen, and sulfur trioxide at 700 K and 2 atm.

   (a) Starting mixture of SO$_2$:O$_2$::1:3, using ideal thermodynamics since only gases are present.

   (b) Increase the temperature to 900 K.

   (c) Use a starting mixture of SO$_2$:O$_2$::1:1/2 at 900 K.

   (d) Discuss the similarities and differences of parts (a)–(c).

4.10$_2$ Ethylene is make by heating ethane at a high temperature. The input to the reactor is 100% ethane.

   (a) Determine the equilibrium composition of ethane, ethylene, and hydrogen at 1100 K and 1 atm using Aspen Plus:

$$C_2H_6 \Leftrightarrow C_2H_4 + H_2 \tag{4.24}$$

   (b) Add in a second reaction and increase the temperature to 1367K and determine the equilibrium composition:

$$C_2H_4 \Leftrightarrow C_2H_2 + H_2 \tag{4.25}$$

Since only gases are present and the pressure is low, an ideal gas is appropriate; also, use the RK–Soave equation of state for comparison.

4.11$_2$ In a steam reformer, steam is added to methane and two (or more) reactions take place:

$$CH_4 + H_2O \Leftrightarrow CO + 3H_2 \text{ endothermic}$$
$$CO + H_2O \Leftrightarrow CO_2 + H_2 \text{ exothermic} \tag{4.26}$$

The composition out of the reformer is based mainly on chemical equilibrium (Marsh et al., 1994, p. 165). Determine the equilibrium composition under the following conditions using Aspen Plus. First, (a) 750°C, 14 bar, steam/methane =

3; then, (b) 650°C, 14 bar, steam/methane = 2. Since the pressure is over 10 atm, use the RK–Soave equation of state.

4.12₁ Solve Problem 4.10 with Excel when the equilibrium constant is 1.76 at 1100 K.

4.13₂ Use MATLAB to solve Problem 4.10 with both reactions at 1367 K. The two equilibrium constants are 37.6 and 0.769.

4.14₂ Use MATLAB to solve Problem 4.11 when the equilibrium constants are 47.17 atm and 1.309 atm.

Numerical Problems

4.15₁ Solve Eq. (4.16) using the Newton–Raphson method.

4.16₁ Solve Problem 4.1 using the Newton–Raphson method.

5

MASS BALANCES WITH RECYCLE STREAMS

Engineers who design and run chemical plants use mass balances to decide which processes are cost-effective. Chemical engineering students are taught in their first course to look at the entire system as well as the individual pieces of equipment. If you know the mass flows of a chemical process, including the input flows and output flows, you can estimate the economic viability of that process by adding up the selling price of the products, and subtracting the cost of the raw materials and waste treatment. If that number is not positive, the process is not cost-effective. You want to find that out before you have spent too much time designing it. Then, you can try to make it economically viable, perhaps by using another chemical reaction pathway, or a new separation scheme for purifying the product. Often, the raw material cost is 70% of the cost of the product, which makes the mass balance very useful. Also, mass balances allow you to find the energy costs in the process, provided you can calculate the enthalpy of each stream. Since thermodynamics allows you to calculate the enthalpy, you can estimate the energy cost.

Most processes involve a recycle stream. The reason is that all the reactants do not react, and businesses cannot afford to throw the rest away. Furthermore, any leftovers have to be disposed of in an environmentally friendly manner, which costs money. Thus, engineers take the unreacted reactants and put them back in the start of the process and try again. This makes the mass balances a bit more complicated, and it leads to iterative methods of solution, which are described in this chapter. Both Excel and Aspen Plus are used to solve mass balances with recycle streams. Situations in which the energy balances affect the mass balance are treated in Chapters 6 and 7, because these are best done using a process simulator such as Aspen Plus.

Introduction to Chemical Engineering Computing, Updated Second Edition. Bruce A. Finlayson.
© 2014 John Wiley & Sons, Inc. Published 2014 by John Wiley & Sons, Inc.

Instructional Objectives: After working through this chapter, you will have

1. Learned how to model processes using mass balances only, in Excel and Aspen, including reactors and separators.
2. Obtained experience with the programs that solve nonlinear equations: how to include them in the calculation, their effect on convergence to a solution, and which options work best in specific cases.

MATHEMATICAL FORMULATION

The mathematical formulation of mass balances is really quite easy. This section develops the equations for three simple units, which you then combine in myriad ways. Once you can do the mass balance for a simple unit, with the help of Excel you can do the mass balance for a process combining those simple units. Equations are given here for a mixer, separator, and reactor, and then those equations are used in Excel to perform mass balances on processes.

A mixer (called MIXR) adds all the flows in and transfers them to the output stream, as illustrated in Figure 5.1. The mass (or mole) balance is that the mass out equals the sum of the masses in. This is true for each chemical component, and hence for the total mass, too. It is more common to make mole balances, since chemical reactions occur using moles, but the conversion from mass to moles and back again is easy. The mass balance equations for the MIXR unit are

$$n_i^{\text{out}} = \sum_{j=1}^{\text{NSTRM}} n_i^j \tag{5.1}$$

where n_i^j is the flow rate of component "i" in the jth stream, and n_i^{out} is the flow rate of component "i" in the outlet stream. If you know the flow rates in, you can easily calculate the flow rates out.

The second unit is a separator, called SEPR. There is one input stream and two (or more) output streams, as illustrated in Figure 5.2. You can specify the fraction of each component going into the unit that goes out the overhead stream. (Note that you need to have some conventions here; clearly identify to which output stream the fractional flow refers.) Call

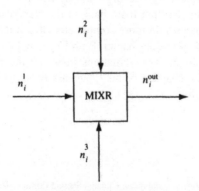

FIGURE 5.1 MIXR unit.

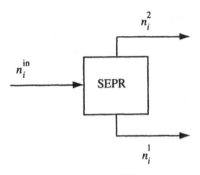

FIGURE 5.2 SEPR unit.

that fraction sf_i, and call it the split fraction going to the overhead stream. The mass balance for each component "i" is

$$n_i^2 = n_i^{in} sf_i, \quad n_i^1 = n_i^{in}(1 - sf_i) \tag{5.2}$$

If you know the split fractions of each component, and the flow rate in, you can easily calculate the flow rates out of stream 1 and 2 for each component.

The final unit is a reactor, called REAC, illustrated in Figure 5.3. This time there is only one input stream and one output stream, but a reaction occurs. Naturally, you need to specify the reaction and the conversion in some way. You can specify the reaction using the stoichiometric equation for the reaction

$$a\mathrm{A} + b\mathrm{B} \rightarrow c\mathrm{C} + d\mathrm{D} \tag{5.3}$$

If N moles of A react, then $(b/a)N$ moles of B also react and form $(c/a)N$ moles of C and $(d/a)N$ moles of D. You specify the amount reacting by giving a conversion for A (i.e., 50% of the A reacts). The conversion is called α. You do have to be careful to specify the conversion based on the limiting reagent, which in the illustration is A. If there were not enough B to react with A, then the conversion of A is limited. If it is too high, it would give a negative concentration of B. The equation for each of the components is

$$n_\mathrm{A}^{out} = n_\mathrm{A}^{in} - n_\mathrm{A}^{in}\alpha, \quad n_\mathrm{B}^{out} = n_\mathrm{B}^{in} - \frac{b}{a}n_\mathrm{A}^{in}\alpha$$

$$n_\mathrm{C}^{out} = n_\mathrm{C}^{in} + \frac{c}{a}n_\mathrm{A}^{in}\alpha, \quad n_\mathrm{D}^{out} = n_\mathrm{D}^{in} + \frac{d}{a}n_\mathrm{A}^{in}\alpha \tag{5.4}$$

Note in particular that the equations for components B, C, and D depend upon the amount of A entering, since that is how the conversion is specified.

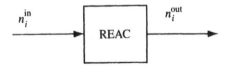

FIGURE 5.3 REAC unit.

All these equations can be programmed easily in Excel. It is the combination of them to represent the mass balances in a process with recycle that becomes challenging. Here are some examples in which the individual units are combined in an Excel spreadsheet to make mass balances for an entire (but simple) process.

EXAMPLE WITHOUT RECYCLE

A stream in a refinery is at 100 psia and 75°F and contains the following:

	lb mol/h
Propane	100
i-Butane	300
n-Butane	500
i-Pentane	400
n-Pentane	500
Total	1800

Your task is to separate this stream into five streams, each of which is a relatively pure stream of one component.

Step 1 You need to decide what physical mechanism to use (distillation, absorption, membranes, etc.) and the operating parameters of the equipment. For the first example, you can use the SEPR unit above as a simplistic model of any of these units. The chemicals are listed in order of their boiling points with the lowest boiling points first. Thus, choose distillation as the preferred separation method and the distillation train (see Figure 5.4) will distill them off one by one.

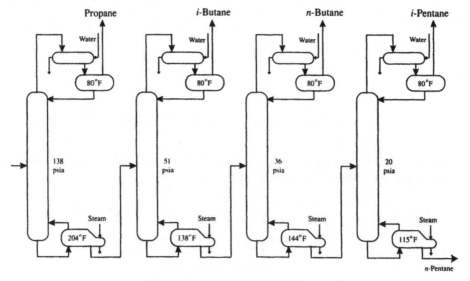

FIGURE 5.4   Distillation train.

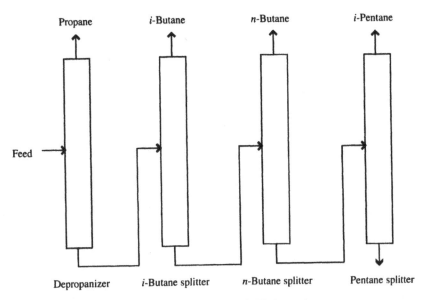

FIGURE 5.5   Simplistic distillation train.

Step 2 You can simplify this by taking out all the energy exchange units, as shown in Figure 5.5.

Step 3 Next, you prepare an even simpler flow sheet, using specified split fractions, as shown in Figure 5.6. In Chapters 6 and 7 you will learn to estimate those split fractions and to model the distillation towers more exactly.

Each unit, SEPR-x, involves a simple mass balance in which you specify the fraction of the feed that you want to go out the overhead and out the bottom stream. You do this for each species. The calculation proceeds from left to right, and the results are shown in Table 5.1.

Step 4 No iteration is necessary in this example. The final step is to check the results. You can add up the input and output streams for each component in each unit and determine that the mass of each species is conserved. You can also check to see that you used the appropriate split fractions. Then, and only then, should you accept the results. This was

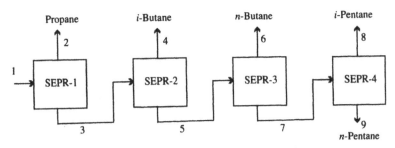

FIGURE 5.6   Distillation train modeled using simple SEPR units.

TABLE 5.1   Recovery Fractions in Distillation Train

Stream No.	1	2	3	4	5	6	7	8	9
Split Fraction		SEPR-1		SEPR-2		SEPR-3		SEPR-4	
C3		0.99		1		1		1	
i-C4		0.005		0.995		1		1	
n-C4		0		0.005		0.995		1	
i-C5		0		0		0.005		0.995	
n-C5		0		0		0		0.005	
				lb mol/h					
C3	100	99	1	1	0	0	0	0	0
i-C4	300	1.5	298.5	297	1.5	1.5	0	0	0
n-C4	500	0	500	2.5	497.5	495	2.5	2.5	0
i-C5	400	0	400	0	400	2	398	396	2
n-C5	500	0	500	0	500	0	500	2.5	497.5

a very simple example, so easy that you could have done it on the back of an envelope. However, if the simple distillation units were replaced by real distillation columns, in which vapor–liquid data were used, the calculations would have been much more difficult and time consuming.

EXAMPLE WITH RECYCLE; COMPARISON OF SEQUENTIAL AND SIMULTANEOUS SOLUTION METHODS

Consider a simple process in which the reactor can only convert 40% of the feed to it (perhaps due to equilibrium constraints), but the separation of reactant A and product B is complete and the unreacted A is recycled. The process is illustrated in Figure 5.7.

Step 1   One way to solve this problem is to do a sequential solution. You start with the feed, and solve the mass balance for the mixer, labeled MIXR. Since you do not yet know the amount of A in stream S5, assume it is zero and go on. Then there is one mole of A fed to the reactor, 40% of it reacts, and the unreacted part is recycled into stream S5.

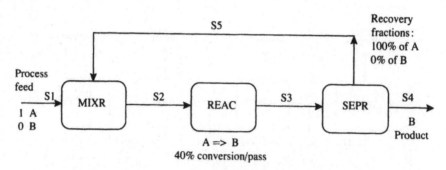

FIGURE 5.7   Process with recycle and 40% conversion per pass.

TABLE 5.2   Mass Balance with Recycle, Solved Using Sequential Method

Iteration Number	Feed, S1	Recycle (Guess), S5	Into Reactor, S2	Out of Reactor, S3	Recycle (at End of Iteration), S5
1	1	0	1	0.6	0.6
2	1	0.6	1.6	0.96	0.96
3	1	0.96	1.96	1.176	1.176
4	1	1.176	2.176	1.3056	1.3056
5	1	1.3056	2.3056	1.38336	1.38336
10	1	1.4848835	2.4848835	1.4909301	1.4909301
15	1	1.4988245	2.4988245	1.4992947	1.4992947
20	1	1.4999086	2.4999086	1.4999452	1.4999452

Step 2 Now repeat the process. This time when doing a mass balance on the mixer, you have a value in stream S5, so you can use it. You continue this process, working through the process over and over. After 20 iterations, the value of A in the recycle when you start an iteration (1.4999086) is close to the value when you end the iteration (1.4999452). You can then stop. The solution is shown in Table 5.2; note that some of the rows have been hidden (see the book website).

Many process simulators use this procedure because it is quite easy to implement. You only need to have a module or subroutine for a mixer, a reactor, and a separator, and the equations for each of these are quite simple. One problem that arises, though, is that the procedure takes many iterations to converge if the conversion is low. And, if there are interlocking recycle streams, convergence may not occur at all. However, it is a quite good scheme, and you can apply it using a spreadsheet.

The other solution technique finds an algebraic solution. Look at Table 5.3. Assign a variable name, R, to the unknown amount of A in stream S5. Work through the same process once; when you finish, the amount of A in the recycle (which you called R) must be equal to the amount of A coming out of the last unit and going into the recycle stream. The equation representing that is

$$R = 0.6^*(1 + R) \tag{5.5}$$

Solving Eq. (5.5) gives $R = 1.5$.

The solution is the same as that found with the sequential method, but you obtained it in one iteration. The only problem is that you had to do it analytically, which usually is not possible. However, you can write the mass balance equations in a general form and create a large set of linear or nonlinear equations to be solved simultaneously on the computer. The difficulty lies in the complexity of the units. The mixer, simple reactor, and simple separator lead to simple equations, but heat exchangers, distillation units, and plug flow reactors do

TABLE 5.3   Mass Balance with Recycle, Solved Using Simultaneous Method

Feed	Recycle (Guess)	Into Reactor	Out of Reactor	Recycle (at End of Iteration)
1	R	1 + R	0.6*(1 + R)	0.6*(1 + R)

not. In return for your work, though, you get enhanced convergence properties, since the simultaneous method will converge when the sequential method does not. In a later chapter you can examine some of the tricks chemical engineers use to make the sequential method work more often. First, let us learn to solve problems with recycle using Excel.

EXAMPLE OF PROCESS SIMULATION USING EXCEL FOR SIMPLE MASS BALANCES

The next example illustrates the use of Excel to solve mass balances for a process consisting of a feed stream, a mixer in which the feed stream is mixed with the recycle stream, a reactor, followed by a separator where the product is removed and the reactants are recycled. In later examples and problems, there will be inerts and purge streams, and so on, but this problem uses a stoichiometric feed. The reactor is limited by chemical equilibrium considerations, which complicates the solution.

The process takes hydrogen and nitrogen (in a 3:1) ratio to make ammonia. The reactor is limited by equilibrium; you will prepare the spreadsheet in stages to aid in troubleshooting. Thus, first prepare a spreadsheet as shown in Figure 5.8, using 25% conversion per pass in the reactor.

Step 1 The mixer takes streams 1 and 6 and adds them and puts them into stream 2. For example, D14 = C14+H14.

Step 2 The nitrogen reacting is given by E14 = −$E$19*D14. Cell E15 is set to three times Cell E14 (hydrogen uses three times the nitrogen reacting), and Cell E16 is set to negative two times the nitrogen in Cell E14. In this column, the signs are negative for reactants and positive for products. The spreadsheet clearly shows that the stoichiometry is correct.

Step 3 Stream 5 takes 0.5% of the nitrogen and hydrogen and 98% of the ammonia: G14 = 0.005*F14; G15 = 0.005*F15; G16 = 0.98*F16.

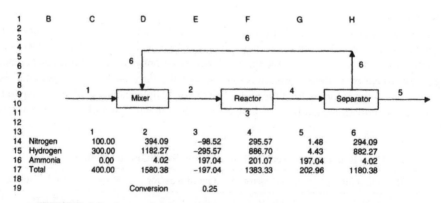

FIGURE 5.8 Ammonia process with 25% conversion per pass in the reactor.

Step 4 The rest goes out stream 6: H14 = F14−G14, and so on. You can easily check the mole balance around the separator.

Step 5 Turn on the iteration feature for the circular reference and the problem is solved.

EXAMPLE OF PROCESS SIMULATION USING ASPEN PLUS FOR SIMPLE MASS BALANCES

To perform this simulation in Aspen, without calculating any energy effects, open Aspen Plus and go to the Navigation Pane. Open the Setup box on the left and choose Calculation Options. Unclick the box that says Perform heat balance calculations. When filling out the information in Aspen Plus, you may have to put in temperatures and pressures, but they will not be used until you authorize the heat balance calculations. For this process, choose a Mixer, Stoichiometric Reactor (RStoic), and Component Separator (SEP). Insert the information on the feed to the process. For the reactor, choose that block, then choose Reactions, and then New. Define the stoichiometry of the reaction and the conversion of a specified component as shown in Figure 5.9 using the Reaction tab. For the Component Separator, choose an output stream and provide the fraction of each component going out that stream. Then run the problem, getting the results shown in Table 5.4. Naturally, they are the same as found using Excel. Problem 5.9 provides a more complicated example of a process with interlocking recycle streams. It was used early in the days of process calculations to test convergence techniques.

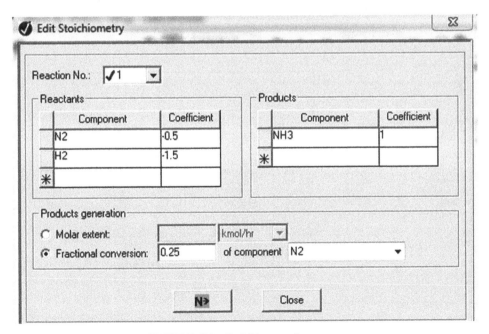

FIGURE 5.9 Stoichiometry for reactor.

TABLE 5.4   Ammonia Process with 25% Conversion per Pass in the Reactor

	1	2	3	4	5
Temperature (°C)	25		25	25	25
Pressure (bar)	1.013	1.013	1.013	1.013	1.013
Vapor fraction					
Mole flow (kmol/h)	400	1580.376	1383.332	202.956	1180.376
Mass flow (kg/h)	3406.112	13491.587	13491.587	3406.112	10085.475
Volume flow					
Enthalpy (Gcal/h)					
Mole flow (kmol/h)					
N_2	100	394.089	295.567	1.478	294.089
H_2	300	1182.266	886.7	4.433	882.266
NH_3	0	4.021	201.066	197.044	4.021

EXAMPLE OF PROCESS SIMULATION WITH EXCEL INCLUDING CHEMICAL REACTION EQUILIBRIA

Next, the reactor conversion is changed to be the equilibrium conversion, which may not be 25%. The equilibrium equation is

$$\frac{1}{2}N_2 + \frac{3}{2}H_2 \Leftrightarrow NH_3, \quad K_p = \frac{p_{NH_3}}{p_{N_2}^{1/2} p_{H_2}^{3/2}} = \frac{y_{NH_3}}{y_{N_2}^{1/2} y_{H_2}^{3/2}} \frac{1}{p} \tag{5.6}$$

The flowsheet and spreadsheet for this case is shown in Figure 5.10. The value of K_p and p are put into Cells F26 and F27, and the mole fractions of the stream coming out of the reactor

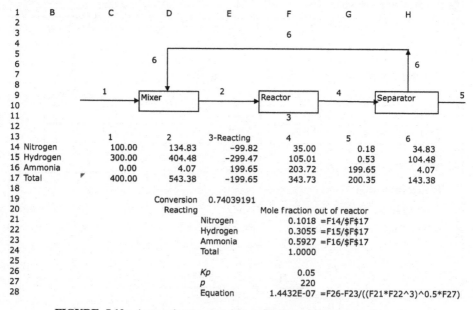

FIGURE 5.10   Ammonia process with equilibrium conversion in the reactor.

TABLE 5.5   Input for Cavett Problem 5.9

		Mole flow (lb mol)	
Temperature (°F)	120		
Pressure (psia)	49	N_2	358.2
Vapor fraction	0.572	CO_2	4965.6
Mole flow (lb mol)	27340.6	H_2S	339.4
Mass flow (lb/h)	1.87E+06	Methane	2995.5
Volume flow (cu⁻)	1.97E+06	Ethane	2395.5
Enthalpy (MME)	−2421.884	Propane	2291
		i-Butane	604.1
		n-Butane	1539.9
		i-Pentane	790.4
		n-Pentane	1129.9
		n-Hexane	1764.7
		n-Heptane	2606.7
		n-Octane	1844.5
		n-Nonane	1669
		n-Decane	831.7
		n-Undeca	1214.5
			27340.6

are obtained by taking the moles of a species in stream 4 and dividing by the total moles in stream 4: F21 = F14/$F$17. Those mole fractions can be used directly in the chemical equilibrium equation, as shown in Cell F28. This is simpler than deriving a chemical equilibrium equation involving the molar flow rates. Notice that as the conversion changes, the mole balance changes. This in turn changes the mole fractions, which no longer satisfy the chemical equilibria equation. Thus, you should use Solver to set F28 to zero by changing cell E19, where the final result is shown. (Note: Goal Seek did not work for this problem.) Goal Seek is available from a third party for the Microsoft operating system and is included in the Macintosh version. The reverse holds for Solver. The solution shown in Figure 5.10 was obtained using the Microsoft operating system and first using the Evolutionary Solver with $0.5 \leq$ conv ≤ 1 to get close (10^{-4}). Then the "GRG Nonlinear" option was used to refine the answer. The only remaining check is to take the final mole fractions and compute the chemical equilibrium equation yourself. If it is satisfied, then you have the correct solution. When using Excel, sometimes the iterations involving a circular reference do not converge. You can use CTRL-Z to reset the spreadsheet or use the undo key in the upper left: ↶ . Another method to avoid difficulties is to save the spreadsheet as a backup before turning on iteration. When "Solver" or "Goal Seek" are being used, and it does not converge, it asks whether you wish to revert to the initial guess; you would choose yes.

Did the Iterations Converge?

In this example, you have two iterations—one because of the circular reference due to the recycle streams and one because of the nonlinear chemical equilibrium equation. Some computers cannot handle both of these complications together. If you need to iterate by hand, follow these steps.

Step 1 In effect you used the spreadsheet to do the molar balances, and you set conv yourself. The spreadsheet shows the value of f(conv) for your choice of conv, but it does the molar balance anyway.

Step 2 If f(conv) is not zero, you can change conv and do it again. Keep this up until f(conv) is small enough to satisfy you. Once you are close, "Solver" should have no problem converging to a tight tolerance. What you are doing is replacing a problem with two iteration loops with a problem in which you supply one of the numbers, and the computer solves the other iteration loop. Then you change your number until the other equation is satisfied.

Extensions

Ideally you would have calculated the phase equilibrium in the separator, too. Then you would have three interacting iterations, and it would be the rare problem that Excel could solve. The difficulty is that during the iterations, the values may be physically unrealistic. Then, the equilibrium relation or the Rachford–Rice equation gives even more unrealistic values. Programs such as Aspen Plus can realize this and take precautionary steps to avoid it. As the flow sheet gets more and more complicated, and involves more and more thermodynamics, the power of Aspen Plus is welcome; see Chapters 6 and 7 for examples.

CHAPTER SUMMARY

In this chapter, you learned how to write the mass balance equations for simple units and then combine them into a complicated process topology. You saw that there were two methods to solve these equations, and Excel used essentially the sequential method because of the circular reference. You learned to handle several complications: recycle streams, chemical reaction equilibrium, phase equilibrium, and purge streams.

CLASS EXERCISES

Divide into teams of two to four students. Work through a spreadsheet (Tables 5.1, 5.2, and 5.4; Figures 5.8 and 5.10), with each person of a team taking a unit, showing where the parameters are for that unit and checking the results. Are they correct? Work through the whole process. Alternatively, your instructor can prepare a flowsheet and data set showing the results, but with some errors for your team to find.

CLASS DISCUSSION (AFTER VIEWING PROBLEM 5.10 ON THE BOOK WEBSITE)

What happened to the mass balances when you introduced a purge stream? (You can run it without carbon dioxide, too.) What happened to the mass balances when vapor–liquid equilibrium was required? Did the ratio of nitrogen to hydrogen in the recycle stream change? Why or why not? What if you had to solve the Rachford–Rice equation in the separator, the chemical equilibrium equation in the reactor, and set the purge fraction to maintain a maximum mole fraction of carbon dioxide in the inlet to the reactor. Could you do that all in Excel? Would it converge? Speculate.

PROBLEMS

Note: All the problems can be solved using Excel or Aspen (with the mass balance only option). The instructor can specify the preferred method, or both for comparison.

5.1₁ The process for making benzene is described below and illustrated in Figure 5.11. The complete process is described in **5.2.** Project below; here the process is simplified. (Toray Industries, Inc. and UOP Process Division, 1975, and Otani et al., 1968.) The process converts toluene (feedrate of 100 mol/h) to benzene and xylenes by disproportionation (also called transalkylation). The reaction and conversion per pass is

$$2C_7H_8 \Leftrightarrow C_6H_6 + C_8H_{10} \qquad 58\% \text{ conversion/pass} \qquad (5.7)$$

toluene   benzene   xylenes

The output from the reactor contains all the components; benzene is removed and then toluene is recycled. In the first splitter, the benzene stream contains 99.5% of the benzene fed to the splitter and 0.1% of the toluene and none of the xylene. In the second splitter, the toluene recycle stream contains all the benzene, 99.5% of the toluene and 0.1% of the xylene fed to the splitter. Perform a mass balance. What is the overall conversion of toluene to xylene?

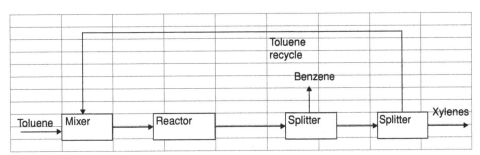

FIGURE 5.11   Simplified benzene process.

5.2   Project. The process for making benzene is illustrated in Figure 5.12, and you are to make the mass balance. (Toray Industries, Inc. and UOP Process Division, 1975, and Otani et al., 1968.) The process converts toluene (feedrate of 100 mol/h) to benzene and xylenes by disproportionation (also called transalkylation). The xylenes also disassociate to form toluene and trimethylbenzene. These reactions occur with a

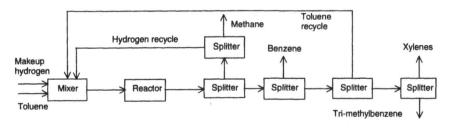

FIGURE 5.12   Benzene process.

catalyst in the presence of hydrogen, with a typical ratio of 7:1 hydrogen to toluene feed to the reactor. The hydrogen also reacts with toluene to form benzene and methane or with xylenes to form benzene and methane. The molar consumption of hydrogen is about 10% of that of toluene.

The reactions and their conversion per pass are

$$2C_7H_8 \Leftrightarrow C_6H_6 + C_8H_{10} \qquad \text{58\% conversion/pass}$$
$$\text{Toluene} \quad \text{Benzene} \quad \text{Xylenes} \tag{5.8}$$

$$2C_8H_{10} \Leftrightarrow C_7H_8 + C_9H_{12} \qquad \text{1\% conversion/pass}$$
$$\text{Xylenes} \quad \text{Toluene} \quad \text{Trimethylbenzene} \tag{5.9}$$

$$C_7H_8 + H_2 \Leftrightarrow C_6H_6 + CH_4 \qquad \text{0.14 percent conversion/pass}$$
$$\text{Toluene} \quad \text{Hydrogen} \quad \text{Benzene} \quad \text{Methane} \tag{5.10}$$

$$C_8H_{10} + 2H_2 \Leftrightarrow C_6H_6 + 2CH_4 \qquad \text{28\% conversion/pass}$$
$$\text{Xylenes} \quad \text{Hydrogen} \quad \text{Benzene} \quad \text{Methane} \tag{5.11}$$

The output from the reactor contains all the components, and the hydrogen and toluene are recycled, the methane is removed from the process, and the benzene and xylenes are removed from the process. The vapor pressures of the components at 20°C are benzene, 75 mm Hg; toluene, 22; xylenes, 4.8–6.6; trimethylbenzene, 1.1. A separation system is used that removes the hydrogen and methane first, then separates the benzene, then the toluene, and then the trimethylbenzene. The hydrogen and methane should be somewhat separated (to recover the hydrogen, which is recycled) and the toluene is recycled. In the separators, use splits of 99.9% of the light component out the top and only 0.1% of the next heavier component. Use splits of 100% for components lighter than the light key and 0% for components heavier than the heavy key. Use "Goal Seek" or "Solver" to adjust the hydrogen makeup feed to insure the proper amount of hydrogen entering the reactor (necessary for the catalyst).

5.3₁ Revise the process modeled in Problem 5.2 by using split fractions of 99% and 1% instead of 99.9% and 0.1% in all the separators. How do the total flows change? Would the equipment have to be larger? Would it cost more? Is the separation cost less? Answer these questions qualitatively now; when you finish your chemical engineering studies you will be able to answer them quantitatively.

5.4₁ Ethyl chloride is manufactured in an integrated process; see Figure 5.13 (Stirling, 1984, p. 72). Complete the mass balance for this process. Ethane reacts with chlorine to make ethyl chloride and hydrogen chloride, and ethylene reacts with hydrogen chloride to form ethyl chloride:

$$C_2H_6 + Cl_2 \Leftrightarrow C_2H_5Cl + HCl \tag{5.12}$$

$$C_2H_4 + HCl \Leftrightarrow C_2H_5Cl \tag{5.13}$$

The process is fed with three streams: ethane, ethylene, and chlorine. The feed streams contain 100 mol/h of each: ethane, ethylene, chlorine. The ethane/ethylene stream also

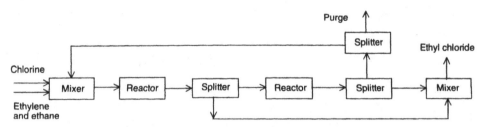

FIGURE 5.13 Ethyl chloride process.

contains 1.0% acetylene and carbon dioxide. (For this problem, just use 1.0% carbon dioxide.) The feed streams are mixed with an ethylene recycle stream and go to the first reactor (chlorination reactor) where the ethane reacts with chlorine with a 95% conversion per pass. The product stream is cooled and ethyl chloride is condensed and separated. Assume that all the ethane and ethyl chloride goes out in the condensate stream. The gases go to another reactor (hydrochlorination reactor) where the reaction with ethylene takes place with a 50% conversion per pass. The product stream is cooled to condense the ethyl chloride, and the gases (predominately ethylene and chlorine) are recycled. A purge or bleed stream takes off a fraction of the recycle stream (use 1%).

5.5 Project. Vinyl chloride monomer is the raw material to make polyvinyl chloride (PVC), which is produced in great quantities. When PVC was invented (as a highly elastic polymer by a University of Washington graduate, Waldo L. Semon, Patent No. 1,929,435) the vinyl chloride was made by reacting acetylene with hydrogen chloride. A process that uses cheaper raw materials is now used (Cowfer and Gorensek, 1997), and a simplification of it is given in Figure 5.14. Complete the mass balance for this process.

The ethylene (100 mol/h) and chlorine (103 mol/h) stream is fed to a direct chlorination reactor where the reaction

$$C_2H_4 + Cl_2 \Leftrightarrow ClC_2H_4Cl \tag{5.14}$$

takes place with essentially 100% conversion of the limiting reagent (ethylene). The selectivity to ethylenedichloride is 99% with the main by-product (and

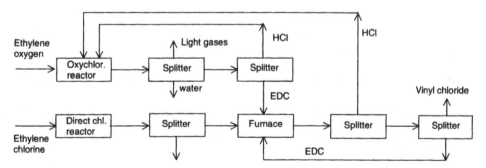

FIGURE 5.14 Vinyl chloride process.

the only one considered here) as 1,1,2-trichloroethane, formed in the following reaction:

$$C_2H_4+2Cl_2 \Leftrightarrow C_2H_3Cl_3 + HCl \tag{5.15}$$

The ethylene and oxygen stream fed to the oxychlorination reactor is 50% ethylene and 100 mol/h total. It is mixed with a hydrogen chloride recycle stream (which has some chlorine) and sent to an oxychlorination reactor, where the reaction

$$C_2H_4 + 2HCl+\tfrac{1}{2}O_2 \Leftrightarrow ClC_2H_4Cl + H_2O \tag{5.16}$$

takes place with a conversion of 96% for ethylene and ethylenedichloride selectivities of 95%. Here you can assume the byproduct is all 1,1,2-trichloroethane.

The output from the direct chlorination reactor is sent to a distillation tower where the heavy components are distilled off. The light components are sent to a furnace, where a pyrolysis reaction takes place:

$$2C_2H_4Cl_2 \Leftrightarrow 2C_2H_3Cl + 2HCl \tag{5.17}$$

The feed to the furnace must be very pure (99.5% ethylenedichloride) to achieve good cracking. The output is quenched. The liquid is the vinyl chloride product and the vapor is hydrogen chloride, which is recycled to the oxychlorination reactor.

The feed to the oxychlorination reactor is 100 mol/h ethylene and 50 mol/h oxygen. The output is cooled to remove the liquid (mostly water). A vent is used to remove light gases. The remainder goes to a distillation column where ethylenedichloride is the heavy component (and is sent to the furnace) and the light components are recycled to the oxychlorination reactor.

5.6₁ Modify Problem 5.5 by using air instead of oxygen as feed to the oxychlorination reactor. Make a mass balance of the revised process.

5.7₂ Synthetic ethanol is made by vapor-phase hydration of ethylene as shown in Figure 5.15. Make a mass balance of this process using Excel:

$$C_2H_4 + H_2O \rightarrow C_2H_5OH \tag{5.18}$$

Water and ethylene is mixed with a recycle stream and sent to a reactor where the reaction in Eq. (5.21) takes place (5% conversion of ethylene per pass). The ethylene feed is 100 mol/h and is 97% ethylene but also contains acetylene (2.9%)

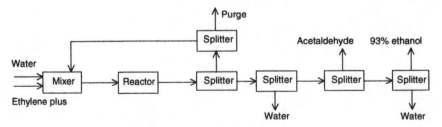

FIGURE 5.15 Ethanol process.

and inert gases (0.1%). The acetylene reacts with water, too, forming acetaldehyde (50% conversion of acetylene per pass):

$$C_2H_2 + H_2O \rightarrow CH_3CHO \tag{5.19}$$

The reactor effluent is cooled to remove the liquids (ethanol, acetaldehyde, water). The gases (ethylene, acetylene, and inert gases) are recycled, but a purge stream is needed to remove the inert gases.

Assume perfect splits. Adjust the fraction purged to keep the ratio of inert gas to ethylene in the stream fed to the reactor at 0.04, and feed enough water to the process to make the molar ratio of water to ethylene 0.6 in the stream into the reactor. Separate the liquids into relatively pure component streams, first removing 60% of the water, then the acetaldehyde, and then purifying the ethanol to 93% ethanol by removing water. In the acetaldehyde separation, remove 99% of the acetaldehyde and 1% of ethanol and water. (Hint: This problem has two constraints. In Excel you can use "Goal Seek" or "Solver" to satisfy one of them and adjust the other variable yourself, followed by another use of "Goal Seek" or "Solver." Repeat this process until both constraints are satisfied.)

5.8₂ Consider a process to make acetaldehyde by dehydrogenation of ethanol over a silver catalyst (Aguiló and Penrod, 1976), as illustrated in Figure 5.16. The ethanol feed rate is 350 mol/h.

The following design constraints must be satisfied:

1. Ethanol decomposes to form acetaldehyde and hydrogen:

$$C_2H_5OH \rightarrow C_2H_3OH + H_2 \tag{5.20}$$

2. The equilibrium limit is about 35% conversion per pass, so that a recycle stream is necessary.

3. Acetaldehyde reacts to form a number of side-products, including ethyl acetate, acetic acid, and butanol; only ethyl acetate is considered here:

$$C_2H_3OH \rightarrow \tfrac{1}{2}C_2H_5 - CO - OCH_3 \tag{5.21}$$

4. The conversion per pass is about 2% for the reaction in Eq. (5.21).

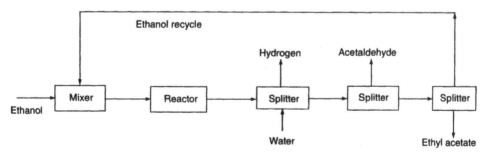

FIGURE 5.16 Acetaldehyde production by dehydrogenation of ethanol.

5. The first splitter must separate the reactor effluent to remove hydrogen; 150 mol/h of water is added.

6. The second splitter must separate the acetaldehyde.

7. The third splitter must remove the ethyl acetate and water and recycle the ethanol.

Make a mass balance with the following separations:

1. Ninety-nine percent of the hydrogen and 0.1% of the other components goes out the overhead of the first splitter.

2. Ninety-nine percent of the acetaldehyde, 100% of the hydrogen, and 0.5% of the other components goes out the overhead of the second splitter.

3. Ninety-nine percent of the ethanol and 1% of the other components goes out the overhead of the third splitter.

5.9₂ Cavett Problem. Chemical plants oftentimes have recycle streams; it can sometimes be difficult to solve the mass and energy balances because iteration is required. As discussed in Chapter 7, tear streams can be identified; you then insert your expected value of that stream and begin the computations. Hopefully, your guess is good

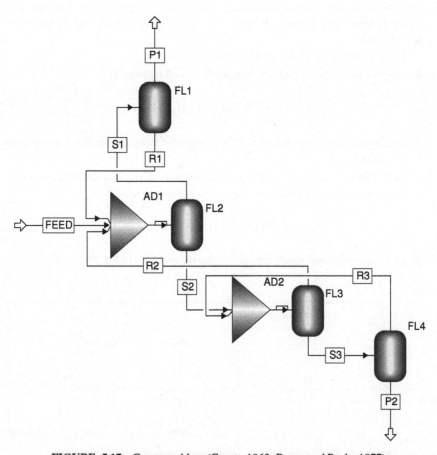

FIGURE 5.17   Cavett problem (Cavett, 1963; Rosen and Pauls, 1977).

enough that the calculations converge as they proceed from unit to unit, and then around again. One of the early problems used to test convergence schemes is called the Cavett problem after the person who formulated it.

The flowsheet is shown in Figure 5.17. There are four flash units operated at the following conditions:

FL1	100°F, 800 psi
FL2	120°F, 270 psi
FL3	96°F, 49 psi
FL4	85°F, 13 psi

The feed rate to the first unit is given in Table 5.5. These chemicals are typical of those in a Refinery, so the Chao–Seader thermodynamics is appropriate. Perform a study of the convergence methods in Aspen Plus: Wegstein, Direct, Broyden, and Newton. Solve the problem with each one, noting the number of iterations, the maximum error/tolerance, and whether the program said it converged. Between each choice of convergence methods, use Run/Reinitialize so that the problem starts from scratch at the beginning.

6

THERMODYNAMICS AND SIMULATION OF MASS TRANSFER EQUIPMENT

Mass transfer is one subject that is unique to chemical engineering. Typical mass transfer problems include diffusion out of a polymer to provide controlled release of a medicine, diffusion inside a porous catalyst where a desired reaction occurs, or a large absorption column where one chemical is transferred from the liquid phase to the gas phase (or vice versa). The models of these phenomena involve multicomponent mixtures and create some tough numerical problems.

In Chapters 3 and 5, you have already seen separations achieved with flash units, where the pressure is lowered and the vapor and liquid output have different compositions. Such separation works when the chemicals have widely differing boiling points or are otherwise easy to separate. When that is not the case, engineers must use distillation. Other phenomena used for separation include absorption, liquid–liquid extraction (moving a chemical from one liquid phase to another one, with immiscible liquids), and adsorption (used for identification in chromatography and for separation in chemical plants). This chapter provides simple examples of some of those processes. Your courses in mass transfer, separations, thermodynamics, and design study these processes in greater detail. In addition, mass transfer involving diffusion and flow (like in microfluidic devices) is covered in Chapters 10 and 11.

When modeling mass transfer equipment, there are two key points to remember: (1) thermodynamics is important and (2) convergence is difficult. The corollary is that you have to compare your thermodynamic predictions with experimental data. Also, you may start with ideal thermodynamics and get a solution. This solution can then be used as the initial guess when the thermodynamic model is more realistic. Process simulators do not always work, because from iteration to iteration the conditions may be unphysical, so you need to be flexible about how you approach a problem.

Introduction to Chemical Engineering Computing, Updated Second Edition. Bruce A. Finlayson.
© 2014 John Wiley & Sons, Inc. Published 2014 by John Wiley & Sons, Inc.

Instructional Objectives: After working through this chapter, you will have

1. Learned to make reasonable choices of thermodynamic models.
2. Obtained experience modeling distillation columns, using both short-cut methods and plate-to-plate calculations.
3. Learned to model packed bed absorption columns.
4. Learned to use Aspen Plus with sensitivity blocks and design specification blocks so that the program can vary parameters to meet your objective.

THERMODYNAMICS

Before doing process calculations, it is necessary to choose a model to describe the thermodynamic properties of the system (called Physical Property Methods in Aspen). The chosen method must account for the forces acting on the molecules of the mixture. It is important to distinguish between specific classes: neutral and nonpolar molecules (such as alkanes), polar molecules (such as water and alcohols), and ionic systems. The forces between these molecules include electrostatic forces between charged particles in ionic systems or permanent dipoles. These develop because, for example, the two ends of the molecule may have opposite signs even though the molecule is neutral overall. It may also happen because atoms are sharing electrons (as in the HCl molecule, where chlorine has seven electrons and shares one with hydrogen, but the electrons in the hydrogen orbit are affected more strongly by the seven electrons in the chlorine orbit. Dipole–dipole interactions are proportional to the square of the dipole moment of each species, but dipole–dipole forces fall off much faster with separation distance than Coulomb forces (Koretsky, 2004). Induced dipoles can occur if the mixture is placed in an electric field, in which case induction forces are important. The normal forces of attraction (dispersion forces) and repulsion are always there, and in some systems loose chemical bonds can be formed (hydrogen bonds). Statistical thermodynamics (Sandler, 2011) is useful for developing these models.

The virial equation of state for a vapor

$$\frac{p\hat{v}}{RT} = 1 + \frac{B(T)}{\hat{v}} + \frac{C(T)}{\hat{v}^2} + \cdots \tag{6.1}$$

can be derived in a number of ways from statistical mechanics, and the numerical values of the virial coefficients can be obtained as integrals over the various intermolecular forces. As the pressure is increased, more and more terms are needed in the expansion. It is possible to develop mixing rules for vapor mixtures, too, based on the type of forces included. This suffices for vapors, but the virial equation of state for liquids converges too slowly. It is then necessary to use radial distribution functions, deduced from scattering experiments, statistical mechanical theory of integral equations (with numerical solutions), or molecular level computer simulations using Monte Carlo methods or molecular dynamics (Sandler, 2011). Although the model parameters are not usually determined a priori, the structure of equations is used with adjustable constants that are chosen to fit experimental data. This point emphasizes again that your choice of thermodynamic model must be validated in some way, and comparison with experimental data is the gold standard.

Some of the options for liquids are (Sandler, 2006) as follows:

Option One: If the liquid solution is ideal, the activity coefficient is 1.0.

Option Two: The Margules one-constant models are for liquid mixtures with chemicals that have similar sizes, shapes, and chemical nature. The two-constant Margules equation expands that for dissimilar molecules.

Option Three: The van Laar model (and regular solution theory by Scatchard–Hildebrand) assumes that the mole fraction around a molecule is the same as the mole fraction of the mixture. But it does account for the fact that the interaction of molecule a with a, and b with b, is different from the interaction of molecule a with b. It works best for nonpolar compounds (like hydrocarbons) and compounds with similar molecular weights, and it may work for weakly polar compounds. The Van Laar options are less recommended in Aspen Plus; they are simpler to use than the others, but less successful in general. In Aspen Plus the ease of use is immaterial since someone else has created the program.

Option Four: The Wilson equation removes the restriction in the van Laar model and permits the mole fraction around the molecule to depend on the size and energy of the molecule, not just the mole fraction of the mixture (this is also true for the nonrandom two liquid (NRTL) and UNIFAC models discussed later). It is more robust, and it works for polar compounds and can handle azeotropes, but not two liquid phases. It works for alcohols and alkanes and requires pair-interaction parameters, which must be determined from data. It uses molar volumes of the different chemicals, so when they are not equal, this is a good method. It also uses binary interaction energies, and the parameters determined by fitting binary experimental data are also used in multicomponent mixtures.

Option Five: The NRTL equation is even more robust, and it works for polar compounds and can handle azeotropes as well as two liquid phases. This method includes the effect of nonrandomness and short-range order as well as differences due to molecular orientation.

Option Six: The UNIQUAC model is perhaps the best of the correlative equations. It is based on statistical mechanics and uses information on molecular volumes and surface areas, which makes it applicable to molecules of different sizes. The activity coefficients are calculated as the sum of two contributions. The first is a size and shape contribution with parameters generally obtained by considering each molecule to be a collection of functional groups (e.g., $-CH_3$ and $-CH_2OH$), and adding up the volumes and surface areas of the groups. All this information is in the Aspen Plus database, and the calculation is done by Aspen Plus. The second contribution in the UNIQUAC model is from the interactions between the molecules. The two parameters for each binary pair of molecules in this contribution are obtained by fitting experimental data. The UNIQUAC model works for polar compounds and can handle azeotropes as well as two liquid phases.

Option Seven: The UNIFAC model is a completely predictive model that is especially useful when no experimental data are available. It is based on the UNIQUAC model, but carries the idea of group contributions further by calculating the values of parameters in the interaction part of the UNIQUAC model from functional group-functional group interactions (unlike from molecule–molecule interactions in UNIQUAC). The functional group interaction parameters, obtained by correlating thousands of sets of data, are in Aspen Plus. As a general rule, UNIFAC predictions are less accurate than UNIQUAC and other models discussed earlier that have parameters obtained by correlating data for each system of interest, but it is valuable for predictions when no data are available. The UNIFAC model works for polar compounds and can handle azeotropes as well as two liquid phases.

Option Eight: The Electrolyte NRTL is generally applicable to electrolyte systems. It is commonly used for acid gas systems because inorganic amines, inorganic carbonates, and other ions are involved.

What method should you use? Aspen Plus recommends using Peng–Robinson or Redlich–Kwong–Soave for hydrocarbons and light gases under moderate conditions. These are Peng–ROB and RK–Soave in Aspen Plus. The equation of state will be used for all thermodynamic properties of both gases and liquids except for liquid density. To see details of the equations of state and the list of properties needed to define them (which Aspen Plus already has included), click on the help menu and type the abbreviation into the search window.

If the mixture contains polar components, then it is necessary to use Wilson, NRTL, or UNIFAC property methods for the liquid. These can be combined with either an ideal gas (WILSON, NRTL, or UNIQ-2) or Redlich–Kwong (WILS–RK, NRTL–RK, UNIFAC, or UNIQ–RK). This works for low-to-moderate pressures. As a rough rule of thumb, chemicals with dipole moments of 0.2 D and below can be considered nonpolar. If the pressure is higher (say greater than 10 atm) then the recommended methods are Redlich–Kwong or Peng–Robinson with the Wong–Sandler (1992) mixing rules (RKSWS or PRWS). Figure 6.1 is a simplified decision tree that applies to most of the examples and problems in this book. If the chemicals are polar and electrolyte, one must use a special model for electrolytes, ELECNRTL.

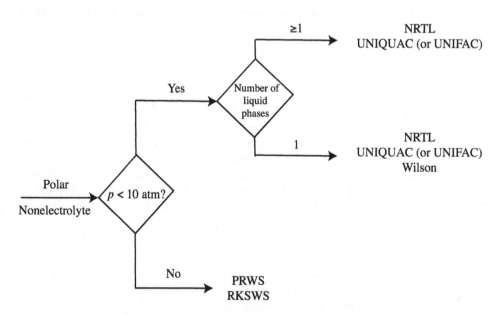

FIGURE 6.1   Decision tree for choosing thermodynamic models.

TABLE 6.1   Methods for Choosing Thermodynamic Models

Home | Methods Assistant
Home | NIST Thermodynamic Data Engine
Home | Run | Troubleshoot | In Search Window: Guidelines for Choosing
    Guidelines for Choosing a Property Method
    Guidelines for Choosing a Property Method for Polar, Nonelectrolyte Systems
    Guidelines for Choosing an Activity Coefficient Property Method

Aspen Plus has hundreds of choices for thermodynamic models. Although the guidelines given earlier will suffice in many cases, there may be times when the comparison to experimental data is not good. Then the features in Aspen Plus can be used to explore other options as summarized in Table 6.1.

Guidelines for Choosing

A convenient decision tree is available in Aspen Plus; enter Guidelines for Choosing a Property Method into the Search window, and figures as in Figure 6.1 will appear. The decisions you have to make are the ones shown in Figure 6.1 as well as: Real or pseudocomponents?; Interaction parameters available?; Vapor phase association?; and Degree of polymerization.

Properties Environment | Home | Methods Selection Assistant

Another aid is the Methods Assistant, under Properties Environment | Home; see Figure 6.2. If you select "Specify component type" you get a window listing of four groups: Chemical,

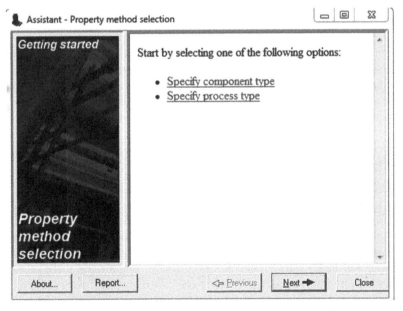

FIGURE 6.2   Property method selection in Aspen Properties.

TABLE 6.2   Equations of State Recommended for Different Processes

Oil and gas production (OIL GAS): Peng–Robinson with Boston–Mathias alpha function (PR–BM) or Redlich–Kwong–Soave with Boston–Mathias alpha function (RKS–BM).

Refinery—medium pressure (REFINERY): Peng–Robinson, Redlich–Kwong–Soave; Chao–Seader, Grayson are older ones developed specifically for refineries.

Refinery—hydrogen-rich applications (REFINERY): Peng–Robinson, Redlich–Kwong–Soave. Grayson is an older one developed specifically for refineries.

Gas processing—hydrocarbon separations (GASPROC): Peng–Robinson with Boston–Mathias alpha function (PR–BM), Redlich–Kwong–Soave with Boston–Mathias alpha function (RKS–BM), Peng–Robinson, Redlich–Kwong–Soave.

Gas processing—acid gas absorption (CHEMICAL AND ELECTROLY): Electrolyte NRTL.

Petrochemicals—aromatics and ether production (PETCHEM): Wilson, NRTL, UNIQUAC.

Chemicals—phenol plants (CHEMICAL): Wilson, NRTL, UNIQUAC.

Chemicals—ammonia plant (CHEMICAL): Peng–Robinson, Redlich–Kwong–Soave, SR–Polar (Schwartzentruber–Renon).

Chemicals—inorganic chemicals (CHEMICAL AND ELECTROLY): Electrolyte NRTL.

Coal processing—combustion (COALPROC): Peng–Robinson with Boston–Mathias alpha function (PR–BM), Redlich–Kwong–Soave with Boston–Mathias alpha function (RKS–BM), or the combustion databank.

Hydrocarbon, Special, and Refrigerants. (1) If you choose Chemical System, the first question is whether the pressure is above 10 bars. If so, you are directed to use equations of state with advanced mixing rules, such as the Wong–Sandler method described earlier. If not, the program recommends using NRTL, Wilson, UNIQUAC, or UNIFAC for liquid activity methods; several cautions are provided for specific systems that can give trouble (carboxylic acids, electrolytes, HF, etc.). (2) If you choose Hydrocarbon system, then additional choices involve whether petroleum assays and pseudocomponents are involved, and whether vacuum conditions are used. Several reasonable choices of thermodynamic models are then displayed. Even then, you should try each of those choices and compare the results to experimental data. While that may not be possible when learning to use Aspen Plus, if you and your company put money on the table, you better test the thermodynamics.

It is also worthwhile to see what methods are recommended for the type of process you are modeling; in the screen shown in Figure 6.2, choose Specific process type. A partial list of processes is shown in Table 6.2. If one chooses Chemical, a new screen gives hints for specific chemical systems as well as a summary of help commands.

Thermodynamic Models

To get general information about the equations of state and the liquid activity coefficient property methods, search on "Liquid Activity Coefficient Property Methods" or on the abbreviation of the equation of state or choose Equations of state. Table 6.3 shows some of the combinations of equations of state and liquid thermodynamic models. This list highlights the most-used equations of state. With all these choices, and limited knowledge of your system, you will likely want to use the recommended options and make predictions of vapor–liquid equilibrium using Aspen Plus to compare those predictions with experimental data. Chapter 3 presents an example of such a comparison for the ethanol–water system.

TABLE 6.3 Pairs of Equations of State and Liquid Activity Coefficient Models (Abbreviated List)

Property Method	Gamma Model	Equation of State
NRTL	NRTL	Ideal gas law
NRTL–RK	NRTL	Redlick–Kwong
UNIFAC	UNIFAC	Redlich–Kwong
UNIFAC–LL	UNIFAC	Redlich–Kwong
UNIQ-2	UNIQUAC	Ideal gas law
UNIQ–RK	UNIQUAC	Redlich–Kwong
WILS-2	Wilson	Ideal gas law
WILS–RK	Wilson	Redlich–Kwong

Property Method	Equation of State	Except for
PENG–ROB	Peng–Robinson	Liquid molar volume
RKSWS	Redlich–Kwong	UNIFAC for excess Helmholz free energy
PRWS	Peng–Robinson	Wong–Sandler mixing rules

EXAMPLE: MULTICOMPONENT DISTILLATION WITH SHORTCUT METHODS

Consider a single-distillation column as shown in Figure 6.3. Short-cut methods have been developed that allow you to see the splits that can be achieved at different pressures, with different reflux ratios, and with a different number of stages. It is often useful to do a short-cut distillation calculation first, before doing the more rigorous plate-to-plate calculations, because the preliminary calculations give you an idea of what is reasonable. The methods are based on the relative volatility of the chemicals, and require you to identify

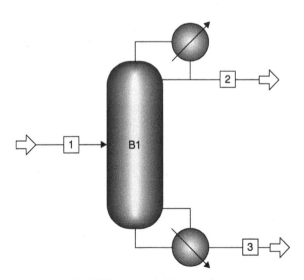

FIGURE 6.3 Distillation column.

TABLE 6.4   Identification of Light and Heavy Key

Chemical	Boiling Point (°C) at 1 atm
Propane	−42.1
i-Butane	−11.9
n-Butane	−0.5
i-Pentane	27.9
n-Pentane	36.1

two components that will be essentially separated (to the extent you specify). If you line up all the components in the order of their boiling points, and draw a line between two of them, the more volatile component is called the light key and the less volatile component is called the heavy key.

For this example, you choose the first column in the process shown in Figure 5.4. In Aspen Plus, you use the module DSTWU for the shortcut method; you also use RK–Soave as the physical property method, because it is a good one for hydrocarbons. The feed is 100 lb mol/h of propane, 300 lb mol/h of i-butane, and the other chemicals as listed in Table 5.1, at 138 psia and 75°F. The column operates at 138 psia with a reflux ratio of 10 (a wild guess initially, confirmed because the column worked). Remember, the minimum number of stages goes together with infinite reflux, so if your column does not work, increase the reflux ratio.

First, you specify the split you would like. To do this, you need the concept of a light key and heavy key. First, line the chemicals up in order of their boiling points, from low to high. (Note: this illustration assumes that the thermodynamics are ideal, so Raoult's Law applies.) Because you are using distillation, if most of the propane is to go out the top stream, a little of the i-butane will, too; very small fractions of the other species will go out the top stream. You want most of the i-butane to go out the bottom. Thus, you draw a line between propane and i-butane in Table 6.4. The component above the line is the light key and the component below the line is the heavy key. The distillation column is based on those two components, and the other components are split according to thermodynamics as described later. In this case, propane is the light key, and you want 99% of it to go out the top stream; i-butane is the heavy key, and you want 1% of it to go out the top stream.

In a shortcut method, correlations are used to relate the reflux ratio, minimum reflux ratio, number of stages, minimum number of stages and the relative volatilities. The equations are available on the book website, but here it suffices to say that the results are based on correlations but are often quite good, especially for hydrocarbons.

You can use Aspen Plus to solve this problem using the DSTWU block.

Step 1   The flowsheet is shown in Figure 6.3, and you enter the components and feed conditions in the usual manner (see other examples in Chapters 2–5 and Appendix C). Using the Home | Methods Assistant, process = refinery gives suggestions for the thermodynamic model: BK10, Chao–Seader, and PENG-ROB. The thermodynamics model chosen was the PENG-ROB. The parts specific to the distillation column are the block parameters. Shown in Figure 6.4 is the screen where you select the block parameters. You choose the light and heavy keys (propane and i-butane), the splits desired (99% and 1%), the pressure of the column (138 psia), a total condenser, and the desired number of stages (10).

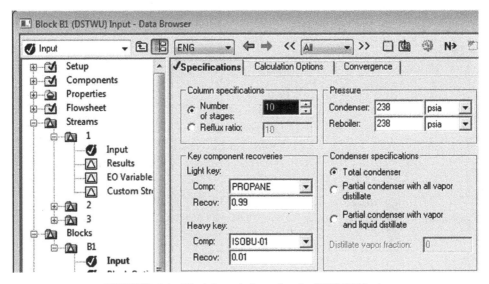

FIGURE 6.4   Block input information for DSTWU block.

Step 2 Figure 6.5 shows the solution with stream information. You can check parts of this output. Look first at the mole balances. Indeed, 99% of the propane and 1% of the *i*-butane went out the top stream. The other components are heavier, and you would expect them to be split in such a way that only a small fraction goes out through the overhead stream; this is indeed the case. You also get the bubble points of the distillate and bottom product (not shown): 82.0°F for the distillate and 200.9°F for the bottom product. These are close to the values shown in Chapter 5, which should give you confidence in the thermodynamics used in the calculations.

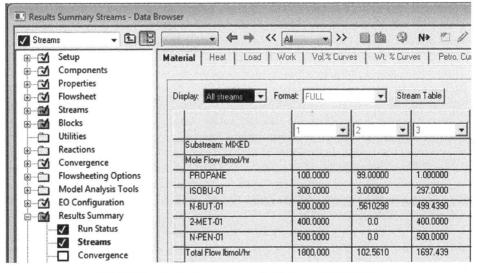

FIGURE 6.5   Multicomponent distillation with DSTWU.

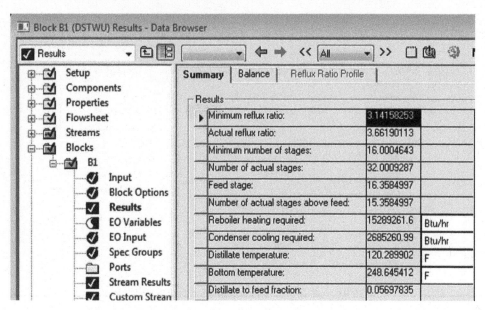

FIGURE 6.6   Block results for DSTWU block.

Step 3 You also want to know the parameters of the distillation column, which are obtained by choosing the block for the column and then clicking the Results tab, giving Figure 6.6. Note that the minimum number of stages is 16, the minimum reflux ratio is 3.14, the actual reflux ratio is 3.66, and the actual number of stages is 32. In this case, your guess of 10 stages was too small and the program modified it for you. If you had entered 32 stages, the program would use exactly 32 stages. You could have specified the reflux ratio instead, and Aspen Plus would compute the number of stages. The feed is on the 16th stage (from the top).

Note that the ratio of reflux ratio to minimum reflux ratio is 3.66/3.14 = 1.17, which is a common and economical one. However, during the energy crisis in the 1970s, engineers found that a ratio of 1.1 was more economical, although the distillation tower was harder to control. This is one of the choices (number of stages or reflux ratio) that you will make as a chemical engineer that affects the economics of the process. The cost of the column and the heating and cooling cost must be examined to find the minimum cost. In this case, the reboiler duty is 15.3×10^6 Btu/h and the condenser duty is 2.7×10^6 Btu/h. Since heating is more expensive than cooling, the major operating cost is the cost of the steam to heat the reboiler. If energy is selling for $6 per million Btu, you would have an operating cost for this column of $61 per hour, or about $540,000 per year.

Your choice of thermodynamic model does make a difference. Table 6.5 shows results from three simulations, all with exactly 26 stages, using different models. The results differ, but since Aspen Plus recommends the Chao–Seader and RKS–BM models, we choose them rather than the RK–Soave.

Difficulties can arise. If you misidentify the light and heavy key (say reversing them), you may get a negative number of stages. If you identify a light and heavy key that has another component that boils between them, that other component will be split in some way, which may not meet your desired specifications. If the thermodynamics is nonideal and the boiling points are not sufficient to guarantee the volatility (i.e., activity coefficients

TABLE 6.5   Comparison of Different Thermodynamic Models

Thermodynamic Model	RK–Soave	RKS–BM	Refinery/Chao–Seader
Minimum number of stages	10.9	16	15.6
Actual number of stages	22	32	31
Minimum reflux ratio	1.0	3.14	2.85
Actual reflux ratio	1.3	3.66	3.34

are important), then you may specify the light and heavy key correctly according to their boiling points, but the simulation may act as if they were not specified correctly.

MULTICOMPONENT DISTILLATION WITH RIGOROUS PLATE-TO-PLATE METHODS

The shortcut distillation method gives reasonable answers to straightforward problems, but distillations can have multiple feed streams or multiple product streams. There might even be azeotropes within the column. To handle complications like that, and to get better accuracy, you will want to use a rigorous plate-to-plate method that takes a vapor stream and liquid stream, equilibrates them, and sends off a vapor and liquid stream with different compositions. This is what happens on each plate, or tray, on the column. Such a method is a large computational problem, and it requires specifying the number of plates, or trays, or stages, in advance. A good starting point is the number of stages suggested by the shortcut method.

The next simulation is for the same column, but using the RadFrac block in Aspen Plus. The feed is the same, the pressure is 138 psia, and the PENG-ROB property method is used. This example uses 32 stages, and you run Aspen Plus to see what the split is. (Note that you cannot easily set the split and find the number of stages or reflux ratio needed to achieve it.) Set the reflux ratio to 3.7 and enter the feed on the 13th stage.

The output is shown in Table 6.6, and only 62% of the propane goes out the top stream; along with 5% of the *i*-butane. The condenser is at 99.5°F and has a heat duty of

TABLE 6.6   Stream Table for Multicomponent Distillation with RadFrac, Reflux Ratio = 3.7

	1	2	3
Temperature (°F)	75	99.5	174.2
Pressure (psia)	138	138	138
Vapor fraction	0	0	0
Mole flow (lb mol/h)	1800	100	1700
Mass flow (lb/h)	110232.9	4937.957	105294.9
Volume flow (cuft/h)	3051.494	157.948	3303.065
Enthalpy (MM Btu/h)	−120.833	−5.67	−108.691
Mole flow (lb mol/h)			
Propane	100	62.336	37.664
i-Butane	300	14.829	285.171
n-Butane	500	3.064	496.936
i-Pentane	400	19.771	380.229
n-Pentane	500	0	500

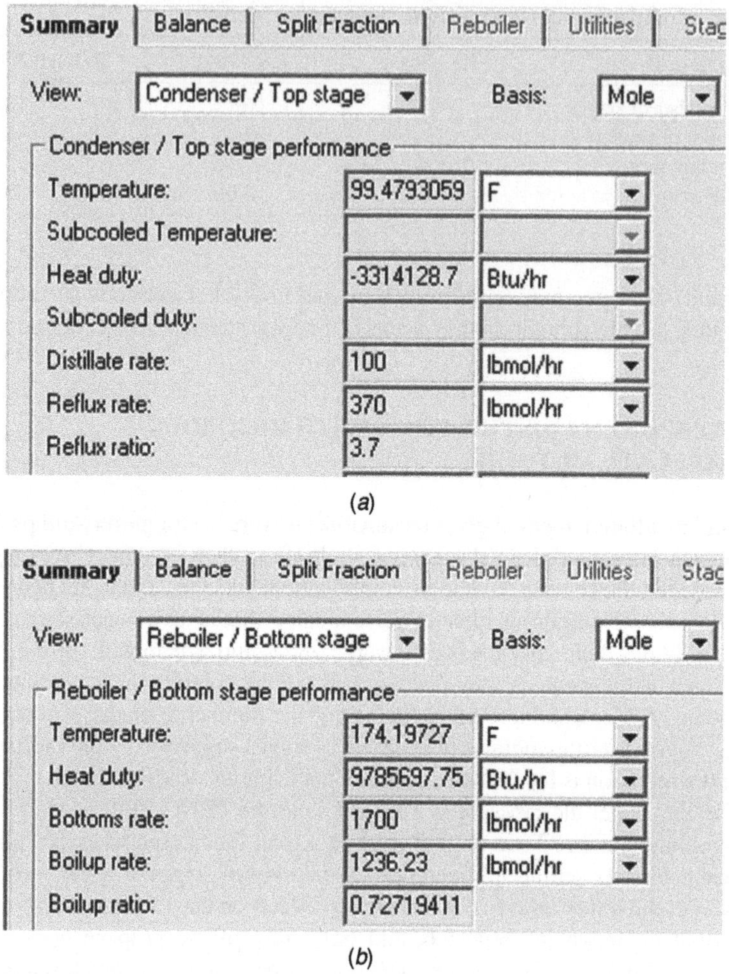

(a)

(b)

FIGURE 6.7   Multicomponent distillation with RadFrac, reflux ratio = 3.7. (a) Condenser results; (b) reboiler results.

-3.3×10^6 Btu/h, and the reboiler is at 174.2°F and has a heat duty of 9.8×10^6 Btu/h (see Figure 6.7). To see the flow rates on each stage, choose the block, then Profiles. Above the feed, the liquid flow rate varies between 330 and 470 lb mol/h, and below it, the flow rate varies between 2920 and 2940 lb mol/h. Although DSTWU assumes the same number of moles on each plate or tray, RadFrac calculates what actually occurs.

Since the desired separation is not achieved, you must run the problem again with more stages (40) and a higher reflux ratio (11). The results are in Table 6.7. The separation is slightly better than required (99.2%). This, of course, changes the temperatures of the condenser and reboiler because they are at the bubble points of the mixtures, and these change with composition. The condenser is at 77°F with a heat duty of -7.0×10^6 Btu/h, and the reboiler is at 177°F and with a heat duty of 14.3×10^6 Btu/h. Above the feed stream, the liquid flow rate varies between 920 and 1200 lb mol/h, and below it the flow rate is about 3520 lb mol/h.

TABLE 6.7   Multicomponent Distillation with RadFrac, Reflux Ratio = 11

	1	2	3
Temperature (°F)	75	77.3	177
Pressure (psia)	138	138	138
Vapor fraction	0	0	0
Mole flow (lb mol/h)	1800	100	1700
Mass flow (lb/h)	110232.9	4421.785	105811.1
Volume flow (cuft/h)	3051.494	143.763	3327.539
Enthalpy (MM Btu/h)	−120.833	−5.206	−109.003
Mole flow (lb mol/h)			
Propane	100	99.135	0.865
i-Butane	300	0.363	299.637
n-Butane	500	0.018	499.982
i-Pentane	400	0.484	399.516
n-Pentane	500	0	500

EXAMPLE: PACKED BED ABSORPTION

If a gas contains a contaminant you would like to remove, one way is to bring the gas into contact with a liquid that can absorb it; this is often done in an absorption column, as illustrated in Figure 6.8. Here, you take an air stream containing acetone at a concentration of 2%; you want to lower its concentration to 0.4% by using water in a column. In Aspen

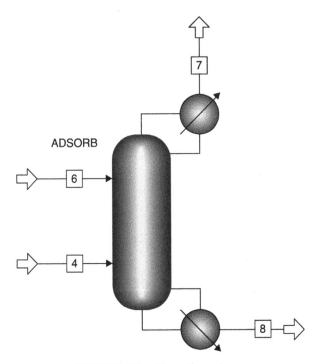

FIGURE 6.8   Absorption column.

Plus, you can use RadFrac. To get this form of the picture for RadFrac, open up the options for the RadFrac block by click on the downward pointing arrow. The only difference from distillation towers is that you will not have a condenser or reboiler, and there are two inputs, one at each end. Before making that model, you have to decide on an appropriate thermodynamic model. Figure 6.9 shows experimental data, along with the predictions of Aspen Plus using the WILS-2 thermodynamic model, which is Wilson with an ideal

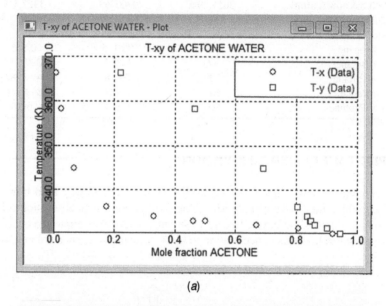

(a)

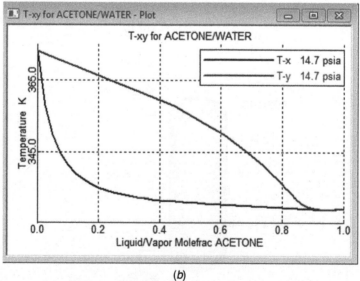

(b)

FIGURE 6.9 Vapor–liquid diagram for acetone–water: (a) experimental data at 1.01×10^5 Pa (Othmer and Morley, 1946); (b) calculations at 1 atm using WILS-2 thermodynamics.

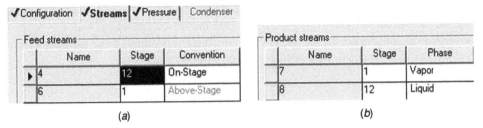

FIGURE 6.10   Absorption column parameters in Aspen Plus. (*a*) Feed streams; (*b*) product streams.

gas. This is one of the recommended models suggested by Aspen using the Properties Environment | Methods Assistant . . . for "Specify component type" and "chemical system," not high pressure. The experimental data were obtained by choosing the blue and red NIST TDE button and looking at *Txy* data. The one plotted is VLE036 due to Othmer and Morley (1946), and reasonable agreement exists for this very nonlinear system.

Use a 12-stage column and have the air stream enter on stage 12 and the water stream enters on stage 1, as shown in Figure 6.10. (The stages are numbered from the top.) The feed streams are both taken as 80°F and 14.7 psia. Table 6.8 shows the feed rates of the air and water stream.

Acetone has been transferred from the air stream to the water stream, and the mole fraction of the air stream leaving is 0.406%. Note also that the air, which was dry entering, is now wet, with water vapor forming about 3.5% of the gas stream. A small amount of air is dissolved in the water, too. In a complete system, of course, the water effluent is taken to another column where the acetone is removed and the water is recycled. Then the water stream is saturated with air and contains some acetone as well.

Absorber/stripper combinations are used to remove a chemical from a stream. For example, carbon dioxide is often absorbed into monoethanolamine (MEA); the steam then is sent to a stripping column at a different pressure and the carbon dioxide is removed out the top. The MEA is sent back to the absorber to be used again. Such units are ubiquitous in refineries, but they are difficult to simulate because of the recycle and the fact that the

TABLE 6.8   Acetone Removal Stream Table

Substream: Mixed	4	6	7	8
Mole flow (lb mol/h)				
Acetone	3.527396	0	0.7288636	2.798533
Water	0	176.3698	6.117179	170.2526
Air	172.8424	0	172.596	0.2463896
Total flow (lb mol/h)	176.3698	176.3698	179.4421	173.2976
Total flow (lb/h)	5208.816	3177.352	5149.347	3236.821
Total flow (cuft/h)	69485.12	51.28876	70657.21	52.37885
Temperature (°F)	80	80	79.70766	59.21119
Pressure (psia)	14.7	14.7	14.7	14.7
Vapor fraction	1	0	1	0
Liquid fraction	0	1	0	3

absorber itself may not converge. Examples are given in Aspen Examples (see Appendix C), and most of the absorbers using amines involve rate-based versions of RadFrac. A good strategy is to not connect the recycle stream, but adjust conditions so that the output from the stripper is about equal to the absorber amine input stream.

EXAMPLE: GAS PLANT PRODUCT SEPARATION

Consider the gas plant product separation process described in Figure 5.4. The pressures in the units have been chosen so the overhead streams can be cooled with cooling water at 80°F. Thus, the designer looked at the vapor pressure of each component before fixing the pressures. The temperatures of the reboilers are basically at the bubble point of the bottom stream, and this will be an output of the computer program.

Use Aspen Plus and model each distillation column using the DSTWU model as shown in Figure 6.11. First, specify the split of key components: for the light component you want 99% out the top and for the heavy component you want 1% out the top. The other components will be split according to the shortcut method [The Fenske equation, Eq. (6.3) on the book website.]

Choose the thermodynamic property method as Refinery/Chao–Seader. Choose the pressures of the columns to agree with the gas plant in Figure 5.4. Each column has 26 stages, and the temperatures shown in Table 6.9 are the result of thermodynamics predicted with Aspen Plus. These are very close to those listed in Figure 5.4.

Aspen Plus also gives information such as the number of stages, minimum reflux ratio, and heat duties in the reboiler and condenser. With the information shown engineers can calculate the capital cost of the equipment (sometimes using rules of thumb to account for instrumentation, pumps, valves, etc.). You can also calculate at least some of the operating cost, the raw material cost and product value, and the cooling and heating cost. For distillation towers at normal temperatures, the heating cost per unit of energy is about 10 times the cooling cost when using cooling water. If refrigeration is used to cool the condensers, that energy cost is even more expensive that the cost of steam. Experienced designers know that, and that is why the pressures are chosen so that cooling water can be used in the condensers. The conditions of the columns are given in Table 6.10.

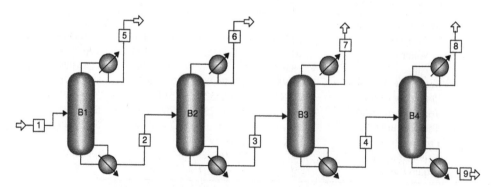

FIGURE 6.11   Gas plant separation in a distillation train.

TABLE 6.9 Results from Gas Plant Separation

	1	2	3	4	5	6	7	8	9
Temperature (°F)	75	200.9	138.7	144.6	82	79.9	80.7	99.9	116
Pressure (psia)	138	138	51	36	138	51	36	20	20
Vapor fraction	0	0	0	0	0	1	0	0	0
Mole flow (lb mol/h)	1800	1697.76	1397.73	900.594	102.24	300.03	497.136	401.985	498.609
Mass flow (lb/h)	115843.62	111289.75	93865.01	64908.713	4553.877	17424.737	28956.298	28933.903	35974.809
Volume flow (cuft/h)	3126.01	3495.803	2655.601	1796.606	148.734	508.141	811.661	768.537	961.268
Enthalpy (MM Btu/h)	−125.049	−111.116	−96.233	−65.507	−5.347	−19.913	−31.512	−30.483	−36.397
Mole flow (lb mol/h)									
Propane	100	1	0	0	99	1	0	0	0
i-Butane	300	297	2.97	0.001	3	294.03	2.969	0.001	0
n-Butane	500	499.76	494.762	4.948	0.24	4.998	489.815	4.948	0
i-Pentane	400	400	400	396	0	0	4	392.04	3.96
n-Pentane	500	500	499.998	499.646	0	0.002	0.352	4.996	494.649

TABLE 6.10   Conditions in Distillation Columns in Gas Plant Separation

Column	B1	B2	B3	B4
Minimum reflux ratio	2.93	9.5	1.9	7.9
Actual reflux ratio	3.43	10.8	2.2	8.9
Minimum number of stages	12.9	31.1	11.3	36.8
Number of actual stages	26	62.3	26	73.5
Feed stage	13.7	33.4	13.6	39.2
Reboiler heating required	11.5×10^6	23.8×10^6	13.4×10^6	40.0×10^6
Condenser cooling required	2.9×10^6	28.8×10^6	14.2×10^6	41.3×10^6

EXAMPLE: WATER GAS SHIFT EQUILIBRIUM REACTOR WITH SENSITIVITY BLOCK AND DESIGN SPECIFICATION BLOCK

Next, we consider an equilibrium reactor and let the temperature change. We have solved for an adiabatic reactor (Table 4.4), so next solve for one with heating, then do a sensitivity calculation, and finally solve it with a design specification. In Chapter 8, the same problem is solved using rate equations.

1. *Reactor with Heating:* Open the same file used to produce Table 4.4. Change the inlet temperature to 25°C; in the reactor setup, set the temperature to 226.85°C; see Figure 6.12. The solution gives a mole fraction 0.079 for CO and H_2O, as was found for Table 4.4. But if we look at the Reactor Results we find we need 0.00499 Gcal/h, or 0.0209 GJ/h. This much energy must be supplied to keep the outlet temperature at 226.85°C.

2. *Sensitivity Block:* Change the reactor setup to use heat duty instead of temperature in its specifications. Although the heat duty will be changed in the sensitivity calculation, it is a good idea to give it a prospective value, here 0.001 Gcal/h, as shown in Figure 6.13. Then open Model Analysis Tools | Sensitivity and define a new analysis called HEAT. Open the Input under that and define two flowsheet variables, FRACTION and TEMP as shown in Figure 6.14. These are defined by putting in the name

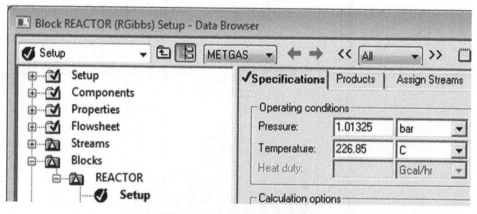

FIGURE 6.12   Reactor with heating.

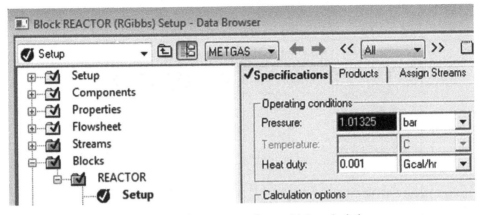

FIGURE 6.13 Reactor setup for sensitivity calculations.

and double clicking it. A variable definition window comes up; for FRACTION, the window is shown in Figure 6.15. This is the mole fraction of CO in the exit stream, S2. TEMP is the temperature of stream S2, which is desired for the sensitivity table. Next, click on the Vary tab and fill in the window as shown in Figure 6.16. By placing your cursor on each arrow and looking down the selections, you can obtain what is shown there. The calculation will be done using heat duties from 0.001 to 0.008 Gcal/h. Finally, click on the Tabulate tab and choose FRACTION for column 1 and TEMP for column 2 of the Table. The parameter being varied will also be listed in the table. The table is shown in Figure 6.17; it can also be copied and pasted into Excel. Now we have a solution for a variety of outlet temperatures along with the heat duty needed to come to that temperature. This takes into account both the heat capacity of the various substances as well as the heat of reaction

3. *Design Specifications:* The next simulation shows how to set a design specification. In this case, we will choose the desired mole fraction of CO in the output stream

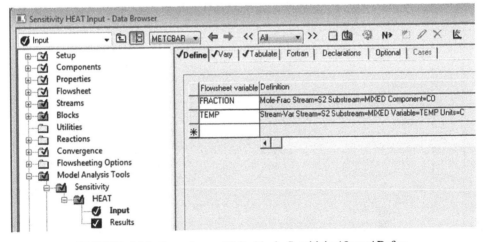

FIGURE 6.14 Input for sensitivity block; Sensitivity | Input | Define.

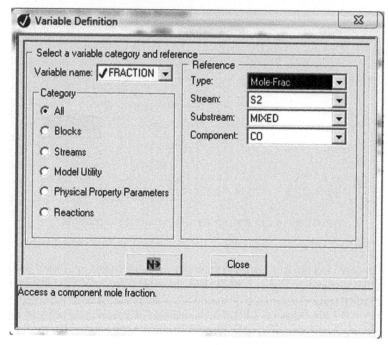

FIGURE 6.15   Variable definition for sensitivity block (click on variable name in Sensitivity | Input | Define).

S2 and let the program figure out what the heating duty is for that outcome. This is done using the Flowsheeting Options/Design Spec option. Click on Design Spec and create a new one, called DS-1. Define the Input | Vary as shown in Figure 6.18 in the same way used for the sensitivity analysis (b)in the preceding text. The specifications are set in the Spec tab, as shown in Figure 6.19: FRACTION is to be set to 0.07, with a tolerance of 0.001. The parameter to be varied is the same as in the sensitivity

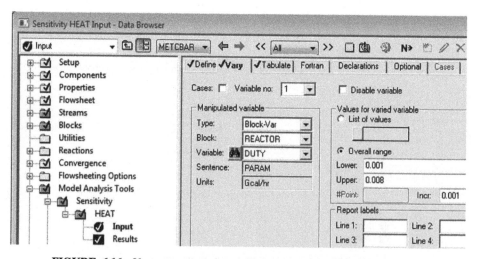

FIGURE 6.16   Vary a parameter in sensitivity block; Sensitivity | Input | Vary.

Summary | Define Variable |

Row/Case	Status	VARY 1 REACTOR PARAM DUTY GCAL/HR	FRACTION	TEMP C
1	OK	0.001	0.00190519	32.5895068
2	OK	0.002	0.00656869	88.5342668
3	OK	0.003	0.01472105	138.682297
4	OK	0.004	0.02590863	184.42408
5	OK	0.005	0.03938760	227.074904
6	OK	0.006	0.05445823	267.678202
7	OK	0.007	0.07053977	307.000736
8	OK	0.008	0.08717223	345.613885

FIGURE 6.17   Table from sensitivity calculations; Sensitivity I HEAT I Results I Summary.

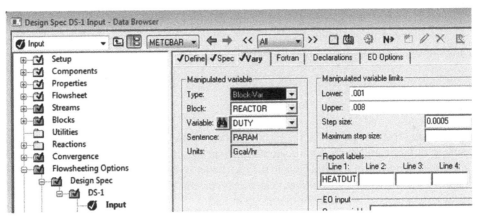

FIGURE 6.18   Design specification input.

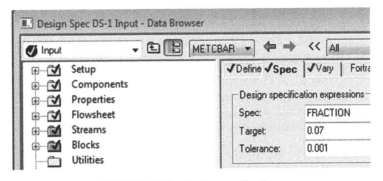

FIGURE 6.19   Design specifications.

Variable	Initial value	Final value	Units
MANIPULATED	0.001	0.00696608	GCAL/HR
HEATRATE	0.001	0.00696608	GCAL/HR
FRACTION	0.00190519	0.06998306	

FIGURE 6.20 Results using the design specification block.

example, namely the heat duty. The manipulated variable limits are from 0.001 to 0.008 with a step size of 0.0005. The results under Design Spec are shown in Figure 6.20. The stream S2 has a mole fraction of 0.070 and a temperature of 305.7, using heat duty of 0.006966 Gcal/h.

CHAPTER SUMMARY

This chapter introduces you to the many thermodynamic models available in Aspen Plus. Aspen Plus was used to solve a variety of distillation problems, with either short-cut methods (DSTWU) or plate-to-plate methods (RadFrac). You also learned how to solve gas absorption problems using Aspen Plus. Finally, you learned how to use the sensitivity option and design spec option in Aspen Plus.

CLASS EXERCISE

Divide into groups of two to four students. Work through each of the examples, with each person on the team taking a unit, showing where the parameters are for that block and checking the results. (1) Tables 6.6 and 6.7; (2) Table 6.8 and Figure 6.8; (3) Tables 6.9 and Figure 6.11. Alternatively, your instructor can prepare a flowsheet and dataset showing the results, but put in some errors for our group to find.

PROBLEMS (Using Aspen Plus)

6.1₁ Model column B2 in Figure 6.11 using a shortcut, DSTWU block with the conditions of the inlet stream as shown in Table 6.9, stream 2, but with a pressure of 51 psia.

6.2₁ Model column B2 in Figure 6.11 using a detailed model, RadFrac. (Hint: use the results from problem 6.1 to choose key operating parameters.) Prepare a report comparing the output when using DSTWU and RadFrac to model the distillation column. What information was needed for each? How do the results compare?

6.3₂ (a) Model column B3 in Figure 6.11 using a shortcut, DSTWU block with the conditions of the inlet stream as shown in Table 6.9, stream 3, but with a pressure of 36 psia. (b) Then model the same column using RadFrac. How do the results compare?

6.4₂ (a) Model column B4 in Figure 6.11 using a shortcut, DSTWU block with the conditions of the inlet stream as shown in Table 6.9, stream 4, but with a pressure of 20 psia. (b) Then model the same column using RadFrac. How do the results compare?

6.5₂ Model Example 3 in Perry's Chemical Engineering Handbook (Perry and Green, 2008, pp. 13–35, Figures 13–37) using (a) DSTWU and (b) RadFrac.

6.6₁ Results for the gas plant are given in Tables 6.9 and 6.10. Which separations are the most expensive? If you were the designer, where would you want to spend your time?

6.7₁ Carbon dioxide from a fermentation process contains one mole percent ethyl alcohol. The alcohol needs to be removed by contact with water at 35°C and 1 atm. The gas flow rate is 400 lb mol/h and the water stream is 620 lb mol/h and contains 0.02 mol% alcohol. Determine the compositions out of the absorption column if you model it with 10 stages using RADFRAC.

6.8₁ Natural gas contains mostly methane but also small amounts of other chemicals that are valuable in themselves. In a refrigerated absorption process, the natural gas is cooled to −40°F at 865 psia and sent to an absorber, which uses dodecane as the absorbing media, with a flow rate of 10,000 lb mol/h. Determine the fraction of each of the chemicals absorbed in a tower with 30 stages with the input shown in Table 6.11.

TABLE 6.11   Input for Problem 6.8

Component	Flow Rate (lb mol/h)
Methane	700,000
Ethane	27,000
Propane	11,000
i-Butane	3,200
n-Butane	2,800
i-Pentane	1,200
n-Pentane	770
Hexane	820

6.9₂ Figure 5.15 shows an ethanol process. The last unit is a distillation tower to remove water from a mixture of water and ethanol. Note that the mixture forms an azeotrope. The feed stream is 100 lb mol/h of a 50–50 molar mixture of water and ethanol at 80°F and 1 atm.

  1. Simulate a distillation tower to create the azeotropic mixture (93% ethanol). Use the WILS-2 option for the thermodynamic model (see Figures 3.5 and 3.6 for the data). Model the column using the DSTWU short-cut model with recovery fractions of 0.85 and 0.15 for ethanol and water, respectively. Start with 10 stages and let DSTWU figure out how many stages are needed.

   2. Use RadFrac with the number of stages and reflux rate set by the output from the DSTWU model.

6.10₂ An air stream with flow rate of 14,515 kmol/h and composition of 0.781 nitrogen, 0.210 oxygen, and 0.009 argon is to be distilled at low temperature and high pressure. Take the input to the distillation column at −138.1°C and 32 bar.

   1. Experimental data (T–xy diagram) at −150.5°C can be obtained using the NIST TDE system. Verify that the PENG-ROB equation of state does a good job of representing this data.

   2. Using Aspen Plus, simulate a distillation column using the shortcut module, DSTWU. For thermodynamics, use the Peng–Robinson option, as suggested by Aspen. Use 35 stages, nitrogen as the light key, recovery (out top) of 0.85, oxygen as the heavy key, recovery (out top) of 0.05. What is the minimum reflux ratio? What is the actual reflux ratio? What is the minimum number of stages? All these quantities are useful when using a more complicated distillation model. Since this process is at a very low temperature, the most expensive part of the process is the refrigeration, that is, the heat duty of the condenser, so note that.

   3. Form a group and have each person examine the effect of changing one of the operational choices: number of stages, reflux ratio, fraction of light key and heavy key, and thermodynamic choice, such as NRTL–RK, and report the results. These guidelines can then be used when simulating the entire gasification plant.

6.11₂ Solve the same Problem 6.10 using the RADFRAC module in Aspen Plus. Use 50 stages, feed on stage 30, distillate rate of 10,000 kmol/h, and a reflux ratio of 30. Sometimes it helps to get a solution by starting out with a higher reflux ratio, or more stages, and then changing them to the ones you want. How does the separation compare with the shortcut method in Problem 6.10? Is one column sufficient to obtain a relatively pure oxygen stream? What could you do to improve the purity of the oxygen stream? Remember that the cost of refrigeration is a significant cost in this process, and that goes up with reflux ratio.

7

PROCESS SIMULATION

Process simulation is used to determine the size of equipment in a chemical plant, the amount of energy needed, the overall yield, and the magnitude of the waste streams. Because the results of process simulation depend upon thermodynamics and transport processes, the mathematical models are complicated and would be time-consuming to solve without a computer. This chapter illustrates the use of a process simulator, Aspen Plus, to model a plant to make ammonia.

The problems solved in Chapters 5 and 6 are simple problems with many numerical parameters specified. You may have wondered where those numbers came from. In a real case, of course, you will have to make design choices and discover their impact. In chemical engineering, as in real life, these choices have consequences. Thus, you must make mass and energy balances that take into account the thermodynamics of chemical reaction equilibria and vapor–liquid equilibria as well as heat transfer, mass transfer, and fluid flow. To do this properly requires lots of data, and the process simulators provide excellent databases. Chapters 2–4 discussed some of the ways thermodynamic properties are calculated. This chapter uses Aspen Plus exclusively. You will have to make choices of thermodynamic models and operating parameters, but this will help you learn the field of chemical engineering. When you complete this chapter, you may not be a certified expert in using Aspen Plus, but you will be capable of actually simulating a process that could make money.

First, the chapter lists the possible unit operations in the Aspen Plus Model Library, because the process is a connected set of the units. Then an example process is illustrated that makes ammonia from nitrogen and hydrogen. You will be able to get both the mass balances and the energy balances for the process. With this information you can figure the size of most of the equipment needed, and hence its cost. You can also figure the operating cost for heating, cooling, compression, and other tasks. The process involves a recycle stream, too, which means you will need to iterate. The chapter ends with a description of

Introduction to Chemical Engineering Computing, Updated Second Edition. Bruce A. Finlayson.

factors that may control whether you get a solution at all, such as convergence acceleration techniques and tear streams.

Instructional Objectives: After working through this chapter, you will have

1. Become familiar with models of different types of equipment in Aspen Plus.
2. Learned to draw a flow sheet, converting physical units into one or more unit modules.
3. Learned which parameters must be specified and which ones are solved for.
4. Understood the vagaries of getting the whole simulation to converge, with large recycle streams and multiple interacting recycle streams.
5. Learned to test the results to verify convergence.
6. Learned to figure costs and optimize the process and control the process to meet specifications.
7. Learned how to add simple user-defined programs in FORTRAN to unit models (in Appendix C).

MODEL LIBRARY

When engineers design a process, they choose different modules to represent the different units. You can view these by clicking on the icons at the bottom of the Aspen Plus screen. If the icons are not displayed, choose the View | Model Palette menu option. When you click on a unit, the words shown at the bottom of the screen give a brief description of the module. Some of the module choices are described in Table 7.1. In Aspen Plus, click on the downward pointing arrow to see different configurations of each module.

Your task is to combine these units with connecting streams into a process, which can then be simulated on the computer.

EXAMPLE: AMMONIA PROCESS

Consider the ammonia process, as illustrated in Figure 7.1. The feed to the process is nitrogen, hydrogen, and a small amount of carbon dioxide (left over from the process to make the hydrogen). The process feed is mixed with a recycle stream, heated to the reactor temperature, and sent to the reactor. The feed to the reactor is roughly a 3:1 mixture of hydrogen and nitrogen, with some ammonia, too. The reaction is limited by equilibrium considerations, so the conversion is not known. You must determine it by solving the equilibrium equation (using RGibbs). In this example, the pressure is high because that favors the reaction. The temperature is also high in order to get a fast reaction, even though this limits the conversion in the reactor. The output from the reactor is cooled and sent to a vapor–liquid separator (this example uses Flash2), where the temperature is lowered. Thus, most of the ammonia condenses and is removed as a liquid. The vapor is recycled and is sent through a compressor to get the pressure back to the desired pressure. A small part of the recycle stream is bled off as a purge stream to prevent the buildup of impurities. To model the ammonia process, we will (1) develop the model (your engineering judgment is necessary), (2) solve it (not a trivial step), (3) examine the results, (4) test the thermodynamic model (which insures your solution is real), (5) introduce utility costs (and raw material costs if desired) to deduce operating expenses, and (6) save the results.

TABLE 7.1 Selected Major Blocks Available in Aspen Plus

1. Mixers/splitters
   a. Mixer—combine several streams into one stream
   b. FSplit—split one stream into two or more streams based on split fraction, flow, flow of a component, etc.
   c. SSplit—divides a stream into substreams
2. Separators
   a. Flash2—rigorous vapor–liquid split or vapor–liquid–liquid split
   b. Flash3—rigorous vapor–liquid–liquid split with three outlet streams
   c. Decanter—separate two liquid phases
   d. Sep—use split fractions of every component
   e. Sep2—separation based on flows and purities
3. Heat exchangers
   a. Heater—heaters, coolers, condensers, fired heaters
   b. HeatX—co- and counter-current heat exchangers
   c. MHeatX—multistream heat exchanger, especially used in LNG or as a cold box
   d. HXFlux—heat transfer calculation model, including furnace
4. Columns
   a. DSTWU—shortcut distillation using Winn–Underwood–Gilliland equations and correlations
   b. Distl—Edmister shortcut distillation
   c. RadFrac—rigorous two phase and three phase, absorber, stripper, distillation columns using stages
   d. Extract—liquid–liquid extraction
   e. MultiFrac, SCFrac, PetroFRAC, BatchSEP—specialized units for specific applications
5. Reactors
   a. RStoic—stoichiometric fractional conversion or extent of reaction
   b. RYield—nonstoichiometric—based on yield distribution
   c. REquil—rigorous equilibrium using Gibbs free energy and user-specified reactions
   d. RGibbs—rigorous equilibrium using Gibbs free energy minimization with determination of all possible reactions, including multiphase ones
   e. RCSTR—continuous stirred-tank reactor, specify volume
   f. RPlug—plug flow reactor, specify length and diameter of tube
   g. RBatch—batch reactor, specify cycle times
   h. The RCSTR, Rplug, and RBatch reactors require choosing a reaction set and building up a reaction rate expression.
6. Pressure Changes
   a. Pump
   b. Compr—compressor
   c. MCompr—multistage compressor
   d. Valve
   e. Pipe—with constant diameter and rise
   f. Pipeline—with multiple segments of pipe
7. Manipulators—including DesignSpec Block and Calculator Block
8. Solids, multiple choices—Crystall, Crusher, Cyclone, Filter, CCD (washer), etc.
9. User models—you program these in FORTRAN to take input temperature, pressure, and flow rate of each component and calculate conditions of outlet streams
10. Conceptual design—interface to Aspen Distillation Synthesis (in Columns in version 8.0)

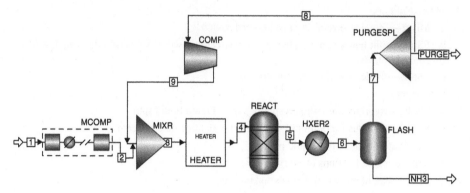

FIGURE 7.1   Ammonia process.

Development of the Model

In this example, you will take an input stream to the process (at 80°F and 300 psia) of

Nitrogen	100 lb mol/h
Hydrogen	300 lb mol/h
Ammonia	0
Carbon dioxide	1 lb mol/h

The inlet stream is compressed to 4000 psi with a multistage isentropic compressor. The stream is mixed with the recycle stream and heated to 900°F, the reactor temperature. In the reactor, there is a pressure drop of 30 psi, and the temperature and component flow rates are set by minimizing the Gibbs free energy under conditions of no heat duty. The outlet is cooled to 80°F and the liquid–vapor phases are separated. The vapor phase goes to recycle, and 0.01% of it is used as purge. A recycle compressor then compresses the rest from 3970 to 4000 psia. In a real process, the heat transfer to preheat the feed to the reactor uses the effluent from the reactor, usually inside the same vessel. In process simulators, though, it is useful to begin as shown in Figure 7.1 to help convergence. Nonrandom two liquid (NRTL) thermodynamics was chosen and is justified a posteriori.

The identification of components is shown in Figure 7.2.

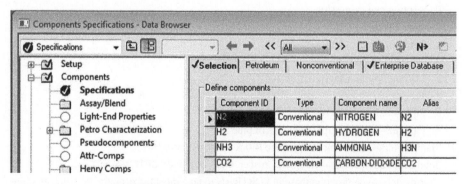

FIGURE 7.2   Components for ammonia process.

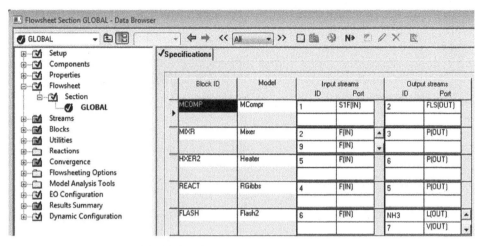

FIGURE 7.3   Partial flowsheet summary.

It is useful to check the connections between the units by using the Flowsheet option | Sections | GLOBAL, and open all the "+" to get Figure 7.3. Any streams that were inadvertently not connected to a unit will show up here because the Section box will be red.

The operating parameters for the MCOMP compressor is shown in Figure 7.4. Both compressors have a discharge pressure of 4000 psia, and the recycle compressor only has one stage. The operating parameters for the fired heater are 900°F and 4000 psia (see Blocks | HEATER | Input). The data for the heat exchanger is similar, with pressure of 3370 psia. You can see the reactor specifications in Figure 7.5.

The specifications for the vapor–liquid separator and simple splitter are shown in Figure 7.6. In Aspen Plus, you can specify different temperature and pressure in the outlet than in the inlet. This may be useful in a first simulation when you are not as interested in the actual heating and cooling required, but this example keeps the unit conditions the same as the stream entering the unit so no heat transfer or compression/expansion occurs. The outlet conditions for the MIXR block are 900°F and 4000 psia.

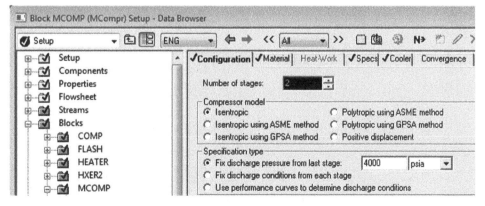

FIGURE 7.4   Specifications for compressor unit MCOMP.

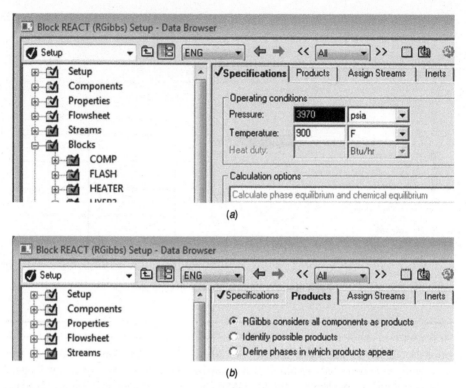

(a)

(b)

FIGURE 7.5 Specifications for the reactor, unit REACT: (*a*) operating conditions of reactor; (*b*) products of the reaction.

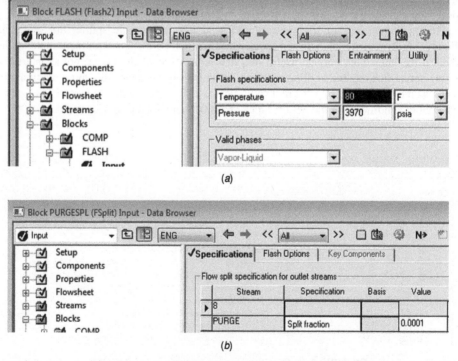

(a)

(b)

FIGURE 7.6 Specifications for the separator and simple splitter, units FLASH, and PURGESPL: (*a*) FLASH specifications; (*b*) PURGESPL specifications.

Solution of the Model

Now that the process is fully specified, you can solve it. Because there is a recycle stream, begin the calculations with the first unit, the compressor. The output from that unit is combined with the recycle stream, which initially is zero. Follow the stream through the rest of the process until it computes the recycle stream. This recycle stream is now different from its previous value. Thus, the program does another iteration, and it continues until there is no further change. Figure 7.7 shows some of the information Aspen Plus provides about the convergence.

Figure 7.7a indicates that the sequential module is used for iteration. Unit MCOMP is calculated only once, since the exact input to that unit is known. Then units HEATER through MIXR are calculated in turn. Thus, convergence is achieved when the input to the MIXR unit agrees with the output from it. If they are the same, convergence has been achieved. If not, more iterations are necessary. This simulation used the Wegstein method of accelerating convergence. Figure 7.7a indicates that the iterations have not fully converged after 30 iterations, and the mass balance error is in Block MIXR. It gives the total discrepancy as 1.3029 in vs. 1.3033 out, for a relative difference of 0.03 percent. This difference is so small that you could accept the results. After the 30 Wegstein iterations the flow rates for N_2, H_2, NH_3, and CO_2 coming into (out of) the MIXR were 269.499 (269.599), 838.580 (838.514), 56.562 (56.582), and 3.122 (3.123) lb mol/h, respectively. If you change to the Direct method (Convergence/Conv Options), it converges about the same. After the Wegstein method, if you change to the Newton method there is a very slight improvement (0.023%), after 147 passes. See these by looking at the MIXR Block and Streams (Figure 7.7b).

The final mass and energy balance are shown in the stream table, Table 7.2. By checking the mass balances around each unit, you can see that they balance. The overall conversion (100 moles of nitrogen to 197.8 moles of ammonia) is reasonable. The rest of the raw materials are lost. Most of the carbon dioxide goes out in the ammonia so perhaps the purge stream was not necessary.

Examination of Results

You can learn several things by looking at "Results" under each "Block." First, look at Figure 7.8, which is for the MCOMP unit, Results I Profile. Two stages were specified in MCOMP because the compression ratio was big: 4000/300 = 13.3. For ideal conditions, you can use thermodynamics to show that the optimal configuration has equal compression ratios with interstage cooling back to the inlet temperature, because this results in less total horsepower. The square root of 13.3 is 3.65, which is the compression ratio used. The temperature out of the first stage was set at 300°F since the stream eventually needs to be heated before going into the reactor, and the pressure after one stage is 1095 psia.

In the Results menu of the fired heater and heat exchanger, the heat duties of each exchanger are given, which permits estimation of the capital cost. The heat duties can be combined with the enthalpies of the process streams to verify an energy balance. Notice also that the ratio of hydrogen to nitrogen in the recycle stream is not 3:1. This is because thermodynamics dictates that the solubility of hydrogen and nitrogen in the ammonia is different, and hydrogen builds up. In the reactor, nitrogen is the limiting reagent, and 73% of the nitrogen fed to the reactor reacts. Thus, the conversion per pass is 73%, based on nitrogen. If you look at the product, though, 197.8 lb mol of ammonia are made from 100 lb mol

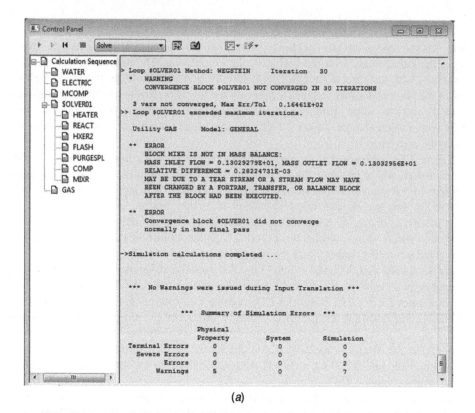

(a)

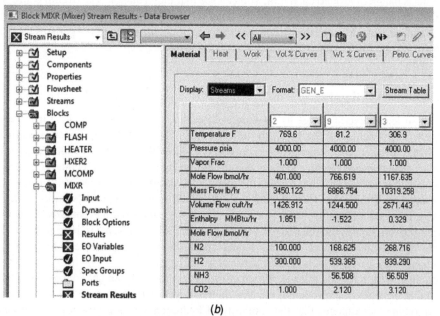

(b)

FIGURE 7.7   Convergence of the mass balance after 30 iterations with Wegstein method and then the Newton method: (a) information about iterations; (b) flow rates at beginning of one iteration (stream 3) and at the end of that iteration (streams 2 and 9).

TABLE 7.2 Mass and Energy Balance for Ammonia Process

	1	2	3	4	5	6	7	8	9	NH$_3$	Purge
Temperature (°F)	80	769.6	306.9	900	900	80	80	80	81.2	80	80
Pressure (psia)	300	4000	4000	4000	3970	3970	3970	3970	4000	3970	3970
Vapor fraction	1	1	1	1	1	0.791	1	1	1	0	1
Mole flow (lb mol/h)	401	401	1167.635	1167.635	969.797	969.797	766.696	766.619	766.619	203.101	0.077
Mass flow (lb/h)	3450.122	3450.122	10319.26	10319.25	10319.25	10319.25	6867.441	6866.754	6866.754	3451.806	0.687
Volume flow (cuft/h)	7809.907	1426.912	2671.443	4568.212	3812.826	1345.256	1250.078	1249.953	1244.5	95.178	0.125
Enthalpy (MM Btu/h)	−0.162	1.851	0.329	5.474	0.813	−7.351	−1.529	−1.529	−1.522	−5.823	0
Mole flow (lb mol/h)											
N$_2$	100	100	268.716	268.715	169.796	169.796	168.642	168.625	168.625	1.154	0.017
H$_2$	300	300	839.29	839.291	542.534	542.534	539.419	539.365	539.365	3.115	0.054
NH$_3$	0	0	56.509	56.509	254.346	254.346	56.514	56.508	56.508	197.832	0.006
CO$_2$	1	1	3.12	3.12	3.12	3.12	2.12	2.12	2.12	1	0
Mole fraction											
N$_2$	0.249	0.249	0.23	0.23	0.175	0.175	0.22	0.22	0.22	0.006	0.22
H$_2$	0.748	0.748	0.719	0.719	0.559	0.559	0.704	0.704	0.704	0.015	0.704
NH$_3$	0	0	0.048	0.048	0.262	0.262	0.074	0.074	0.074	0.974	0.074
CO$_2$	0.002	0.002	0.003	0.003	0.003	0.003	0.003	0.003	0.003	0.005	0.003

	Summary	Balance	**Profile**	Coolers	Stage Curves	Wheel Curves	Utilities

Compressor profile

	Stage	Temperature	Pressure	Pressure ratio	Indicated power	Brake horsepower	Head developed	Volumetric flow
		F ▼	psia ▼		hp ▼	hp ▼	ft-lbf/lb ▼	cuft/hr ▼
▶	1	415.652892	1095.44512	3.65148372	374.748956	374.748956	154847.315	7809.90694
	2	769.585318	4000	3.65148372	545.795492	545.795492	225524.222	3077.22587

FIGURE 7.8   Compressor specifications, unit MCOMP.

of nitrogen and 300 lb mol of hydrogen. Because the raw materials can be converted into a maximum of 200 lb mol, the overall yield of the process is 197.8/200 or 98.9%.

Even more information can be obtained using the Home | Summary | Report menu. This gives details about individual blocks. The Home | Summary | History menu gives information about the convergence and what calculations have been done. The Home | Summary | Input lists the parameters you have set.

Testing the Thermodynamic Model

The thermodynamics can be tested using the results from Table 7.2. The reactor effluent is in equilibrium. Take the mole fractions of stream 5 and compute the K-value:

$$K_p = \frac{y_{NH_3}}{(y_{N_2} y_{H_2}^3)^{1/2}} \frac{1}{p} \tag{7.1}$$

With values of $y_{N_2} = 0.175$, $y_{H_2} = 0.559$, $y_{NH_3} = 0.262$, $p = 270$ atm, the K-value is 0.0056. Stephenson (1966) gives a measured value of 0.006 at these conditions. Thus, the chemical reaction equilibrium is satisfactory when using the PENG-ROB thermodynamic model.

Another test of the thermodynamics is whether the solubility of hydrogen and nitrogen in the liquid ammonia is correct. Looking at the exit to the flash unit, the amount of hydrogen is 3.115 lb mol for 197.8 lb mol ammonia. Solubilities are usually expressed in cc (at STP) per gram of liquid. In this case, those numbers are 20.7 and 7.7 for hydrogen and nitrogen, respectively. Data by Wiebe and Tremeame (1933, 1934) give values at 270 atm (interpolated between 200 and 400 atm) of 26.4 and 27.6. Thus, the hydrogen is close, but the nitrogen is off. This is a very difficult system to simulate; many other equations of state did not lead to a convergent result. Vapor–liquid equilibria is less useful in this case since the amount of hydrogen and nitrogen in the ammonia is so small.

Utility Costs

The cost of raw materials is usually the major operating cost, but following that in importance is the cost of utilities. Aspen Plus provides a convenient way to obtain the cost of utilities in the entire process. You begin by opening the Utilities folder: Home | Summary | Utilities. There are several possible forms of utilities; choose New, select water and fill out the form: energy price of $0.6/MMBtu, Heating/Cooling Value of 45 Btu/lb, Inlet and Outlet Temperatures of 75 and 120°F. Since water is being used as a coolant, it is also necessary to select the Consistency Check/Ignore. (Otherwise errors will be indicated, although there are not any errors.) Then choose gas and fill out the form: $4/MMBtu. Choose electricity and assign a price of 5 cents/kwh.

The next step is to tell the computer which blocks are going to use utilities. Select Block HXER (the cooler after the reactor), and select WATER. For the Block COMP select ELECTRICITY. For the Block MCOMP, select Setup and then the Specs Tab. At the bottom you can specify the utilities as ELECTRICITY. Since a counter current heat exchanger is used, no gas is used to heat the feed to the reactor. Now the utilities have been specified to Aspen Plus.

After the mass and energy balances are computed, the program calculates how much cooling water and natural gas is needed, and its cost. See Results | Summary | Operating Costs in the Navigation Pane, as shown in Figure 7.9. The cost of the natural gas is almost one-half million dollars per year. By combining the heater and cooler before and after the reactor, this cost can be reduced. The cost of electricity for the compressors is $300,000 per year.

This completes the example. The steps are as follows:

Step 1 Development of the model, including a provisional choice of thermodynamic model and equipment specifications.

Step 2 Solution of the model by doing the calculations iteratively. Convergence acceleration techniques are sometimes necessary.

Step 3 Examination of results.

Step 4 Testing the thermodynamic model (may be done as step 2).

Step 5 Obtain the utility costs.

Step 6 Summarizing the results using View/Report, View/History, and View/Input Summary, save the stream information to Excel and the flow diagram as a figure.

Operating Cost Summary	**Utility Cost Summary**			
Utility summary				
Utility ID:		ELECTRIC	GAS	WATER
▶ Utility type:		ELECTRICITY	GAS	WATER
Costing rate:	$/hr	34.4239545		4.89863751
Mass flow:	lb/hr			181431.029
Duty:	Btu/hr	2349188.17		8164396.26
Heating/Cooling value:	Btu/lb		23903	45

FIGURE 7.9 Summary of utility cost.

Greenhouse Gas Emissions

Aspen Plus will also calculate the greenhouse gas emissions for a process and the cost of the carbon tax. In Setup | Calculation Options, choose the standard you wish to use; here we use what Aspen calls USEPA (2009). Also set the carbon tax; for $50 per ton, put in 2.5 cents/lb. Then when the simulation is finished, choose Results/CO_2 Emission/ and you get the summary. For the ammonia process treated here there is carbon dioxide going in and out, which balance, and the carbon tax is extremely small and due to round off error in the iterations. For other processes, the carbon tax will be more important.

CONVERGENCE HINTS

To understand the convergence process used in most simulators, consider the process shown in Figure 7.10. Chapter 5 discusses the sequential and simultaneous methods. The sequential method is simple, but it might not converge. Thus, convergence acceleration techniques are used to improve the chances of convergence; two of those are described here [Wegstein (1958) and Broyden (1965, 1970) methods (Perry and Green (2008)]. Simultaneous methods are called equation oriented (EO) methods in Aspen Plus. Take a single component for illustration; in Figure 7.10, you would know the flow rate in the feed stream. The mass balance on the mixer takes the flow rate of streams 1 and 6 and adds them to become stream 2. The reactor changes the amount, and then the separator separates the stream into two streams. One of those is the recycle stream. Usually, you do not know stream 6 when you start the calculation; in the first iteration you often take it as zero, or some other assumed value. After reaching the separator, you have a new possible value for stream 6. Write this process as an equation:

$$x = f(x) \tag{7.2}$$

In this form, x is the flow rate of the chemical in the recycle stream, and $f(x)$ is the value of the flow rate of the chemical in the output of the separator, obtained after doing the calculation once around the loop. You have solved the problem when these two values agree. To make this an iterative process, call x^n the value of x after n iterations.

$$x^{n+1} = f(x^n) \tag{7.3}$$

This iterative process is simple to apply, since the function $f(x)$ is really a series of functions evaluated one after the other, and each step is a mass balance on one simple unit.

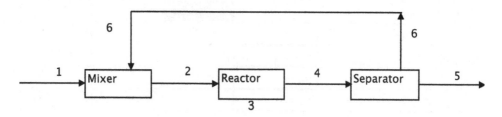

FIGURE 7.10   Process with recycle stream.

It may not converge, however. In order to speed up the convergence, the Wegstein method is very effective. The method calculates a provision value of $\hat{x}^{n+1}$ and then uses the last two iterates to calculate a "modified" value, which is then used as the new value in place of $\hat{x}^{n+1}$:

$$\hat{x}^{n+1} = f(x^n), \quad x^{n+1} = qx^n + (1 - q)\hat{x}^{n+1}, \quad \text{where}$$

$$q = -\frac{b}{1 - b}, \quad b = \frac{x^n - \hat{x}^{n+1}}{x^{n-1} - x^n} \tag{7.4}$$

Wegstein (1958) found that the iterations converged differently depending on the value of q.

$q < 0$	Convergence is monotone
$0 < q < 0.5$	Convergence is oscillatory
$0.5 < q < 1$	Direct method diverges in oscillatory manner, Wegstein method converges
$1 < q$	Diverges monotonically

The derivation of the Wegstein method begins with Eq. (7.2), written as a new function $F(x)$ as shown in Eq. (7.5).

$$F(x) = x - f(x) \tag{7.5}$$

Apply the secant method to this function. The secant method is the Newton method with a numerical version of the derivative based on the last two iterations (see Appendix E):

$$0 = F(x^n) + \frac{F(x^{n-1}) - F(x^n)}{x^{n-1} - x^n} \left[(x^*)^{n+1} - x^n\right] \tag{7.6}$$

But you also know that

$$F(x^n) = x^n - f(x^n) = x^n - x^{n+1}$$
$$F(x^{n-1}) = x^{n-1} - f(x^{n-1}) = x^{n-1} - x^n \tag{7.7}$$

Combining these equations gives

$$(x^*)^{n+1} = \frac{x^{n-1}x^{n+1} - (x^n)^2}{x^{n-1} - 2x^n + x^{n+1}} \tag{7.8}$$

which is the same as the Wegstein method, Eq. (7.4). Sometimes convergence is enhanced by applying Wegstein only every so many iterations rather than each iteration. The Wegstein iteration method is the default method in Aspen Plus, but there are others, including the direct method (Eq. 7.3), Broyden's method, and Newton's method. While Wegstein's method works on one variable at a time, Broyden's method works on all the variables at once using a matrix that is adjusted iteration by iteration. In the direct method, Eq. (7.3) is used as it stands. In the ammonia process, when using the Wegstein method, the lack of convergence was apparent in the MIXR unit, even though this is the last unit calculated. This is because after the calculation x^{n+1} is replaced by $(x^*)^{n+1}$ using Eq. (7.8). Thus, it is important to note the order of calculation in the iterations.

TABLE 7.3 Overall Iteration Errors for Ammonia Process Using Wegstein and Direct Methods

Iteration Number	Error-Wegstein	Error-Direct
1	1.00E4	1.00E4
2	1.00E4	1.00E4
3	1.47E4	6.72E3
4	−2.50E3	−3.22E3
5	−8.79E2	−2.58E3
10	5.58E2	−6.48E2
15	−5.62E0	−9.40E1
20	3.31E1	3.06
25	2.60E1	9.70
30	1.64E1	11.1

Another term used when discussing iteration methods is the tear stream. This is the stream that is guessed (or set to zero) when stepping through the process the first time. In Figure 7.10, if you compute in the following order: mixer, reactor, separator, then stream 6 is the stream you must assume and check—it is the tear stream. If you compute in the order: reactor, separator, mixer, then stream 2 is the tear stream. You apply the Wegstein iteration method to the tear stream. Of course, if the calculations are simple, as would be the case if all physical properties are specified and there are no energy balances, then you can use a spreadsheet. It may be so fast that you do not care whether 10, 100, or 1000 iterations are needed.

Sometimes it is convenient to have the calculations made one unit at a time. This can be done by clicking on the Step button (the open triangle in Figure C.2b in Appendix C) or CTRL-F5. It may also be necessary to reinitialize, that is, set the variables to their values at the start of the simulation. This is useful if the iterations do not converge and the answer is not reasonable. Just rerunning the problem will start from the unreasonable answer. To put the variables back to their beginning value, use the reinitialize button (see Figure C.2b in Appendix C). The progress of iterations can be seen clearly if you expand the Results I Streams to include all the streams and then click the Step button over and over.

For the ammonia process, the errors at each iteration (when starting from zero) are shown in Table 7.3. The errors decrease relatively rapidly but then decrease more slowly. Using the Newton method after Wegstein provided a converged solution.

OPTIMIZATION

Aspen has the capability of optimizing a function that you define. An example is given here that is very simple, but it illustrates the process. Take a single unit, the reactor in the ammonia process. The reactor RGibbs is used and the production of ammonia is maximized with respect to the reactor pressure. In this example, the reactor is unit REAC, with input stream 1 and output stream 2. The input is 100 lb mol/h nitrogen and 300 lb mol/h hydrogen at 900°F and 270 atm.

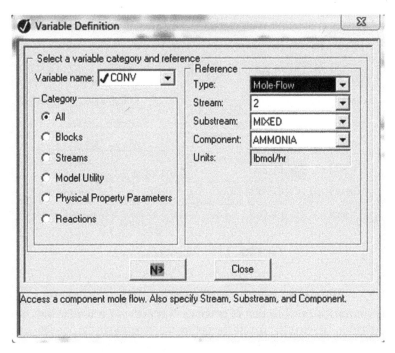

FIGURE 7.11 Definition of the variable.

Step 1 In the Navigation Pane, open Model Analysis Tools and the Optimization folder. Select New in the Object manager; change the name, if desired, and click OK. Here, it is CONV.

Step 2 Open the Optimization CONV. Initially this window has no entries; select New and define a variable name that will be involved in the optimization (any name is acceptable). Here, it is CONV.

Step 3 A window appears to define CONV. Put CONV into the Flowsheet variable and click Edit. Complete it as shown in Figure 7.11. Select in turn the type, Mole-Flow; Stream: 2 (the output stream of this simple example); Component: Ammonia.

Step 4 Click on the Objectives and Constraints Tab; you wish to maximize CONV.

Step 5 Next click on the Vary tab in order to define which variable to change. Figure 7.12 shows the choice of Block-Var for Block B1 (the reactor in this simple example), and indicates that the pressure is to be varied, between the limits of 2000 and 6000 psia.

Step 6 Solve the problem by running the program. The final results are shown in Figure 7.13; select Results Summary/Streams to see this.

Note that you only get the final results, but do not know how sensitive the result is to the reactor pressure. In this case, though, the pressure is at the maximum allowed in

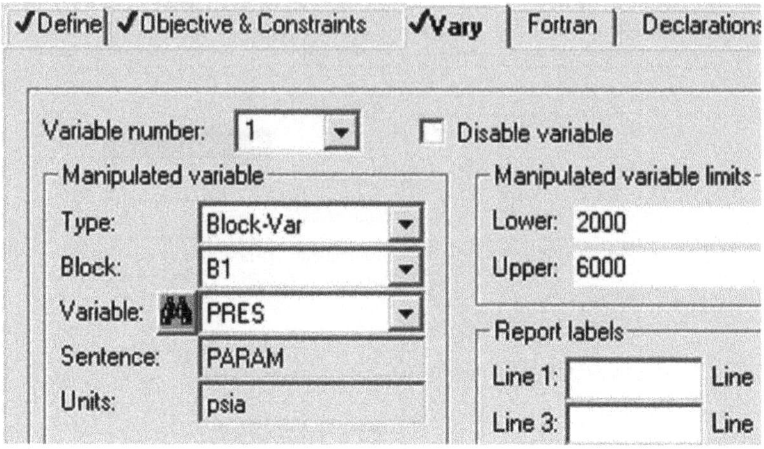

| ✓Define| ✓Objective & Constraints | ✓Vary | Fortran | Declarations |

FIGURE 7.12   Parameters to vary.

Figure 7.12, as expected. You can use the Sensitivity option in the Model Analysis Tools to find the conversion as a function of pressure. It operates in a similar way, but provides more information. To see this, take out the optimization option in the previous example and replace it with the Sensitivity option. The screens are similar to Figures 7.11 through 7.12. In Sensitivity/CONV/Input, choose the Tabulate Tab. Add New, CONV as the second column. After running the model, select Modal Analysis Tools | Sensitivity | CONV | Results. Then one gets the ammonia output for a variety of pressures as shown in Figure 7.14.

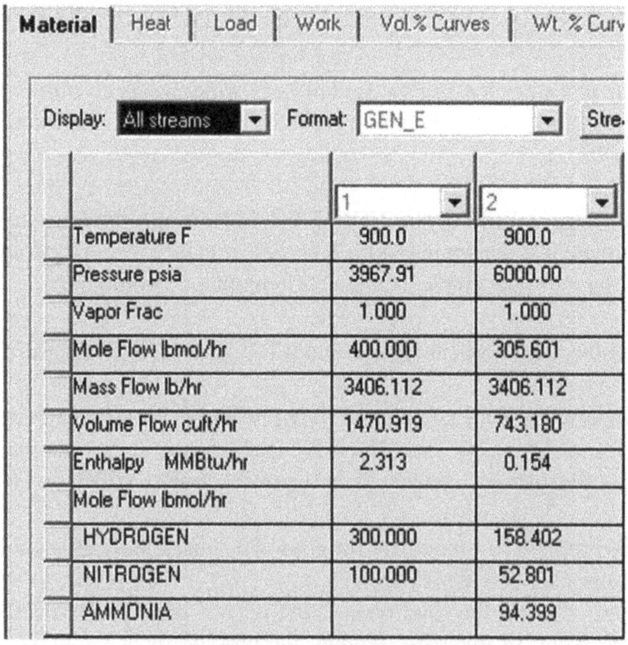

	1	2
Temperature F	900.0	900.0
Pressure psia	3967.91	6000.00
Vapor Frac	1.000	1.000
Mole Flow lbmol/hr	400.000	305.601
Mass Flow lb/hr	3406.112	3406.112
Volume Flow cuft/hr	1470.919	743.180
Enthalpy MMBtu/hr	2.313	0.154
Mole Flow lbmol/hr		
HYDROGEN	300.000	158.402
NITROGEN	100.000	52.801
AMMONIA		94.399

FIGURE 7.13   Optimization results.

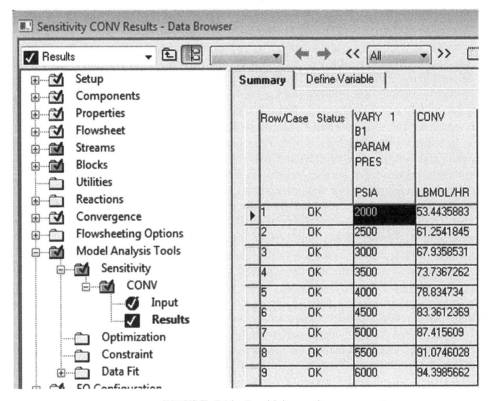

FIGURE 7.14 Sensitivity results.

INTEGRATED GASIFICATION COMBINED CYCLE

Prosperity frequently depends on the availability of cheap energy. Two countries, United States and China, have significant deposits of coal within their respective borders (Chu, 2009). One-half the electricity in the United States is generated by burning coal (National Academy of Engineering, 2008). Yet burning coal causes large emissions of carbon dioxide, a greenhouse gas. Thus, both countries are the largest emitters of carbon dioxide in the world. There are ways to change that, and one of them is to use integrated gasification combined cycle (IGCC). The technology is not new—Eastman Chemical has designed, built and operated IGCC processes for many years (since 1983) (Hess, 2006). But new advances in part of the process make them cheaper, and if there is a penalty for emitting carbon to the atmosphere, they become especially attractive compared with just burning the coal.

The steps in the process are simple: burn the coal with limited oxygen so that you make syngas, with lots of hydrogen and carbon monoxide. In fact, it is better to separate the air stream beforehand into an oxygen stream and nitrogen stream, since that reduces costs later for the emission control—the amount of nitrogen is huge so if it is not there the process is smaller. The air is separated in an Air Separation Unit (ASU). The coal and oxygen are combined in a gasifier to make the syngas stream. This stream contains some solid material (particulates), including flyash, which must be removed. Next, the water gas shift reactor makes more hydrogen; in the process the sulfur is converted to hydrogen sulfide, which is then removed. The carbon dioxide is removed in an acid gas removal process (this happens every day in a refinery, so it is not new). It can be stored underground in a Carbon Capture and Sequestration step. What is left is a hydrogen-rich fuel that is sent to a gas turbine where it burns. After combustion, the hydrogen is in the form of water. During this whole process, some units are very hot and need to be cooled. The cooling fluid (starting out as water, turning into steam) used to do that is sent to a steam turbine where additional energy is extracted. This step adds about 7–10 percentage points to the efficiency.

In addition to the flyash and sulfur compounds, there are emissions of NO_x, particulate matter and mercury to be considered. Coals come in different varieties, so that the process will be tuned somewhat to meet the requirements of a specific type of coal. For example, the coal in India often has a high ash content (Prasad and Gautam, 2004), and this affects the temperature in the gasifier and the flyash removal step. The sulfur content may very from 0.5 to 7 wt.%; ash content may vary from 1 to 40 wt.%; chlorine content may vary by a factor of 20 (Cooper, 2010). Furthermore, a similar process can be used with biofuels as the input instead of coal. Thus, it is worthwhile to learn to design an IGCC process.

The material fed to the process—air and coal—basically goes straight through the process with no recycle. Thus, many of the single units can be examined individually without reference to other parts of the process, and that is done in Chapters 2–6. It is important to model the thermodynamics correctly, and this aspect is considered in Chapters 2–6, also. While the raw materials and resulting products do not involve recycle streams, the process is tied together with heat exchangers, since the water/steam is recycled. A detailed report of this process and others is available (National Energy Technology Laboratory, 2007). Here, we utilize the process in Aspen for Problems 7.9–7.10.

CELLULOSE TO ETHANOL

Ethanol has been made from biomass since the time of the pharaohs in Egypt. Early in the twenty-first century there was a push to develop an industry doing this in the United States. Part of the push was for energy independence—to have a fuel that did not have to be imported. Farmers groups were all for making ethanol from corn, the government subsidized it and established fuel standards that required some ethanol, but chemical engineers were wary. They knew some of the problems: unless done carefully, the process used more energy than was in the final product. Still, there was less carbon dioxide emitted to the atmosphere in a complete life cycle since the growing of the biomass used carbon dioxide to offset the carbon dioxide formed during combustion of ethanol. It was also found that growing corn required a great deal of water, 2.3–8.7 million liters per megawatt hour, versus 90–190 liters per megawatt hour for petroleum extraction and oil refining (Service, 2009). Later, the use of corn for ethanol caused food prices to increase.

Another option was being looked at too: making ethanol from cellulosic feedstocks, the stalks, leaves, and wood of plants. Another twist was to make "drop-in" fuels instead of ethanol. The idea was to develop chemical processes that made hydrocarbons that are similar to those already used in power plants, trains, and automobiles (Anonymous, 2010; Weekman, 2010). This avoids the problem of having to replace the entire gasoline pipeline system in the United States, since ethanol corrodes the current pipelines. It also allows fuels for aviation and diesel, where ethanol cannot be used. Enzymes can be developed to tweak the reactions to make the most desirable hydrocarbons and to make more of them, thus increasing the overall efficiency.

The production of drop-in fuels (Anonymous, 2010) uses enzymes and bacteria to turn sugar into straight-chain alcohols, C12–C16, which are the main component of diesel fuel. Furthermore, if butanol is made rather than ethanol, the corrosion problems are reduced since butanol does not absorb water as readily as ethanol and it has four carbon atoms for each oxygen atom, rather than the two in ethanol. This is where the energy comes from. While this is also a small market, the US Navy projects that by 2020 half of their fuel will be from renewable sources (Anonymous, 2010).

There are a variety of processes that can use biomass: gasification, pyrolysis, and fermentation. Grains and cellulosic material is usually processed using hydrolysis and fermentation. Vegetable oils and algae use a process of esterification and pyrolysis, whereas wood and stalks can be gasified (Regalbuto, 2009; Weekman, 2010). We focus here on fermentation, primarily of cellulosic materials.

The raw material for cellulosic feedstocks is quite varied: trees, corn stover, and switch-grass, among others. The economics of moving that material dictate that a plant can economically collect material from a radius of about 50 miles (Ladisch et al., 2010). Thus, we immediately lose one of the hallmarks of chemical engineering processes—the economy of scale. The next problem is that different sources of feedstocks have different chemical composition, so that each process must be tuned for the feedstock being used. Sometimes enzymes are engineered to handle a specific feedstock, too. The feedstock contains variable amounts of cellulose (33–44%), xylan (15–21%), arabinan, mannan, and galactan (collec-tively 0–5%), acetyl (2–5%), lignin (18–29%), ash (1–10%), and other material. Xylan is a complex polysaccharide found in plant cell walls. Arabinan is a monosaccharide, mannan is a plant polysaccharide, and galactan is a polymer of galactose found in hemicellulose.

The process involves a number of steps in series:

1. *Feedstock preparation*: First, the feedstock is prepared to eliminate dirt and metals.
2. *Pretreament*: Then it goes to a pretreatment process that cooks the material in hot water or steam, acid or base, in order to open the plant cell wall to expose the cellulose to enzymes that hydrolyze these macromolecules.
3. Hydrolysis produces both five-carbon and six-carbon sugars.
4. *Fermentation*: Then the stream goes to fermentation units where the sugars are made into ethanol.
5. Distillation and drying processes provide dry ethanol.

While the process is mostly straight through, there is a significant energy load, and this is reduced via many recycle streams involving water and steam and solids for combustion. In fact, there is a great deal of water circulating throughout the process, which adds to the expense. The lignin can be used as a boiler fuel, for example. Ladisch et al. (2010) estimate that the maximum yield of ethanol from corn stover is 95 gal/dry ton biomass. Using the density of ethanol (6.58 lb/gal), this gives an ethanol product of only 625 lb/dry ton, or only 31%. Since 40% of the feedstock is water, that means that of the weight of wet feedstock, only about 18% by weight leaves as product. This limitation makes the development of enzymes particularly important so that the process can be as efficient as possible.

Here, a process developed by the National Renewable Energy Laboratory and the Harris Group will be used to illustrate many features (Aden et al., 2002; Aspen, 2011). The mass balances can easily be made using Excel, and they can be done in Aspen, too, using the mass-balance-only option. To do the energy balances in Excel would be too tedious, since there are so many components. To do energy balances in Aspen requires that one develop the appropriate thermodynamics for each chemical, including the solids. Frequently, this is done using user-generated information, which must then be inserted into the Aspen simulation. For this reason, here we only consider the mass balances.

The report is Aden et al. (2002). The Aspen Model is at http://devafdc.nrel.gov/biogeneral/Aspen_Models/. Download the BKP file (the Aspen backup file), the input file, and the spreadsheet.

CHAPTER SUMMARY

This chapter listed many of the possible units in the Model Library of Aspen Plus. The ammonia process illustrated the procedures (and computer windows) you used to set the process conditions and examine the results. The thermodynamics choices can be verified by comparison with data reported in the literature. Sometimes the calculations do not converge, and then use methods for accelerating convergence: Wegstein, direct, Broyden, Newton. Some experimentation may be necessary and your choices should be guided by an understanding of how the iterations proceed through the process and from iteration step to step. Two processes were introduced: IGCC and ethanol from cellulose. While the problems for those processes are simplified, they are processes currently being worked on by chemical engineers.

CLASS EXERCISE

Divide into teams of two to four students. Work through a spreadsheet (Table 7.2), with each person of a team taking a unit, showing where the parameters are for that unit and checking the results. Are they correct? Does the mass balance? Does the energy balance? Work through the whole process. Alternatively, your instructor can prepare a flowsheet and dataset showing the results, but with some errors for your team to find.

PROBLEMS

In Problems 7.2–7.8 report the Greenhouse Gas Emissions, too.

7.1₁ Derive Eq. (7.8) from Eqs. (7.6) and (7.7). Show that Eq. (7.8) is the same as Eq. (7.4).

7.2₁ Simulate the benzene process (Problem 5.1) using Aspen Plus. Take the feed at room temperature and 1 atm. Compress it to 35 atm. Preheat the feed to the reactor to 550°C and cool the effluent. Model the reactors as RStoic reactors, and model the separations using distillation towers. You will have to decide on the number of stages and reflux ratio, and using DSTWU first might be useful.

7.3 **Project.** Simulate the benzene process (Problem 5.2) using Aspen Plus. Take the feed at room temperature and 1 atm. Compress it to 35 atm. Preheat the feed to the reactor to 550°C and cool the effluent. Model the reactors as RStoic reactors, and keep the hydrogen/methane separations as simple splitters; model the other separations using distillation towers. You will have to decide on the number of stages and reflux ratio, and using DSTWU first might be useful.

7.4 Project. Simulate the ethyl chloride process (Problem 5.4) using Aspen Plus. The feed streams are at room temperature and 20 psia and they are compressed to 95 psia. Preheat the feed to the reactor to 800°F and cool the effluent. Model the reactors as RStoic reactors. Take the effluent from the first reactor, cool it, and send the liquid to the product stream. Compress the gases to 265 psia, and heat them to 350°F (if needed). Lower the pressure and cool the effluent from the second reactor and separate the liquid and vapor. Recycle the vapor and send the liquid to the product stream. You will have to decide on the cooling temperature to use in the vapor–liquid separators.

7.5 Project. Simulate the vinyl chloride process (Problem 5.5) using Aspen Plus. Take the feed at room temperature and 20 psia. Operate the direct chlorination reactor at 65°C and 560 kPa. A distillation column removes the trichloroethane and the rest of the stream is sent to the furnace. Heat the stream to 1500°F, so pyrolysis takes place. Cool the effluent from the furnace, and recycle the vapor (mostly HCl). Send the liquid (vinyl chloride and ethylenedichloride) to a distillation column for separation. You will have to decide on the number of stages and reflux ratio, and using DSTWU first might be useful. The oxychlorination reactor operates at 200°C and 1400 kPa. Wash the effluent from the oxychlorination reactor with water to remove contaminants and then remove the water and light gases in a distillation column. The ethylenedichloride must be dry (<10 ppm) before going to the furnace. The stream then goes to a distillation column where the ethylenedichloride is purified ($>99.5\%$) before it is sent to the furnace. Since the furnace operates at 560 kPa, the pressure is reduced before reaching the furnace.

7.6₁ How does the process in Problem 7.5 change if the furnace is run at the high pressure (1400 kPa)?

7.7 Project. Simulate the ethanol process (Problem 5.7) using Aspen Plus. The feed streams are at one atmosphere and room temperature, but the reactor operates at 960 psia and 570°F. Thus, you must heat the reactor feed, and after the reaction occurs you must cool the product. The first splitter is a vapor–liquid separator (you choose the temperature that will separate ethanol from ethylene), and the remaining splitters are distillation columns. You will have to decide on the number of stages and reflux ratio, and using DSTWU first might be useful.

7.8₁ Modify the process in Problem 7.7 to use a counter-current heat exchanger (and trim heater) to heat the reactor feed and cool the reactor product.

7.9₂ Obtain access to the following files either from your administrator or from the area (usually):

Local Disk(C@/Program Files (x86)/AspenTech/Aspen Plus v7.2/GUI/App/IGCC.

Your system may be different, so the x86 and v7.2 are adjustable. There is a pdf file entitled *Aspen Plus IGCC Model*. There is also an Aspen model that can be loaded. The process includes almost 40 chemicals and several subprocesses. The subprocesses are as follows:

Sizing: to prepare the coal
ASU: to separate air into nitrogen and oxygen
GASFR: the gasifiers

CLEANING: to clean the gas

CLAUS: to process the hydrogen sulfide to make elemental sulfur

POWER: to collect all the steam in the process to use it in a turbine or in the process

WGS: the water gas shift reactor and separators

METHANZ: the make methanol

NH₃: to make ammonia

If you open the process and look at the flow sheet, each of these subprocesses is in a square box, and the process units are inside. Thus, there are blocks inside the blocks. The subprocess blocks can be examined by double clicking on the box. If you look at a few of them, you will see what a complicated process it is. Examine the process and answer the following questions:

1. What thermodynamic choices are made? Are they the same in all subprocesses?

2. The process involves solids at the beginning. Look at the components and see how the solids are specified.

3. What is the objective of each subprocess, based on the input and output?

4. What blocks are used? Are they a priori blocks that can be calculated, or must you choose some parameters that you might not know unless you are an expert in the field? (*ex*. A simple splitter (SEP block) will separate components of a stream in the manner you specify, but you need to choose the split. By contrast, a flash unit or distillation column will determine the split based on chemistry and physics without a priori knowledge.

7.10₂ Consider the two-stage water–gas shift reactor for the IGCC process. The flow rate of the syngas is taken proportional to that from the Aspen model (stream SG-A) (kmol/h): nitrogen, 1126; water, 1700; carbon monoxide, 1686; hydrogen, 1134; carbon dioxide, 0. First, compare the mole fraction of this stream (from a coal gasification process) with that from a methane reformer (Problem 4.6). Note particularly the presence of nitrogen here (much less in the methane reformer because air has been separated so that only oxygen is involved) and the mole fraction of hydrogen. Why would there be more hydrogen from methane reforming than from coal gasification? Both reactors are modeled using the RGibbs block. The input to the first reactor is set at 400°C and 20 bar, based on the suggestions of Platon and Wang (2010). The output from the first reactor is cooled to 200°C; steam is added to enhance the conversion. Platon and Wang suggest a steam/dry molar ratio of 0.4 going into the second reactor. Add a steam flow going into the second reactor, too, and adjust it so that the ratio of 0.4 is achieved. This can be done using the Design Spec block.

7.11 **Project.** A one-stage process to form acetaldehyde from ethylene and oxygen is given below. The feed stream is 100 lb mol/h of ethylene and 50 lb mol/h of water at 20 psia and 80°F. The reaction is

$$C_2H_4 + \left(\tfrac{1}{2}\right) O_2 \rightarrow CH_3CHO$$

and the conversion per pass is taken as 95%. The output is cooled to 80°F and the gases are removed in a stripper (use a version of RADFRAC without a condenser or reboiler), the water–acetaldehyde stream is separated in another RADFRAC, and finally, light gases are removed from the acetaldehyde stream. The gases from the stripper are recycled, and 1% of them are purged. (This is not strictly necessary for the process as described, but would be necessary if there were light impurities in the oxygen inlet stream.) You will have to experiment with the number of stages and distillate rates and reflux ratios.

7.12₂ Examine the Aspen example for a biodiesel process. (See Aspen examples in Appendix C for directions to find this.) Identify the main sections of the process. Look at major input/output streams: oil, methanol, FAME (fatty acid methyl ester or biodiesel), and glycerol. What is the mass flow rate of product glycerol to product FAME? Lots of glycerol is generated by the process. Suggest some ways glycerol can be used.

Problems Involving Corn Stover and Ethanol

The process modeled here is a mass balance only version of this process to make ethanol from corn stover. The nomenclature adheres as much as possible to that in the original document. You should download the Aspen version and examine both the pdf report, especially Appendix F (process flow diagrams) and the Aspen model itself. Also see the spreadsheet. See the appendix for how to view sections of the process in Aspen. Also, save the input report, which provides additional details. There is not exact agreement between the pdf document, Appendix F (which shows actual process units), and the Aspen model (where several modules may represent one process unit), but the stream numbers are usually the same. The mass balances on the pdf document are not the same as those in the Excel spreadsheet, because the various documents represent the process at different time during development over several years. Likewise the input file may be different from what is shown in the pdf document. Despite some inconsistencies, all versions of the process are generally the same. The numbers used here to define the process are representative; in many cases, they must be determined by experiment. In particular, the Aspen model has a Process Section 400, which is used to produce enzyme (in the 1999 version); it has been eliminated in the later versions (pdf); instead, the enzyme will be purchased. This is an area of active research, so this permits purchase of the best material at the time.

The process to change corn stover into ethanol involves many different chemicals. To simplify the process, the polymers xylan, arabinan, mannan, and galactan are combined and used simply as xylan. In addition, we have left out a number of components that are small and are not involved in the reactions. In addition, we consider mass balances only. Streams can contain both liquids and solids; thus, it is necessary to learn how to do this in Aspen. While version 7.2 requires the solids be in a separate stream (CISOLIDS), version 7.3 allows many of them to be in the liquid stream (MIXED), here we will use them in a separate stream to illustrate the procedure. Aspen has certain restrictions for a mass-balance-only simulation, which are described in the Aspen index (insert "mass balance only simulations" into the search window). In particular, it is necessary to turn off the calculation of energy balances by unchecking the box "Perform Heat Balance Calculations" in the Simulation options form. Also, it is convenient to develop the flowsheet in sections, which can be plotted and examined by themselves. The method to do that is demonstrated in Appendix C.

100 Process—Feedstock Storage and Handling

Corn stover arrives in bales, 98,039 kg/h at 15% moisture. It is washed, and the dirty water is cleaned in a clarifier and thickener. The washed stover has metals removed magnetically and is then shredded. Some of it must be stored for future use, since the delivery of material is not year-round. The shredded material is 30–40% moisture.

200 Process—Pretreatment and Hydrolyzate Conditioning

The hydrolysis reactions convert most of the hemicellulose in the feedstock to soluble sugars; here, all combined in xylose. Glucan is converted to glucose using sulfuric acid. Acetic acid is also liberated from the hemicellulose hydrolysis. Following the pretreatment reactor, the stream is flash cooled, which vaporizes a large amount of water and some acetic acid. The solids are washed and pressed to separate the liquid portion, which contains sulfuric acid. Lime is added to the liquid to raise the pH to 10. After a delay, the stream is neutralized and gypsum precipitates. The gypsum is filtered out and the hydrolyzate is mixed with the solids (cellulose) and water before sending on to the next area.

Section A201. The mass-balance-only process begins with the input to the 200 process, stream 105 as indicated in Table 7.4. The first part is A201, the prehydrolysis section. The flow rate is about 40% water by mass; some of it is a liquid and some is a solid. 2104 kg/h

TABLE 7.4 Input to Process (kg/h)

kg/h	105 From Feed Handling	Molecular Weight
Liquids		
Ethanol	22	46.0684
Water	45,002	18.0153
Glucose		180.1560
Xylose		150.1300
Acetic acid	232	60.0520
Lactic acid		90.0779
Sulfuric acid		98.0785
Carbon dioxide		44.0095
Oxygen		31.9988
Nitrogen		28.0134
Other soluble solids	7,417	193.0000
Total liquids	52,673	
Solids		
Cellulose	31,187	162.1410
Xylan	22,924	132.1147
Lignin	21,806	–
Gypsum		172.1720
CaO		56.0774
Total solids	75,917	
Total flow	128,590	

TABLE 7.5   Reactions Occurring in the Ethanol Process

Reaction Number	Reaction	Fractional Conversion Based on Nonwater Substance
	M202HI	
3	H_2O + Cellulose (Cisolid) $\rightarrow$ Glucose	0.07
6	H_2O + Xylan (Cisolid) $\rightarrow$ Xylose	0.9
	T209	
	H_2SO_4 + CAO (Cisolid) + H_2O $\rightarrow$ Gypsum Conversion based on H_2SO_4	1
	T224	
	H_2SO_4 + CAO (Cisolid) + H_2O $\rightarrow$ Gypsum Conversion based on H_2SO_4	1
	T310	
3	H_2O + Cellulose (Cisolid) $\rightarrow$ Glucose	0.9
	F301-6F	
7	Glucose $\rightarrow$ 2 Ethanol + $2CO_2$	0.9
11	Glucose $\rightarrow$ 3 AACID	0.015
12	Glucose $\rightarrow$ 2 LACID	0.002
16	3 Xylose $\rightarrow$ 5 Ethanol + $5CO_2$	0.8
	F300CONT	
1	Glucose $\rightarrow$ 2 LACID	1
2	3 Xylose $\rightarrow$ 5 LACID	1
	F300F	
7	Glucose $\rightarrow$ 2 Ethanol + $2CO_2$	0.95
11	Glucose $\rightarrow$ 3 AACID	0.015
12	Glucose $\rightarrow$ 2 LACID	0.002
16	3 Xylose $\rightarrow$ 5 Ethanol + $5CO_2$	0.85

of sulfuric acid (stream 212) is combined with 98,256 kg/h of water (stream 211) in a mixer (A201). The dilute sulfuric acid (stream 214) is combined with stream 105 in M202MIX where it is heated to 100°C; while we would not worry now about the energy transfer, the combined stream is 214A. This stream is combined with 11,507 kg/h of low-pressure steam (stream 215) in M202LO, where it is heated to 100°C and the pressure is raised to 39.26 atm. The output (stream 215A) is fed to the reactor (M202HI) along with high-pressure steam (stream 216, 37,234 kg/h). There are several different reactions taking place, and the RStoic module can be used to model the mass balance, as displayed in Table 7.5. Be careful, though, since the balances are on mass and the reactions occur on a mole basis. The output, stream 217, is fed to a flash unit (T203) at 1 atm. For this mass-balance-only model, the split fractions (the fraction of the feed stream going out the top stream 218 as vapor) are ethanol 78%, water 27%, acetic acid 2.9%. This completes the prehydrolysis Section A201.

Section A203. (It is more convenient to consider this section before Section 202.) The Lime Addition Section, A203, takes stream 240 from section 202 and adds lime (stream 227, 2395 kg/h); all the sulfuric acid reacts with the lime in T209 to form gypsum and the ouput is stream 228. This goes to unit T-224, a reacidification tank. Sulfuric acid is added

(1185 kg/h) and all the lime reacts with sulfuric acid to form gypsum. The output is stream 239, which is sent to S-222 (SSplit) to remove the gypsum. The solid stream (229) contains 99.5% of the solids and 0.4% of the liquids; the rest is in the liquid stream (230), which is recycled to Section A202.

Section A202. In the solid–liquid separation section, water is added in stream 252 (125,000 kg/h) and stream 253 (75,000 kg/h with 1000 kg/h ethanol. In the real process, this water comes from the lignin separation process, which is not modeled here. Meanwhile, stream 220 from the previous section is split into two streams in unit S205A (SSplit): stream 221 is 78% of it and stream 225 is the rest. Streams 225 and 252 are combined in S205MIX to form stream 254. Stream 254 is sent to unit S205B (SEP unit) that is a filtration unit; the output is determined from experimental data: stream 256 contains 75% of the water and 90% of the other liquid components. Stream 255A contains all the solid components, plus the rest of the liquid components. Steams 221 and 256 are combined in MIXR to form stream 240, which is sent on to the Lime Addition Section, A203. Stream 255A is combined with air (964 kg/h of oxygen and 3703 kg/h of nitrogen in stream 259) and sent to the S205FLSH unit. The air stream and 404 kg/h water come out of this unit as a gas stream (260) and the remainder is a liquid as stream 255. Stream 255 is combined in unit C202 with stream 253 and stream 230 from the Lime Addition (A203); this is thus a recycle stream. The output is stream 250, which is sent to the Saccharification and Fermentation Section, A300. This completes the section.

A300 Process—Saccharification and Cofermentation

Section A300. Saccharification (or hydrolysis) of cellulose to glucose uses cellulase enzymes, and it works best at high temperatures where the enzyme activity is high. Thus, the saccharification is performed first. The next section of the A300 process uses fermentation to convert the sugars to ethanol. This is done using bacteria as a biocatalyst. The enzymes and bacteria are specially made to enhance production; here, the conversion is taken as from the NREL study, and the conversions were based on experimental data; see pages 27–28 of the report for information about the specific compounds.

In Process Section A300, stream 250 from Process Section A202 is fed to a vessel (T310). This vessel in Aspen actually represents a series of five stirred tanks that are designed to allow the enzymes to grow (over a period of 36 hours). The reactions for T310 are listed in Table 7.5; most of the cellulose is converted to glucose. Ninety percent of the output from T310 is sent directly to the fermenters, FEMMIX, in stream 302. The rest is sent to another reactor unit. The seed train of five reactors is modeled as unit F301-6F (RStoic), with the conversions shown in Table 7.5. The output is flashed (F301-6FL) to remove the carbon dioxide and oxygen formed in the reactors. In addition, 3% of the ethanol, 0.22% of the water, and 0.2% of the acetic acid go out as vapor. The liquid from the flash drum goes on to be combined in the mixer FEMMIX with the previous stream 302.

Three percent of stream 304A is split off to send to a contamination reactor (F300CONT). In that unit the glucose and xylose are all converted to lactic acid with 100% conversion. The other stream (97% of 304A) goes to a fermenter (F300F) where a number of reactions take place (Table 7.5). The output from both reactors is sent to a flash unit to release gases (all the carbon dioxide and oxygen and 2.8% of the ethanol, 0.2% of the water, and 1.1% of the acetic acid. The liquid stream (306) from the flash unit (F300FLSH) goes on to the next section for purification.

A500 Process—Distillation, Dehydration, Scrubber, Evaporator, Solids Separation

Section A500. Stream 306 is fed to a distillation column (D501, SEP) to distill most of the water (and remove the dissolved carbon dioxide). This column acts with a side draw, which is an ethanol–water mixture. For purposes of this model, remove all the carbon dioxide, 0.24% of the ethanol and 0.005% of the water out the top in stream 508. This stream goes to a scrubber. The side draw (stream 510) takes 99% of the ethanol fed to the column, 10% of the water, and 5% of the acetic acid; this stream goes to the rectification process. The remainder goes out the bottom, stream 518. This stream goes to a series of evaporators; these are not modeled here, but some of the output from the evaporator goes to a lignin separation process (A505, not modeled), and some of the output from that process becomes stream 516; it is 97% water. Ultimately, water from the lignin separation plant is recycled as streams 252 and 253, which are specified here.

The rectification process takes stream 510, adds recycle stream 521 from the molecular sieve process, and distills it in unit D502. The overhead from this unit (stream 511) is at the ethanol–water azeotrope, which is 0.826 mole fraction ethanol at 27 psia and 92.4°C. The rest of that stream is water. The bottoms (stream 516) contains everything else, but negligible amounts of ethanol, and it is recycled. Since we are specifying the output stream of D502, we use the SEP2 unit to model this.

The ethanol–water azeotrope, stream 511, is sent to a molecular sieve process (T507-8) for drying. This is an adsorption unit that requires regeneration periodically. During the regeneration, ethanol and water are generated, and this is sent back to the rectification column as stream 521. This stream contains 20% of the ethanol fed to the dehydration unit and 95% of the water in stream 511. The very dry ethanol output (stream 515) is sent to storage. That is the product of the process.

The flow rates into each section are given so that each section can be done as a homework problem. The outputs can then be compared with the other information. The entire process is a Project. Both problems and the project can be done in Excel or in Aspen Plus. For each section, compare the ease of calculation and the transparency of the flowsheet between the two programs. Compare the results with the results of the NREL process.

7.13₁ Simulate Section 201 in (a) Aspen Plus; (b) Excel. Flow rates to the process (kg/h): ethanol, 22; water, 45002; acetic acid, 232; other soluble solids, 7417; cellulose, 31187; xylan, 22924; lignin, 21806. Stream 215 is 11,507 kg/h of steam; stream 216 is 37,234 kg/h of steam.

7.14₁ Simulate Section 202 in (a) Aspen Plus and (b) Excel. Flow rates to Section 202, stream 220: ethanol, 5; water, 137,928; glucose, 2426; xylose, 23445; acetic acid, 225; sulfuric acid, 2104; other soluble solids, 7417; cellulose, 29004; xylan, 2292; lignin, 21806. Stream 230 is ethanol, 5; water, 222430; glucose, 2363; xylose, 22837; acetic acid, 219; other soluble solids, 7225; cellulose, 113; xylan, 9; lignin, 85; gypsum, 37. Stream 252 is 125,000 kg/h of water and stream 253 is 1000 kg/h of ethanol and 75,000 kg/h of water. The air stream is 964 kg/h of oxygen and 3703 kg/h of nitrogen.

7.15₁ Simulate Section 203 in (a) Aspen Plus; (b) Excel. Flow rates to Section 203: ethanol, 5; water, 224092; glucose, 2372; xylose, 22929; acetic acid, 220; sulfuric acid, 2058; other soluble solids, 7254; cellulose, 22623; xylan, 1788; lignin, 17009; the lime addition (stream 227) is 2395 kg/h and the sulfuric acid steam is adjusted to react all the lime.

7.16[1] Simulate Section 300 in (a) Aspen Plus; (b) Excel. Flow rates to Section 300: ethanol, 1005; water, 335862; glucose, 2416; xylose, 23353; acetic acid, 224; sulfuric acid, 46; other soluble solids, 7388; cellulose, 6494; xylan, 513; lignin, 4882; gypsum, 37.

7.17[1] Simulate Section 500 in (a) Aspen Plus; (b) Excel. Flow rates to Section 500: ethanol, 14775; water, 334468; glucose, 259; xylose, 3126; acetic acid, 352; lactic acid, 905; sulfuric acid, 46; other soluble solids, 7388; cellulose, 649; xylan, 513; lignin, 4882; gypsum, 37. The model for unit D502 must achieve a certain concentration in the overhead—the azeotrope with water and ethanol. Use a SEP2 block for this unit. For this block, set the split fraction of the overhead to 1.0 for ethanol, and 0.0 for all the other components except water. The window has a place for a second specification. For the water, set that to a mole fraction of 0.174. Then the amount of water sent to the overhead will be adjusted to give that mole fraction in a mixture of ethanol and water, and the rest of the water, and the other components, will go out the bottom stream.

7.18 **Project.** Do the entire process using either Excel or Aspen Plus. Compare the results with the results of the NREL process. In your report, show an overall mass balance, indicate the amount of water fed to the process and the ratio of water to ethanol produced. Using the stoichiometry in Tables 7.4–7.5, calculate the maximum amount of ethanol that can be produced and the efficiency of your process.

8

CHEMICAL REACTORS

Ethane flowing through a cylindrical tube can be heated to the point that it reacts into hydrogen and ethylene, the precursor to polyethylene. Many reactors, though, require a catalyst to speed up the reaction. The water–gas shift reactor makes hydrogen by using a cylindrical tube that is filled with catalyst and past which the gas flows. One feature that sets chemical engineers apart from other engineering disciplines is their ability to handle chemical reactions, the process of converting one or more chemicals into other chemicals that are more valuable. Thus, chemical engineers need to understand chemical reactions and should be able to model equipment in which the chemical reactions take place—that is, chemical reactors. Sometimes reactors are as simple as a cylindrical tube, with reactants fed in one end and products (and unreacted reactants) coming out the other end. Reactants may be placed in a vessel and stirred to create new chemicals.

Real-life complexities challenge engineers as they design chemical reactors. In some reactors, the reaction rate is slow. When this happens, you need to have a bigger reactor to achieve a specified production rate, and it may be too big to be economical. Then, you look for a catalyst that speeds up the reaction but retains its existence after the reaction. The tube is packed with catalyst, and reactants flow around the catalyst. Oftentimes the reaction is exothermic and the reactor gets hot. The energy must be removed somehow, and heat transfer must be included in the calculations. Sometimes the reaction occurs inside a porous catalyst, perhaps an alumina pellet (Al_2O_3) with catalyst coating the internal surface. Then it may be necessary to include heat and mass transfer effects inside the pellet.

When it is necessary to include these effects—slow reaction rates, catalysts, heat transfer, and mass transfer—it can make an engineering problem extremely difficult to solve. Numerical methods are a must, but even numerical methods may stumble at times. This chapter considers only relatively simple chemical reactors, but to work with these, you must learn to solve ordinary differential equations as initial value problems.

Introduction to Chemical Engineering Computing, Updated Second Edition. Bruce A. Finlayson.
© 2014 John Wiley & Sons, Inc. Published 2014 by John Wiley & Sons, Inc.

This chapter begins with reactor models described in mathematical terms. You will learn how to solve ordinary differential equations by using MATLAB®. Then, after you have mastered this new skill, you will solve several chemical reactor problems by using MATLAB. Sometimes these problems can be solved using Comsol Multiphysics and Aspen Plus, too, as the chapter discusses next. A few problems use Excel to solve ordinary differential equations. The reason is that all the other programs have guaranteed error control. Although it is possible to solve simple ordinary differential equations with Excel, it becomes cumbersome if the problem is difficult, and you would not know the accuracy of your solution without extra effort. Some reactors operating in steady state are expressed as algebraic equations, and both Excel and MATLAB are used to solve those equations.

Instructional Objectives: After working through this chapter, you will be able to

1. Model chemical reactors that are stirred tanks, plug flow reactors, including non-isothermal ones.
2. Solve such problems using MATLAB, Comsol Multiphysics, Aspen Plus, and Excel, for comparison.
3. Solve reactor problems in which there are mole changes and/or mass transfer limitations.
4. Appreciate the need for special integration techniques when the equations are stiff.
5. Introduce complicated reaction rate expressions into Aspen Plus when modeling reactors.

MATHEMATICAL FORMULATION OF REACTOR PROBLEMS

A differential equation for a function that depends on only one variable is called an ordinary differential equation. The independent variable is frequently time, t, but for reactors, it can also be length down a plug flow reactor. An example of an ordinary differential equation is

$$\frac{dy(t)}{dt} = f(y), \quad y(0) = y_0 \tag{8.1}$$

The general solution to the differential equation includes many possibilities; the engineer needs to specify initial conditions to specify which solution is desired. If all conditions are available at one point [as in Eq. (8.1)] then the problem is an initial value problem and can be integrated from that point on. If some of the conditions are available at one point and others at another point, then the ordinary differential equations become two-point boundary value problems (see Chapter 9). Initial value problems as ordinary differential equations arise in control of lumped parameter models, transient models of stirred tank reactors, and generally in models where there is no diffusion of the unknowns.

Example: Plug Flow Reactor and Batch Reactor

Consider a plug flow reactor—a cylindrical tube with reactants flowing in one end and reacting as they flow to the outlet. A mole balance is made here for the case when the operation is steady. Let F be the molar flow rate of a chemical, which changes dF in a small part of the tube; take the volume of that small part to be dV. The rate of reaction is expressed as the moles produced per unit of time per unit of volume. If a chemical specie reacts, then the r for that specie is negative. Thus, the overall equation for a small section of the tube is

$$dF = r\,dV \quad \text{or} \quad \frac{dF}{dV} = r \tag{8.2}$$

The molar flow rate can be related to the velocity in the tube:

$$F = uAc \tag{8.3}$$

In many cases, the velocity and cross-sectional area of the tube are constant. Then you can write

$$dF = uA\,dc, \quad dV = A\,dz \tag{8.4}$$

If you put these into Eq. (8.2) and divide by dz you get

$$u\frac{dc}{dz} = r, \ \text{plug flow reactor} \tag{8.5}$$

Normally, r is a function of c, and this is an ordinary differential equation for $c(z)$. In addition, you need to know the initial value of the concentration, that is,

$$c(0) = c_0 \tag{8.6}$$

If the reactor contains a catalyst, then the reaction rate is expressed as the molar rate of change per mass of catalyst, r'. One must then multiply it by the mass of catalyst per unit volume, but otherwise the situation is the same:

$$dF = \rho_B r'\,dV \quad \text{and} \quad u\frac{dc}{dz} = \rho_B r', \ \text{catalytic plug flow reactor} \tag{8.7}$$

In a batch reactor, let N be the moles of a chemical; the molar change of a chemical is the reaction rate (molar change per unit time per unit volume) times the volume and the time:

$$dN = rV\,dt \tag{8.8}$$

This can be put into a differential equation, too:

$$\frac{dN}{dt} = rV \tag{8.9}$$

If the volume of the batch reactor is constant, then

$$c = \frac{N}{V} \tag{8.10}$$

and the equation becomes

$$\frac{dc}{dt} = r, \ \text{batch reactor} \tag{8.11}$$

This equation is similar to the equation for a plug flow reactor, with the time t taking the place of the variable z/u. You can also introduce catalyst into the batch reactor, in which case the reaction rate r is replaced by $\rho_B r'$. In addition, you need an initial condition to go with Eq. (8.11), $c(t = 0) = c_0$. The structure of the problem is the same as that for a plug flow reactor, and if you can solve one, you can solve the other. So, there is some economy of effort.

Many chemicals are involved, and you need a reaction rate for each reaction occurring. Thus, use $\{c_j\}$ for all the chemical concentrations and $\{r_i\}$ for all the reactions. If the reaction is

$$a\mathrm{A} + b\mathrm{B} \rightarrow c\mathrm{C} + d\mathrm{D} \tag{8.12}$$

then the rate of reaction of B is b/a times the rate of reaction of A, and the rates of reaction of C and D are $-c/a$ and $-d/a$ times the rate of reaction of A. The important point is that the rate of reaction is an algebraic function of the concentrations, temperature, and pressure. Each reaction rate needs to be known specifically for the reaction in question.

Example: Continuous Stirred Tank Reactor

Another type of reactor is known as a continuous stirred-tank reactor (CSTR). This reactor is a vessel with an inlet stream and an outlet stream, and the contents are constantly being stirred. Engineers often assume that this reactor is so well mixed that the concentration is the same everywhere. The mole balance on this reactor is

$$F_{\mathrm{out}} - F_{\mathrm{in}} = rV \tag{8.13}$$

where V is the volume of the reactor. The molar flow rates are related to the volumetric flow rates, Q, as follows:

$$F = Qc \tag{8.14}$$

When the volumetric flow rates are constant, Eq. (8.13) becomes

$$Q(c_{\mathrm{out}} - c_{\mathrm{in}}) = rV, \quad \mathrm{CSTR} \tag{8.15}$$

In a well-mixed reactor (CSTR), the concentration going out the reactor is the same as the concentration in the reactor. Since the rate expression is a function of c, it is a function of c_{out}. Then Eq. (8.15) becomes an algebraic equation in one variable (here) for c_{out}. You can solve this using either Excel or MATLAB.

USING MATLAB TO SOLVE ORDINARY DIFFERENTIAL EQUATIONS

Simple Example

In MATLAB, you define the problem by means of a function, called an m-file. You then tell MATLAB to solve the differential equation. Thus, you need to prepare an m-file that defines the equation and then call the subroutine "ode45" to do the integration. This method may seem mysterious at first because you call a subroutine, which in turn calls your m-file.

You do not directly call your m-file when solving the differential equation. This process is illustrated using a single, simple differential equation:

$$\frac{dy}{dt} = -10y, \quad y(0) = 1 \tag{8.16}$$

Integrate this equation from $t = 0$ to $t = 1$. The exact solution can be found by quadrature and is

$$y(t) = e^{-10t} \tag{8.17}$$

Step 1 To use MATLAB, you first construct an m-file that defines the equation.

```
% filename rhs.m
function ydot=rhs(t,y)   %compute the rhs of eqns.for any input
                         %values t, y
ydot = -10*y             %compute rhs
```

Step 2 Then to test the function, issue the command

```
q = rhs(0.2,3)
ydot = -30
q = -30
```

When $y = 3$, the value of the right-hand side should be -30, so the m-file is correct. In more complicated cases, you might remove all the ";" from the rhs.m so intermediate results are printed. This simplifies your task when checking the program. Now add ";" to all commands in rhs.m or else you will get lots of output.

Step 3 Next create an m-file that provides the script to run the problem. These commands can also be typed in the command window, but when you have several commands that you will use over and over, it is more convenient to create a script that can run all the commands with one message from you.

```
% filename simple.m
y0(1)=1.;                % set the initial condition to 1.
tspan=[0 1];             % integrate from t = 0 to t = 1
[t,y]=ode45(@rhs,tspan,y0)    % call the routine ode45
plot(t,y,'r')            % plot the solution
xlabel('t')                  % label the x-axis
ylabel('y')                  % label the y-axis
```

The results give a table of values for t and y and Figure 8.1.

```
t = 0   0.0050  0.0100  0.0157 ...  0.9475 0.9738        1.0
y = 1.0 0.9510  0.9044  0.8601 ...  0.000076   0.000059  0.000046
```

Step 4 The validity of your solution depends upon several things. First, you check the m-file "rhs" and determine that it gives the correct result when you insert values for t and y. You always have to do this check to show that your work is correct. You depend upon MATLAB to do its job correctly. You can easily check that it uses the initial and ending times correctly from the output and can see that the initial condition for y is correct, too.

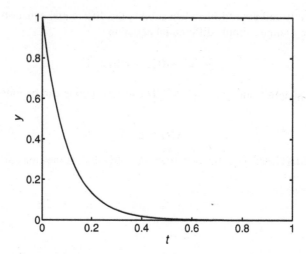

FIGURE 8.1 Solution to Eq. (8.16) using MATLAB.

You can always solve a problem for which you know the solution, and you try to make that problem as much like your problem as possible. In this case, you do not need to use numerical methods for the test, because you have an exact solution. Thus, you can compare the numerical result to the analytical result (they are the same) and determine that MATLAB did its job correctly. In cases without an analytical solution, you check your m-file "rhs" and then depend on MATLAB (and the accuracy checks). Note that in the tabulated values shown earlier, the increments between time points are 0.005 at the start of the calculation and 0.0272 at the end of the calculation. The integration routine adjusts the time step to insure that a specified accuracy is met. You can also decrease the numerical error by setting a parameter (error tolerance) to improve the result (see Appendix B).

Use of the "Global" Command

In the aforementioned example, you solved the problem when the right-hand side was $-10y$, and you simply used the value 10 in the program. There are two other ways to set this value, using the command "global" or by passing k as a parameter. You might want to write this as $-ky$ and solve the problem for several values of k. Because the function only knows what is set inside the function, or what gets transferred to it, you need to learn how to transfer information from the workspace into the function. You will need to make those changes to the code used earlier. If you use the "global" command, put the variable name in both the workspace and the m-file and assign its value in one place or the other. Then that value is accessible from either the m-file or the workspace.

Step 1 Here, you will set the value of k in the workspace, or calling program, and then it will be available in the m-file.

```
% filename simple2.m   %compute the rhs of Eq. (8.16)
global k
k=10;                   % set k to 10; it is now available to any
                        % m-file that has the
```

```
                  % global k statement in it.
y0=1;             % set the initial condition to 1
tspan=[0 1];      % integrate from t = 0 to t = 1
         % call the routine ode45; use the m-file called rhs2,
[t,y]=ode45(@rhs2,tspan,y0)
plot(t,y)  % plot the solution
```

Step 2 Next, add the global command to the m-file, called rhs2. (Note you added a numerical identifier to both the calling program and the m-file to help keep them straight.)

```
% filename rhs2.m
function ydot=rhs(t,y)
                  % compute the rhs of Eq. (8.16) for any input
                  % values t, y
global k             % The m-file can access the numerical
                     % value of k, if
                     % it has been set to global and given a
                     % value elsewhere
disp(k) % display the numerical value of k
              % use this only when testing the function rhs2
ydot = -k*y;   % compute rhs
```

Naturally, you expect the same results, which you can verify by running the new code.

Step 3 Check the code as done before, with one addition. It is helpful the first time you run the code to print the variables inside the function to make sure they are accessible. If not, MATLAB will tell you that the variable is not defined. Alternatively, use the debugging techniques described in Appendix B.

Passing Parameters

Still another way to introduce k into the function is to use it as a parameter in the calling argument.

```
% filename simple3.m
k=10;             % set k to 10
y0=1;             % set the initial condition to 1
tspan=[0 1];      % integrate from t = 0 to t = 1
options = []      % the OPTIONS = [] is a place holder
                  % The symbol [] is made as [] without a space.
                             % call the routine ode45;
                             % use the m-file called rhs3,
[t,y]=ode45(@rhs3,tspan,y0,options,k)
plot(t,y) % plot the solution
```

Now the function is

```
% filename rhs3.m
function ydot = rhs(t,y,k)
% compute the rhs of eqns. for any input values t, y
```

```
disp(k)          % display the numerical value of k; use only for
                 % testing of rhs3
ydot = -k*y;     % compute rhs
```

This program gives the same results as given by simple1.m and simple2.m.

Example: Isothermal Plug Flow Reactor

In this section, you will solve the equations for an isothermal plug flow reactor. The first problem is very simple and is patterned after a problem on the California Professional Engineers License Examination (Fogler, 2010). Here it is modified. You take a reactor in which components A and C are fed in equimolar amounts, and the following reaction takes place:

$$2A \rightarrow B \tag{8.18}$$

You assume that the reaction takes place in the liquid phase and that the volumetric flow rate remains constant even when reaction occurs. The equations are Eq. (8.5) for each species:

$$u \frac{dC_j}{dz} = r_j, \quad j = 1, \ldots, 3 \tag{8.19}$$

where C_j is the molar concentration of the jth species, u is the velocity, z is the distance down the tube from the inlet, and r_j is the rate of reaction of the jth species, in moles per volume time. Here the rate of reaction is taken as second order:

$$\text{Rate of formation of B} = kC_A^2 \tag{8.20}$$

where the units of k are volume per mole time. The equations for all three species are then

$$u \frac{dC_A}{dz} = -2kC_A^2, \quad u \frac{dC_B}{dz} = +kC_A^2, \quad u \frac{dC_C}{dz} = 0 \tag{8.21}$$

At the inlet you take

$$C_A(0) = 2 \text{ kmol/m}^3, \quad C_B(0) = 0, \quad C_C(0) = 2 \text{ kmol/m}^3 \tag{8.22}$$

and we take $u = 0.5$ m/s, $k = 0.3$ m^3/kmol s, and the total reactor length as $z = 2.4$ m.

Step 1 The MATLAB program requires you to write a function that defines the right-hand side. The input parameters to the function are the concentrations of all species. (a) Thus, to solve the problem you use the variables

$$y_1 = C_A, \quad y_2 = C_B, \quad y_3 = C_C \tag{8.23}$$

(b) The function also needs the velocity, u, and the rate constant, k. The distance from the inlet, z, takes the place of time and is the independent variable. (c) The rates of reaction are then evaluated, and the function returns the numerical value of the right-hand side. The code for the function is

```
% rate1.m
% This function gives the right-hand side for a simple
% reactor problem which is isothermal.
function ydot=rate1(VR,y)
% y(1) is CA, y(2) is CB, y(3) is CC
% k = 0.3 and u = 0.5
CA=y(1);
rate = 0.3*CA*CA;
ydot(1) = - 2.*rate/0.5;
ydot(2) = + rate/0.5;
ydot(3) = 0.;
ydot = ydot';
```

Step 2 You test this m-file by calling it with specific values for $y_j, j = 1, 2, 3$ to ensure that it is correct. Using $y(j)$ for y_j, issue the following commands.

```
y(1) = 0.2; y(2) = 0.3; y(3) = 0.4; rate1(0.1,y)
```

You get

```
ans = -0.048, 0.024, 0
```

which agrees with hand calculations. This is a very important step because this is where you add value. MATLAB will integrate whatever equations you give it, right or wrong, and only *you* can ensure that the program has solved the right equations.

Step 3 Next, write a code that serves as the driver. This code must (a) set any constants (here they are just put into the function rate1 for simplicity), (b) set the initial conditions and total reactor length, and (c) call the ode solver.

```
% run_rate1.m
% This is the driver program to solve the simple flow reactor
example.
y0 = [2 0 2]              % set the initial conditions
zspan = [0 2.4]           % set the total volume of the reactor
[z y] = ode45(@rate1,zspan,y0)              % call the ode solver
plot(z,y(:,1),'*-',z,y(:,2),'+-',z,y(:,3),'x-')   % plot the result
xlabel('length (m)')
ylabel('concentrations (kgmol/m^3)')
legend('A', 'B', 'C')
```

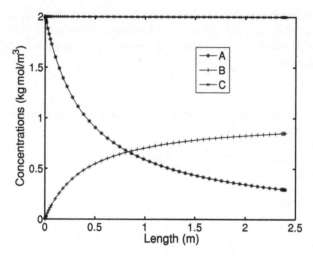

FIGURE 8.2   Different plot of solution to problem posed by Eqs. (8.21)–(8.22).

Step 4 Each curve is labeled with a different symbol in the plot command, and the curves are easy to distinguish (see Figure 8.2). The plot command given earlier will make the three curves have different colors, but those are not visible in this black and white version. Now that you have a validated code, you can vary the parameters to see their effect.

Example: Nonisothermal Plug Flow Reactor

Most reactions have heat effects, and this means that reactor models must be able to model the heat transfer as well as the mass balances. The following example models a simple reactor oxidizing SO_2 to form SO_3 (Young and Finlayson, 1973). After some manipulation, the equations are

$$\frac{dX}{dz} = -50R', \ \frac{dT}{dz} = -4.1(T - T_{surr}) + 1.02 \ \times 10^4 R' \tag{8.24}$$

where the reaction rate is

$$R' = \frac{X[1 - 0.167(1 - X)]^{1/2} - 2.2(1 - X)/K_{eq}}{[k_1 + k_2(1 - X)]^2} \tag{8.25}$$

$$\ln k_1 = -14.96 + 11,070/T, \ \ln k_2 = -1.331 + 2331/T$$
$$\ln K_{eq} = -11.02 + 11,570/T \tag{8.26}$$

with the parameters: $T_{surr} = 673.2$, $T(0) = 673.2$, $X(0) = 1$. The variable X is the concentration of SO_2 divided by the inlet concentration, $1–X$ is the fractional conversion, and T is the temperature in kelvin. The first equation is the mole balance on SO_2, and the second is the energy balance. The first term on the right-hand side of Eq. (8.24) represents cooling at the wall; the second term there is the heat of reaction.

These equations are appropriate when radial and axial dispersion are not important. Axial and radial dispersion are actually important in this case, as discussed by Young and Finlayson (1973), but the simpler case is a good vehicle for seeing how to include temperature effects in the problem. The reaction rate equation shows that equilibrium can be reached (then $R' = 0$), and the equilibrium constant depends on the temperature. Thus, the conversion depends intimately on the temperature.

Step 1 The MATLAB program requires you to write a function (m-file) that calculates the right-hand side of Eq. (8.24), given the input, z, X, and T, which are the axial location in the reactor and the conversion and temperature at that axial location, respectively. Thus, (a) to solve the problem you use the variables

$$y(1) = X, \ y(2) = T \qquad (8.27)$$

The axial location, z, and the variables $y(1)$ and $y(2)$ will be available to the function. (b) You first move $y(1)$ into X and $y(2)$ into T (for clarity in the program) and then (c) evaluate the rate of reaction. (d) Finally, you calculate the right-hand sides of the differential equations and put them in the vector ydot, which is the output from the function.

```
% rateSO2.m for SO2 reaction
function ydot = rateSO2(z,y)
% X is concentration SO2 divided by the inlet concentration
X = y(1);
% T is temperature in degrees K
T = y(2);
k1 = exp(-14.96+11070/T);
k2 = exp(-1.331+2331/T);
Keq = exp(-11.02+11570/T);
term1 = X*sqrt(1-0.167*(1-X));
term2 = 2.2*(1-X)/Keq;
denom = (k1+k2*(1-X))^2;
rate = (term1-term2)/denom;
ydot(1) = -50*rate;
ydot(2) = -4.1*(T-673.2)  + 1.02e4*rate;
ydot = ydot';    % converts ydot to a column vector (required)
```

An alternative for the last three lines is

```
term1 = -50*rate;
term2 = -4.1*(T-673.2)  + 1.02e4*rate;
ydot = [term1;term2]    % ydot is a column vector
```

Step 2 You test this m-file by setting the input variables and calling the m-file.

```
y(1) = 0.2; y(2) = 573.2;
rateSO2(0.1,y)
ans = -0.001148    410.235
```

This agrees with hand calculations, so the m-file is correct. (To make the comparison easier, you can remove the semicolons from rateSO2. After checking, the semicolons are added again.) You *must* make this check to be sure you have solved the desired problem.

Step 3 Next, you write the code as the driver. This code must set the constants, set the initial conditions for conversion and temperature, decide how far down the reactor to integrate, and then call the ode solver.

```
% run_SO2.m
% set the dimensionless initial conditions
y0 = [1 673.2]
% set the integration range
zspan = [0 1]
% call the solver
[z y] = ode45(@rateSO2,zspan,y0)
% plot the result
plot(z,y(:,1),'r-')
xlabel('dimensionless axial position')
legend('fraction converted')
pause
plot(z,y(:,2),'g-')
xlabel('dimensionless axial position')
legend('temperature (K)')
```

The results are shown in Figure 8.3. The concentration and temperature are plotted in separate graphs because they have such different magnitudes. Now you can change the parameters to see their effect. Change the inlet temperature to 800 K and see what happens.

USING COMSOL MULTIPHYSICS TO SOLVE ORDINARY DIFFERENTIAL EQUATIONS

Comsol Multiphysics is treated in detail in Chapters 9–11 and Appendix D, but it can also be used to solve reactor problems. The advantage of Comsol Multiphysics is that you program with a graphical user interface, so computer errors are less likely. It is still necessary to check your work, though. Although the applications in this chapter are all one-dimensional (to compare with MATLAB solutions), it is easy to solve two-dimensional problems, as described in more detail in later chapters. We show here how to solve the same three problems already solved using MATLAB: the simple exponential, Eq. (8.16); the isothermal flow reactor, Eqs. (8.21)–(8.22); and the nonisothermal reactor, Eqs. (8.24)–(8.26).

Simple Example

The problem (8.16) can be restated as

$$\frac{du}{dt} = -10u, \ u(0) = 1, \quad f(u, u_t, u_{tt}, t) \equiv ut + 10^*u = 0 \qquad (8.28)$$

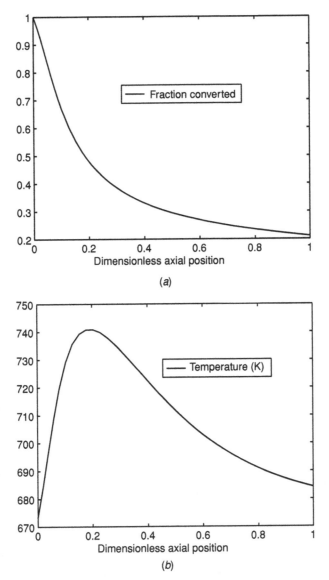

FIGURE 8.3 Solution to problem posed by Eqs. (8.24)–(8.26). (*a*) Fraction converted; (*b*) temperature.

We wish to solve this initial value problem using Comsol Multiphysics; since there are no space dimensions, the initial value problem is designated as a zero-dimension problem—zero space dimensions.

Step 1 (Choose the Problem) Open Comsol Multiphysics and choose "0D" in the Model Wizard and click the next arrow (blue, at the top of the Settings window). Then click Mathematics to expand the folder, expand the "Global ODEs and DAE interface folder,"

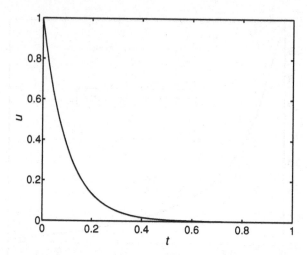

FIGURE 8.4   Solution to Eq. (8.16) using Comsol Multiphysics.

and select "Global ODEs and DAEs(ge)"; click the next arrow. The Settings window will have a list of options; select "Time Dependent" and then click the Finish flag.

Step 2 (Define the Equation)   Open the "Global ODEs" node (click on the triangle before it) and click on "Global Equations." In the settings window, enter *u*, *ut*+*10**u, 1. This says the variable is called *u*, the equation is Eq. (8.28), and the initial condition is 1.0.

Step 3 (Solve the Problem)   Click on the arrow before "Study 1" to expand the node and then click on Step 1: "Time Dependent." Change the Times from range (0:0.1:1) to (0:0.01,1). Right click on "Study 1."

Step 4 (Plot the Solution)   The plot should appear automatically. If not, choose Results under Model Builder; right click to get "1D Plot Group." If that already exists, choose it. Click on "1D Plot Group" and in the Settings window change "Time Selection" from All to "From list" to see the list of *t* values. Click on "Plot Group 1" and set the axis and plot settings. You will get Figure 8.4, which is the same as obtained using MATLAB, Figure 8.1.

Step 5 (Save the Figure)   To save the figure, click on the camera, click on Browse, give the figure a name, click Save, and then OK.

Example: Isothermal Plug Flow Reactor

The problem is solved in the original coordinates [molar flow rates and reactor volume In Eq. (8.2) since those are conveniently included in Comsol Multiphysics:

$$\frac{dF_A}{dV} = -2kC_A^2, \quad \frac{dF_B}{dV} = kC_A^2, \quad \frac{dF_C}{dV} = 0 \tag{8.29}$$

Step 1 (Choose the Problem)   Open Comsol Multiphysics and select "0D" in the Model Wizard; click the next arrow. Then select "Chemical Species Transport/Reaction

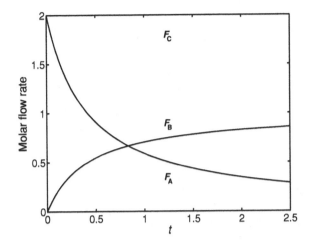

FIGURE 8.5   Solution to problem posed by Eq. (8.21)–(8.22) using Comsol Multiphysics.

Engineering (RE)" and click the next arrow. Finally, choose "Stationary Plug Flow" and click the Finish flag. If you do not have the Reaction Engineering Module, choose the "Global ODEs."

Step 2 (Define the Equation)   Click on "Reaction Engineering" in the Model Builder; in the Settings window set the reactor type to plug flow and make the mixture for a liquid. Keep the flow rate v as 1 as a simplification in this example. Then the molar flow rate $F_A = vC_A = C_A$. Right click "Reaction Engineering" again; this time choose Reaction. Set the chemical reaction to 2A $\Rightarrow$B, irreversible. The program will automatically add Species:A and Species:B to "Reaction Engineering." Set the forward rate constant kf to 0.6; set the reaction rate to kf_1*c_A if it is not done automatically. Go to Species:A and set the reaction rate to $-2*r\_1$. For Species:B set it to $+r\_1$. Right click the "Reaction Engineering" node and add a species, C. The rates of reaction for this species are zero. Open the "Species Feed Stream" node for each species and set the values to F_A = 2, F_B = 0, F_C = 2.

Step 3 (Solve)   Under Study 1, choose "Stationary Plug Flow" and change the range of solution from 0:0.1:1 to 0:0.1:2.5. Solve the problem by choosing "Study 1" and right clicking; choose =.

Step 4 (Plot the Solution)   The plot should appear automatically. If not, choose "Results" under Model Builder; right click "1D Plot Group." If that already exists, choose it. Right click on "1D Plot Group" and choose Global. Then click the plot icon in the Settings panel. You can change the labels for the x- and y-axis here under "Plot Settings." The y-axis label is changed to "Molar flow rate." You will get Figure 8.5.

Step 5 (Save the Figure)   To save the figure, click on the camera, click on Browse, give the figure a name, click Save, and then OK. You see that the solution shown in Figure 8.5 looks the same as obtained using MATLAB, Figure 8.2.

Example: Nonisothermal Plug Flow Reactor

Consider next the problem posed by Eqs. (8.24)–(8.26). Because there are two variables, you need two equations.

Name	Expression
k1	exp(-14.96+11070/T)
k2	exp(-1.331+2331/T)
Keq	exp(-11.02+11570/T)
num	X*(1-0.167*(1-X))^0.5-2.2*(1-X)/Keq
denom	(k1+k2*(1-X))^2
Rate	num/denom

▼ Variables

FIGURE 8.6 Variables.

Step 1 (Choose the Problem) Open Comsol Multiphysics and choose "0D" in the Model Wizard and click the next arrow. Then click Mathematics to expand the folder, expand the "Global ODEs and DAE interface folder," and select "Global ODEs and DAEs(ge)"; click the next arrow. The Settings window will have a list of options; select "Time Dependent" and then click the Finish flag.

Step 2 (Define Equation) Under the "Model 1" node, right click Definitions and select Variables. In the settings window, create the Variables as shown in Figure 8.6. Expand "Global ODEs and DAEs" and click on "Global Equations." Set the two equations as shown in Figure 8.7, which agree with Eq. (8.24). The initial conditions on X and T are set here, too.

Step 3 (Solve) Expand the Study 1 node, choose Time Dependent and note that the range of solution is from 0 to 1, as desired. Right click on Study, $=$.

Step 4 (Plot) The plot should appear automatically. If not, choose Results under Model Builder; right click to get the "1D Plot Group." If that already exists, choose it. Click on the arrow before "1D Plot Group" and choose Global. Then create the expressions to plot as shown in Figure 8.8. The results (X, T, Rate, and K_{eq}) are all scaled so that they appear on Figure 8.9. The "x-Axis Data" is changed to Expression, and set to t, so that the description can be changed to length. Since there are no units in this problem, click on "Plot Group 1" and remove the "(s)" from the x-axis label. It is useful to see what part of the reactor is doing the most work and to see how the equilibrium constant changes with temperature, which changes with axial position.

▼ Global Equations

$$f(u,u_t,u_{tt},t) = 0, \quad u(t_0) = u_0 \quad u_t(t_0) = u_{t0}$$

Name	f(u,ut,utt,t)	Initial value (u_0)	Initial value (u_t0)
X	Xt+50*Rate	1	0
T	Tt+4.1*(T-Tsurr)-Rate*1.02e4	673.2	0

FIGURE 8.7 Equations.

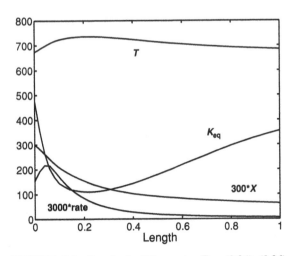

Expression	Unit	Description
300*mod1.X		State variable X (300*mod1.X)
mod1.T		State variable T (mod1.T)
3000*mod1.Rate		Rate of reaction (3000*mod1.Rate)
mod1.Keq		Equilibrium constant (mod1.Keq)

FIGURE 8.8 Expressions to plot to obtain Figure 8.9.

REACTOR PROBLEMS WITH MOLE CHANGES AND VARIABLE DENSITY

The examples shown earlier are for liquid streams or dilute gaseous streams where the density does not change drastically. Sometimes the density does change, though, and such problems are not much harder to solve numerically than their constant-density counterparts. The governing equation is Eq. (8.2), here expressed as a differential equation:

$$\frac{dF_j}{dV_R} = r_j, \quad j = 1, \ldots, 3 \tag{8.30}$$

F_j is the molar flow rate of the jth species, V_R is the volume of reactor (from the inlet), and r_j is the rate of reaction of the jth species. Let us take the same reaction, Eq. (8.18), and reaction rate, Eq. (8.20). Putting these together gives the differential equations to be solved:

$$\frac{dF_A}{dV_R} = -2kC_A^2, \quad \frac{dF_B}{dV_R} = +kC_A^2, \quad \frac{dF_C}{dV_R} = 0 \tag{8.31}$$

The equations are differential equations for the molar flow rates, F_A, F_B, and F_C but the rates of reaction are expressed in concentration, C_A. It is possible to derive the concentration

FIGURE 8.9 Results for SO_2 reactor, Eqs. (8.24)–(8.26).

of each chemical from the molar flow rates, and this section illustrates how to do that in MATLAB and Comsol Multiphysics. In a gas phase, the total concentration is governed by the perfect gas law, although more complicated equations of state are possible:

$$C_T = \frac{p}{RT} \tag{8.32}$$

Thus, at any point in the reactor, if you know the pressure and temperature, you can find C_T. The pressure is determined by fluid mechanics and the pressure drop, and the temperature is determined by the energy equation. Then the concentration of a chemical is the mole fraction times the total concentration, for example

$$C_j = x_j C_T, \quad \text{and the mole fraction is}$$

$$x_j = \frac{F_j}{F_A + F_B + F_C}, \quad j = A, B, \text{ or } C \tag{8.33}$$

If the temperature and pressure are constant, the total concentration is the total concentration at the inlet, which is taken here as 5 kmol/m^3. Thus, the flow rate is

$$F_A + F_B + F_C = \dot{V} C_T \tag{8.34}$$

The problem is summarized here as

$$\frac{dF_A}{dV_R} = -2kC_A^2, \quad \frac{dF_B}{dV_R} = +kC_A^2, \quad \frac{dF_C}{dV_R} = 0$$

$$F_A(0) = 2 \text{ kmol/s}, \quad F_B(0) = 0, \quad F_C(0) = 2 \text{ kmol/s} \tag{8.35}$$

$$C_A = \frac{F_A}{F_A + F_B + F_C} C_T$$

The MATLAB program requires you to write a function (m-file) that calculates the right-hand side, given the input, which are the flow rates of all species. Thus, to solve the problem you would use the variables

$$y(1) = F_A, \quad y(2) = F_B, \quad y(3) = F_C \tag{8.36}$$

The function will have available to it the accumulated reactor volume V_R, and the local value of $y(j), j = 1, \ldots, 3$ at the same reactor volume. You then calculate the concentrations at that reactor volume using Eq. (8.37):

$$C_A = \frac{y(1)}{y(1) + y(2) + y(3)} C_T \tag{8.37}$$

The rates of reaction are then evaluated, and the function returns the numerical value of the right-hand side. You can complete this example by doing Problem 8.7 at the end of the chapter.

The same considerations apply when using Comsol Multiphysics. The variables being solved for are now F_A, F_B, F_C. In Variables/Expressions/ you define equations for new variables ca, cb, and cc representing Eq. (8.37), and then use the same reaction rate expression (in terms of ca). The rest of the solution proceeds as before:

$$ca = CT^* F_A/(F_A + F_B + F_C)$$
$$cb = CT^* F_B/(F_A + F_B + F_C) \qquad (8.38)$$
$$cc = CT^* F_C/(F_A + F_B + F_C)$$

CHEMICAL REACTORS WITH MASS TRANSFER LIMITATIONS

When a reaction occurs on a catalyst particle, it is necessary for the reactants to get to the catalyst. Sometimes this transfer is hindered by mass transfer, and then the concentration near the catalyst is not the same as it is in the bulk stream. This phenomenon is demonstrated with the problem posed in Eqs. (8.21)–(8.22), except that now the rate of reaction is evaluated at a concentration that is near the catalyst. The concentration C_s is the concentration on the surface, expressed as the kmol of a species per volume of the catalyst. In your Chemical Reactor Design courses, you will learn various ways to represent this concentration. The net effect, however, is that Eq. (8.21) is changed to Eq. (8.39):

$$u\frac{dC_A}{dz} = -2k_s C_{A,s}^2, \quad u\frac{dC_B}{dz} = +k_s C_{A,s}^2, \quad u\frac{dC_C}{dz} = 0 \qquad (8.39)$$

You will be solving for $C_A(z)$, but you cannot evaluate the rate of reaction in Eq. (8.39) because you do not know $C_{A,s}$. You need a mass balance relating the rate of mass transfer to the catalyst to the rate of reaction. One form of that is Eq. (8.40), where k_m is a mass transfer coefficient in units of m/s, determined from correlations derived in fluid mechanics and mass transfer courses, a is the surface area exposed per volume of the reactor (m^2/m^3), and k_s is a rate of reaction rate constant (here m^3/kmol s). Other formulations are possible, too:

$$k_m a(C_A - C_{A,s}) = k_s C_{A,s}^2 \qquad (8.40)$$

Looking closely at Eqs. (8.39)–(8.40) you can see that to solve the differential equations in Eq. (8.39) you must solve Eq. (8.40) *at every position z*. Thus, this is a problem that combines ordinary differential equations with nonlinear algebraic equations.

MATLAB easily handles these kinds of problems. Basically, you call a routine to integrate the ordinary differential equations (e.g., ode45). You construct a right-hand side function (m-file) to evaluate the right-hand side. The input variables are z and the three concentrations, and the output variables are the three derivatives. Take C_A and solve Eq. (8.40) using either "fzero" or "fsolve." Then you have $C_{A,s}$ at this location, z. You can evaluate the rates of reaction in Eq. (8.39) and put them in the output from the m-file.

Step 1 You first construct an m-file to evaluate Eq. (8.40).

```
% mass_rxn.m
% this m-file evaluates the mass transfer equations
% with a reaction
function y=mass_rxn(CAs,CA)
global k km
% the first entry is the guess of CAs and
% the second entry is the fluid concentration
y = km*(CA-CAs)-k*CAs*CAs;
```

Step 2 Test this function by using the feval function.

```
global k km
k = 0.3; km = 0.35
feval(@mass_rxn,0.6,0.5)
ans = -0.143
```

This is correct.

Step 3 Next, test the use of the "fzero" command.

```
CAguess = 0.5; CA = 0.5;
OPTIONS=[]
CAs = fzero(@mass_rxn,CAguess,OPTIONS,CA);
```

The result is 0.37771, which is correct.

Step 4 Next, construct an m-file, "rate1_mass.m", to evaluate the right-hand sides of Eq. (8.39). This m-file is almost the same as that shown following Eq. (8.23), except that a global command is used so the parameters can be changed easily. The only significant change to this m-file from that used earlier is the use of "fzero" to find CAs, and then using CAs rather than CA to evaluate the rate of reaction.

```
% rate1_mass.m
% This function gives the right-hand side for a
% reactor problem which is isothermal but has
% mass transfer effects
function ydot=rate1_mass(VR,y)
global k vel km
% y(1) is CA, y(2) is CB, y(3) is CC
CA=y(1);
% solve the mass transfer problem
% the first entry is the guess of CAs and
% the second entry is the fluid concentration
OPTIONS=[];
CAguess = CA;
CAs = fzero(@mass_rxn,CAguess,OPTIONS,CA);
rate = k*CAs*CAs;
ydot(1) = - 2.*rate/vel;
ydot(2) = + rate/vel;
ydot(3) = 0.;
ydot = ydot';
```

Step 5 The last m-file is the one to run the problem, which you do using the value $k_m a = 0.2$. The only changes to the program run_rate1 following Eq. (8.23) are (a) the use of "global" to set the parameters and (b) changing the name of the right-hand side function from rate1 to rate1_mass.

```
% run_rate1_mass.m
% This is the driver program to solve the simple flow reactor
% example.
```

```
global k vel km
k = 0.3; vel = 0.5; km = 0.2;
% set the initial conditions
y0 = [2 0 2]
% set the total volume of the reactor
zspan = [0 2.4]
% call the ode solver
[z y] = ode45(@rate1_mass,zspan,y0)
% plot the result
plot(z,y(:,1),'*-',z,y(:,2),'+-',z,y(:,3),'x-') % or plot(z,y)
xlabel('length (m)')
ylabel('concentrations (kgmol/m^3)')
legend('A', 'B', 'C')
```

Step 6 As a check, you run this program while changing the "rate1_mass.m" program to use

```
%CAs = fzero(@mass_rxn,CAguess,OPTIONS,CA);
rate = k*CA*CA;
```

instead of

```
CAs = fzero(@mass_rxn,CAguess,OPTIONS,CA);
rate = k*CAs*CAs;
```

This has the effect of removing the mass transfer limitation, and the result should be the same as shown in Figure 8.2, and it is the same. The code is returned to its proper form, and the program "run_rate1_mass" calculates the output (see Figure 8.10). With mass transfer resistance included, the outlet concentration of B is 0.61. When there was no mass transfer limitation, the outlet concentration of B was 0.85. Thus, the reactor is not able to produce as much product, and a bigger reactor is required. This same type of

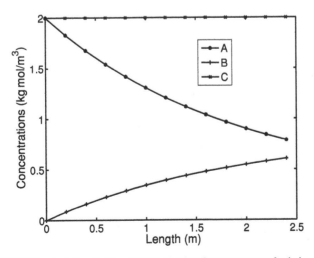

FIGURE 8.10   Solution to Eqs. (8.39)–(8.40) when mass transfer is important.

problem can arise with multiple concentrations or with one concentration and temperature. Then you would need to replace the "fzero" command with the "fsolve" command (from the Optimization Toolbox). You might need to experiment to provide good initial guesses for the concentrations and temperatures, because Eq. (8.40) would be turned into a set of nonlinear algebraic equations, and the solution is very sensitive to the temperature variable.

PLUG FLOW CHEMICAL REACTORS IN ASPEN PLUS

The water–gas shift reaction was studied in Chapters 4 and 6 as an equilibrium reactor. Here the same reactor is modeled in Aspen Plus using a packed bed with a kinetic expression, as a plug flow reactor, but still using a reactor at constant temperature. The reaction is

$$CO + H_2O \rightarrow CO_2 + H_2 \tag{8.41}$$

If the rate of reaction is r (mol/min m^3) then the following equation can be written for each species. The positive value is used for the products and the negative value is used for the reactants:

$$\frac{dF_i}{dV} = \pm r \tag{8.42}$$

The reaction rate is obtained from Choi and Stenger (2003) and Kolavennu et al. (2006):

$$r = k_0 \exp(-E/RT) \left[p_{CO}\, p_{H_2O} - \frac{p_{CO_2}\, p_{H_2}}{k_{eq}} \right] \tag{8.43}$$

The parameters are

$$K_{eq} = \exp\left[\frac{4577.8}{T} - 4.33 \right] \tag{8.44}$$

and

$$k_0 \exp(-E/RT) = 6.195 \times 10^8 \frac{mol}{atm^2 m^3\, min} \left[-\frac{47.53\ kJ/mol}{RT} \right]. \tag{8.45}$$

The reactor is to be isothermal at 450 K and 1 atm (1.013 bar). Thus, the mole fraction is the same as the partial pressure in atmospheres. The feed rate to the reactor is 33 kmol/min (1992 kmol/h) half carbon monoxide and half water. A reactor volume of 0.25 m^3 is considered. Under these conditions the reaction constant is

$$k_0 \exp(-E/RT) = 1882 \frac{mol}{m^3/min} = 0.0314 \frac{kmol}{m^3/s} \tag{8.46}$$

Open the program and use the unit RPLUG in Reactors, with stream in S1, and stream out as S2. Insert the chemicals, and choose Peng–Robinson equation of state since that is suggested as a good one (see Chapter 6). For stream S1, use 16.5 mol/min for carbon monoxide and 16.5 mol/min for water, as a vapor at 1.013 bar (1 atm) and 450 K. In the reactor block, in Setup I Specifications, choose Reactor at constant temperature, at inlet temperature. Under the Configuration tab, choose a length of 25 m and a diameter of 0.1129 m; then the cross-sectional area is 0.01 m^2 and the reactor volume is 0.25 m^3. Under the Reaction tab, create a new reaction set by right clicking the "available reaction sets." It creates a reaction set R-1 and asks you what type of equation you wish to use. Since the reaction kinetics is given in the form of Langmuir–Hinshelwood–Hougen–Watson (LHHW) formulation, choose LHHW. Next, go to the Reactions node in the Navigation Pane and open it: the reaction set R-1 is there. Click on New and create the stoichiometry of the reaction in the pop-up window.

Now comes the difficult part: defining the kinetics with the proper units. Click the kinetic tab. Make the reacting phase a vapor. It is helpful to look at the help display for Langmuir, which is summarized here. The reaction rate expression for a four-component system is in the form

$$\text{Rate} = \frac{\text{(Kinetic factor)(driving force expression)}}{\text{(Adsorption expression)}} \tag{8.47}$$

The form of these is

$$\text{Kinetic factor} = k \left(\frac{T}{T_0}\right)^n \exp\left[-\frac{E}{R}\left(\frac{1}{T} - \frac{1}{T_0}\right)\right] \tag{8.48}$$

We can make this agree with our expression by taking $n = 0$, $T_0 = 450$, $E = 0$ and $k = 0.0314$. Enter those values. The units of k have to be in kmol/s m^3, which are already calculated. Next, consider the driving force expression; click on the Driving Force box. Make the reacting phase vapor and make mole fraction be the concentration basis. If it were something different the value of k would have to be adjusted. The term is going to be

$$\text{Driving force} = k_1 \prod_{i=1}^{N} C_i^{\alpha_i} - k_2 \prod_{i=1}^{N} C_i^{\beta_i} \tag{8.49}$$

For our case we want

$$\text{Driving force} = y_{CO}y_{H_2O} - \frac{y_{CO_2}y_{H_2}}{K_{eq}} \tag{8.50}$$

Since the concentration basis is mole fraction, the Cs in Eq. (8.49) are mole fractions. The first k is 1.0 and the second one is $1/K_{eq}$. The standard form of these ks is

$$\ln k = A + \frac{B}{T} + C \ln T + DT \tag{8.51}$$

Thus, for k_1, we want $A = B = C = D = 0$. We also want for term1 (the first term in driving force) to have $a = 1$ for both CO and H_2O. We achieve this by making sure the "Enter term" says term1, choosing CO and H_2O and setting their coefficients to 1, keeping the others zero, and keeping all four constants as zero as shown in Figure 8.11. Then change "Enter term" to term2, the second term in Eq. (8.49). This time we set the coefficients of CO_2 and H_2 to 1, keeping the others as zero, and setting $k = 1/K_{eq}$. Thus, we set $A = 4.33$ and

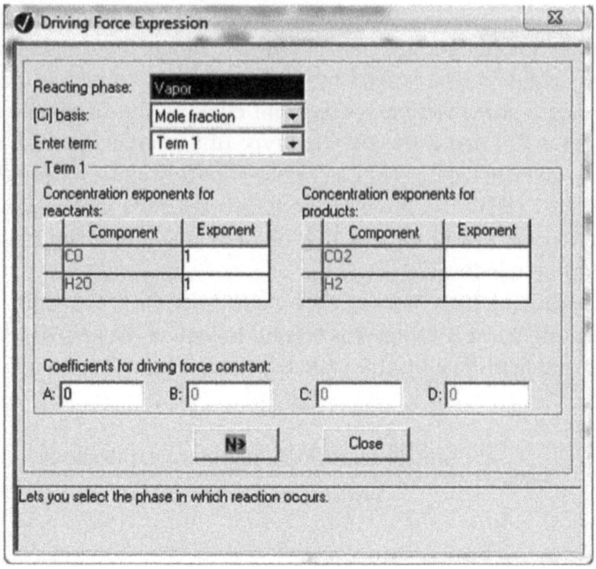

(a)

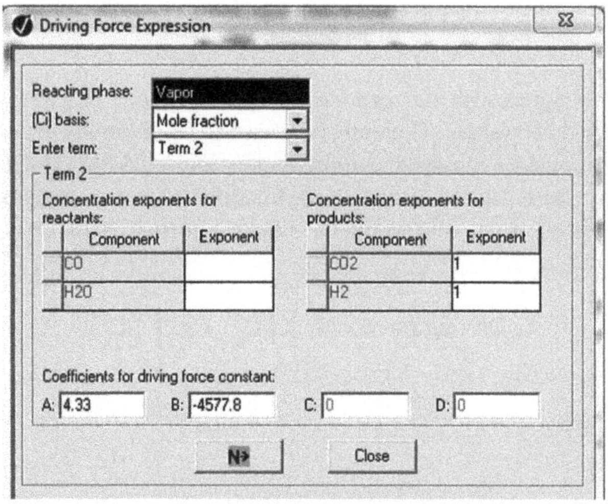

(b)

FIGURE 8.11 Driving force: (*a*) term 1; (*b*) term 2.

$B = -4577.8$. Finally, we set the Adsorption Expression to 1.0. The adsorption term is

$$\text{Adsorption term} = 1 + \prod_{i=1}^{N} k_i C_i^{\alpha_i} \tag{8.52}$$

Since we want this term to be one, we make all the k_i a small number. To do this we activate coefficients for terms 2 through 5 and set $A_i = -1000$. One way to check whether this has all been done correctly is to calculate the rate of reaction at the inlet conditions and compare it with the computed result (as the slope of the mol/min vs. volume curve).

Length	CO	H2O	CO2	H2
meter				
0	0.5	0.5	0	0
2.5	0.2916944	0.2916944	0.2083056	0.2083056
5	0.20608957	0.20608957	0.29391043	0.29391043
7.5	0.15956585	0.15956585	0.34043415	0.34043415
10	0.13039053	0.13039053	0.36960947	0.36960947
12.5	0.11042714	0.11042714	0.38957286	0.38957286
15	0.09594860	0.09594860	0.4040514	0.4040514
17.5	0.08500105	0.08500105	0.41499895	0.41499895
20	0.07645808	0.07645808	0.42354191	0.42354191
22.5	0.06962113	0.06962113	0.43037887	0.43037887
25	0.06403766	0.06403766	0.43596233	0.43596233

(Process Stream | Coolant Stream | Properties | User Variables)

Process stream profiles — View: Molar composition — Substream: MIXED

FIGURE 8.12 Mole fraction profiles in water–gas shift reaction, using Aspen Plus.

Note that in Aspen one does not need to worry about the fact that the total number of moles changes (if it did) down the reactor—it is handled automatically.

The problem can then be solved. The stream2 gives the output conditions (450 K, 1 atm, 0.127 kmol/h of CO), but we also want the profile within the reactor, which is available from the Reactor Block | Profiles. The tabulated values of mole fraction are shown in Figure 8.12. These are changed to conversion, which is plotted in Figure 8.13. These are nominally the same as derived in MATLAB. Aspen Plus has an advantage in that it calculates the energy balance and will work without much additional effort even if the temperature is changing.

CONTINUOUS STIRRED TANK REACTORS

Equation (8.15) gave the mass balance for a CSTR. A similar equation can be written as an energy balance. This example considers a CSTR in which a first-order reaction occurs, but the temperature also changes due to the heat of reaction. The equations to be solved are

$$\frac{Q}{V_R}(1 - c) = c \ \exp[\gamma(1 - 1/T)]$$

$$\frac{Q}{V_R}(1 - T) = -\beta \ c \ \exp[\gamma(1 - 1/T)]$$

(8.53)

The left-hand sides are the flow rate times a concentration or temperature difference between the input and output, divided by the volume. The equations have been normalized by the inlet concentration and temperature. The right-hand sides are the rate of reaction and the rate of energy generation due to reaction, respectively.

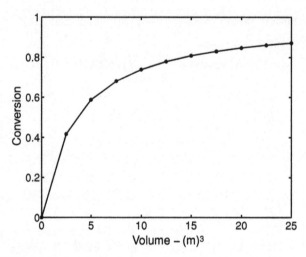

FIGURE 8.13   Conversion in water–gas shift reaction, using Aspen Plus.

The case described by Eq. (8.53) is for an adiabatic reactor. When the reactor is adiabatic, the equations can be combined by multiplying the first equation by β and adding it to the second equation; then the right-hand side vanishes:

$$\beta(1 - c) + (1 - T) = 0 \tag{8.54}$$

This equation can be solved for T:

$$T = 1 + \beta(1 - c) \tag{8.55}$$

Now the mass balance can be considered a single equation in one unknown (c):

$$\frac{Q}{V_R}(1 - c) = c \ \exp[\gamma(1 - 1/\{1 + \beta - \beta c\})] \tag{8.56}$$

Solution Using Excel

The first set of parameters used is $Q/V_R = 8.7$, $\beta = 0.15$, $\gamma = 30$. The simple spreadsheet is shown in Figure 8.14; the cells are named and the names are used in the expressions. You can use either "Goal Seek" or "Solver" to find the value of c that makes the function zero.

beta	0.15	
gamma	30	
flowvel	8.7	
conc	0.731073	
T	1.040339	=1+beta*(1-conc)
rate	2.339659	=conc*EXP(gamma*(1-1/T))
function	0.000006	=flowvel*(1-conc)-rate

FIGURE 8.14   Spreadsheet solving Eq. (8.56).

Solution Using MATLAB

MATLAB allows you to extend the method to more complicated situations when Excel does not work.

Step 1 First, construct an m-file that represents Eq. (8.56).

```
% rate_T
function fn=rate_T(c)
global beta gamma flowvol
T = 1 + beta*(1-c)
rate = c*exp(gamma*(1-1/T))
fn = flowvol*(1-c)-rate
```

Step 2 Test it using the commands

```
>> global beta gamma flowvol
>> beta = 0.15; gamma = 30; flowvol = 8.7;
>> rate_T(0.5)
T = 1.0750
rate = 4.0547
fn = 0.2953
ans = 0.2953
```

The answer is correct.

Step 3 To solve the problem, type the following commands into the command window.

```
>> fzero(@rate_T,0.5)
ans = 0.73107565193641
```

This is the same solution as that obtained using Excel, as expected.

CSTR with Multiple Solutions

For a different set of parameters, the CSTR can have more than one solution. For this problem, the solutions lie between 0 and 1, because the concentration has been normalized by the inlet value, where the normalized concentration is 1.0, and the reaction uses up the material. The solution you get depends upon the initial guess of c. Use Excel to solve the problem when $Q/V_R = 25$, $\beta = 0.25$, keeping $\gamma = 30$. Successive trials led to the results shown in Table 8.1.

TABLE 8.1   Multiple Solutions to Eq. (8.56) When $Q/V_R = 25$, $\beta = 0.25$, $\gamma = 30$

Initial Guess of c	Final Result
0	0.0863
0.5	0.5577
1.0	0.9422

Solutions to Multiple Equations Using MATLAB When two or more variables must be found, as in Eq. (8.53), a solution can be found to make both the equations zero without rearranging to produce Eq. (8.56).

Step 1 You can use the following MATLAB m-file along with the "fsolve" command (available in the Optimization Toolbox).

```
% rate_T2
function fn=rate_T2(y)
global beta gamma flowvol
c = y(1);
T = y(2);
rate = c*exp(gamma*(1-1/T));
fn(1) = flowvol*(1-c)-rate;
fn(2) = flowvol*(1-T)+beta*rate;
```

Step 2 Test the m-file with the following commands.

```
>> global beta gamma flowvol
>> beta = 0.15; gamma = 30; flowvol = 8.7;
>> feval(@rate_T2,[0.5 1.1])
ans = -3.29556351331856    0.27683452699778
```

This answer is correct.

Step 3 Then in the command window, issue the following commands.

```
beta = 0.25; gamma = 30; flowvol = 25;
fsolve(@rate_T2,[0.0 1.25])
```

With this initial guess, the program returns the first answer shown in Table 8.1. With an initial guess of [0.5 1.125], it returns the second answer. With an initial guess of [1 1] it returns the third answer. Such problems would not always converge, and the convergence is harder for larger β and γ. As the convergence becomes harder, it is more and more crucial to provide a reasonable starting guess.

TRANSIENT CONTINUOUS STIRRED TANK REACTORS

Reactors do not always run at steady state. In fact, many pharmaceutical products are made in a batch mode. Such problems are easily solved using the same techniques presented earlier because the plug flow reactor equations are identical to the batch reactor equations. Even CSTRs can be run in a transient mode, and it may be necessary to model a time-dependent CSTR to study the stability of steady solutions. When there is more than one solution, one or more of them will be unstable. Thus, this section considers a time-dependent

CSTR as described by Eq. (8.57):

$$V\frac{dc'}{dt'} = Q(c' - c'_{in}) - Vk_0c'\exp(-E/RT')$$

$$[\phi(\rho C_p)_f + (1-\phi)(\rho C_p)_s]V\frac{dT'}{dt'} = -(\rho C_p)_f Q(T' - T'_{in})$$

$$+ (-\Delta H_{rxn})Vk_0c'\exp(-E/RT') \qquad (8.57)$$

The variables are as follow:

V = reactor volume
Q = volumetric flow rate
c' = concentration
T' = temperature
t' = time
k_0 = reaction rate constant
E = activiation energy
ϕ = void fraction
ρ = density
C_p = heat capacity per mass

where subscript f for fluid, s for solid; $-\Delta H_{rxn}$ is the heat of reaction in energy per mole. The nondimensional form of these equations is

$$\frac{dc}{dt} = (1-c) - cDa\exp[\gamma(1 - 1/T)]$$

$$Le\frac{dT}{dt} = (1-T) + \beta cDa\exp[\gamma(1 - 1/T)] \qquad (8.58)$$

The parameters are defined as

$$c = \frac{c'}{c'_{in}}, \quad T = \frac{T'}{T'_{in}}, \quad t = \frac{Qt'}{V}, \quad Da = \frac{V}{Q}k_0\exp\left(-\frac{E}{RT'_{in}}\right)$$

$$Le = \frac{\phi(\rho C_p)_f + (1-\phi)(\rho C_p)_s}{(\rho C_p)_f}, \quad \beta = \frac{(-\Delta H_{rxn})c'_{in}}{(\rho C_p)_f T'_{in}} \qquad (8.59)$$

The parameter Le is a Lewis number, and it includes the heat capacity of the system. The Da is a Damköhler number and includes the rate of reaction. The parameters are taken as

$$\beta = 0.15, \quad \gamma = 30, \quad Da = 0.115 \; Le = 1080, \quad c(0) = 0.7, \quad T(0) = 1 \qquad (8.60)$$

Step 1 To integrate these equations using MATLAB, you need an m-file to compute the right-hand side, and then an m-file to call "ode45." The right-hand side function is

```
% rate_limit.m
% solve for particle with limit cycle
function ydot=rate_limit(t,y)
```

```
global beta gamma Damk Lewis Tin
c = y(1);
T = y(2);
rate = c*Damk*exp(gamma*(1 - 1/T));
ydot(1) = 1 - c - rate;
ydot(2) = Tin - T + beta*rate;
ydot(2) = ydot(2)/Lewis;
ydot = ydot';
```

Step 2 After checking this code you can run it using the following m-file.

```
% run_rate_limit
global beta gamma Damk Lewis Tin
beta = 0.15; gamma = 30; Lewis = 1080 ; Damk = 0.115
Tin = 1
tic
[t y]=ode45(@rate_limit,[0 2],[ 0.7 1.0])
toc
plot(t,y(:,1),'*-k',t,y(:,2),'o-k')
xlabel('time')
legend('concentration','temperature')
```

The result is shown in Figure 8.15. It looks as if steady state is achieved by the time that $t = 2$. However, this is not true. Integrate to $t = 1000$ and look at the results in Figure 8.16. It still has not reached steady state. The reason is that the temperature responds much more slowly than does the concentration. Thus, the concentration comes to a steady-state value appropriate to the current temperature, but then the temperature keeps changing and the concentration must change to keep up. Notice the very rapid change of c from the initial value of 0.7 to about 0.875 shown in Figure 8.17. This is because the value of $c = 0.7$ was not appropriate for a temperature of 1.0. In mathematical terms, the time response of the

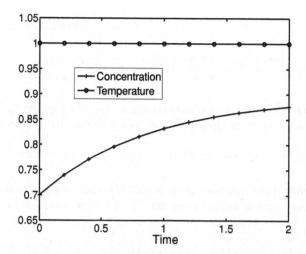

FIGURE 8.15 Transient CSTR, up to $t = 2$, elapsed time $= 0.0061$ s.

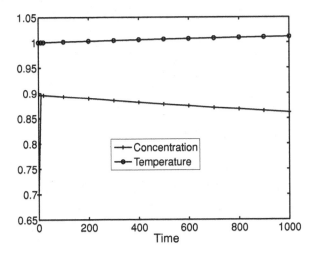

FIGURE 8.16 Transient CSTR, up to $t = 1000$, elapsed time $= 0.19$ s.

two variables is very different (the eigenvalues of the equation are widely separated), and the system is called stiff. See Appendix E for more discussion about stiff equations.

MATLAB has some solvers especially suited to stiff problems. You merely change from "ode45" to "ode15s." When integrating to $t = 1000$, you get Figure 8.17, but it only takes 0.016 seconds. Thus, the solution is 12 times faster. Since steady state still has not been reached, integrate to $t = 40,000$. Figure 8.17 shows that steady state has been reached. Now the computation times are 7.5 s for "ode45" and 0.026 seconds for "ode15s," giving "ode15s" a speed-up factor of over 290. The difference is made more dramatic if you remove the [t y] = in front of the call to ode45 and ode15s. Then the solution is plotted as it is calculated, and the difference in speed is very apparent.

Next change the parameter Le from 1080 to 0.1 and integrate to $t = 100$ using "ode15s."

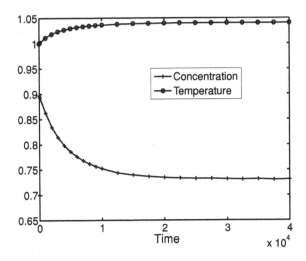

FIGURE 8.17 Transient CSTR, up to $t = 40,000$, elapsed time $= 0.026$ s.

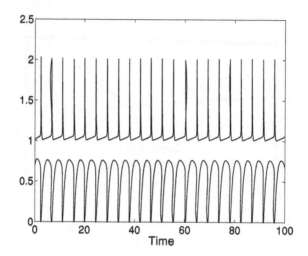

FIGURE 8.18   Transient CSTR, $Le = 0.1$, integration to $t = 100$.

Step 1 The code is shown below.

```
% run_rate_limit
global beta gamma Damk Lewis Tin
beta = 0.15; gamma = 30; Lewis = 0.1; Damk = 0.115; Tin = 1
tic
[t y]=ode15s('rate_limit',[0 100],[ 0.7 1.0])
toc
plot(t,y(:,1),'-k',t,y(:,2),'-k')
xlabel('time')
pause
plot(y(:,1),y(:,2),'-k')
xlabel('concentration')
ylabel('temperature')
```

The solution is shown in Figure 8.18. The pattern seems to be repeated over and over.

If the temperature is plotted versus the concentration at the same time, the repeated pattern is very apparent (see Figure 8.19). This problem exhibits what is called a limit cycle. This is also a nice problem to run without the [t y] = in front of the ode15s; then you see the solution develop before your eyes.

CHAPTER SUMMARY

In this chapter equations were derived for different reactors: plug flow, batch, and CSTR. You learned how to integrate ordinary differential equation using MATLAB, and then applied those skills to reactor problems in plug flow reactors (both isothermal and non-isothermal). You also learned to do the same thing using Comsol Multiphysics, Aspen Plus, and, for simple problems, Excel.

Complications arose when the moles or total concentration changed, or when mass transfer effects were important, and you learned how to incorporate those complications.

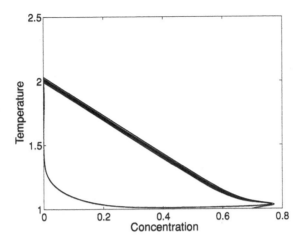

FIGURE 8.19   Limit cycle display of Figure 8.18.

Finally, CSTRs were studied using both MATLAB and Excel in cases where the solution is steady, where multiple steady solutions exist, and you learned to use MATLAB when the solution is time-dependent and the problem is stiff, leading to limit cycles.

PROBLEMS

8.1₁ Solve for the concentration distribution in a plug flow reactor governed by the following equations. (a) Use MATLAB and (b) use Comsol Multiphysics:

$$Q\frac{dc}{dV_R} = -R(c), \quad c(0) = c_{in} \tag{8.61}$$

$$R(c) \equiv \frac{kc}{c+v}, \quad Q = 50, \ c_{in} = 2, \ V = 2000 = \text{total volume},$$

$$k = 0.198, \ v = 3.8. \tag{8.62}$$

8.2₁ Solve for the concentration distribution in a plug flow reactor with the following reaction rate expression and parameters. (a) Use MATLAB and (b) use Comsol Multiphysics:

$$R(c) \equiv \frac{kc}{(1 + K_1 c)^2}, \quad Q = 2.2, c_{in} = 1, \ V = 50 = \text{total volume},$$

$$k = 2, \ K_1 = 4. \tag{8.63}$$

8.3₁ A CSTR is governed by the following equation

$$Q(c - c_{in}) = -VR(c) \tag{8.64}$$

(a) Solve for c when

$$R(c) \equiv \frac{kc}{c+v}, Q = 50, c_{in} = 2, V = 2000, k = 0.198, \psi = 3.8.$$

(b) Solve when

$$R(c) \equiv \frac{kc}{(1 + K_1 c)^2}, \; Q = 2.2, \; c_{\text{in}} = 1, \; V = 50, \; k + 2, \; K_1 = 4.$$

8.4₁ The problem posed by Eqs. (8.21)–(8.22) can be solved for a CSTR, too. In that case, the equation is multiplied by the cross-sectional area of the pipe to obtain

$$u \frac{dC_A}{dz} = -2k_s C_A^2 \rightarrow uA \frac{dC_A}{dAz} = -2kC_A^2 \rightarrow Q \frac{dC_A}{dV_R}$$
$$= -2kC_A^2 \rightarrow Q(C_A - C_{A,\text{in}}) = -2VkC_A^2 \tag{8.65}$$

Take the cross-sectional area as 0.001 m² and solve for the same conditions as in Eq. (8.22).

8.5₁ The growth rate for cells in a bioreactor can follow a variety of growth equations. Monod kinetics is represented by Problem 8.3(a). A substrate-inhibition growth rate is given by (see Shuler (1988))

$$\text{Growth rate} = \frac{\mu_m S}{K_s + S + S^2/K_I} \tag{8.66}$$

where μ_m is the maximum specific growth reaction rate; K_s is the monod constant, K_I is the inhibition constant, and S represents the substrate concentration (the material being reacted). The equation growing cells in a CSTR is

$$Q(S - S_{\text{in}}) = -V \frac{\mu_m S}{K_s + S + S^2/K_I} \tag{8.67}$$

In nondimensional form this is

$$(\sigma - 1) = -Da \frac{\sigma}{\omega + \sigma + \varepsilon \sigma^2},$$

where

$$Da = \frac{V \mu_m}{Q S_{\text{in}}}, \quad \omega = \frac{K_s}{S_{\text{in}}}, \quad \varepsilon = \frac{S_{\text{in}}}{K_I} \tag{8.68}$$

Solve this problem for $Da = 1.19$, $\omega = 3.56 \; 10^{-3}$, $\varepsilon = 2.53$. (Hint: look for multiple solutions.) See Schmidt (2005) and Fogler (2010) for additional information about biological reactors.

8.6₁ Solve the same Problem 5 with the CSTR replaced by a plug flow reactor. The equations are then

$$\frac{d\sigma}{dz} = -Da \frac{\sigma}{\omega + \sigma + \varepsilon \sigma^2}, \; z = 0 \rightarrow 3, \; \sigma(0) = 1 \tag{8.69}$$

8.7₁ Solve Eq. (8.35) using MATLAB for a total volume of 1.2 m³, $k = 0.3$ m³/(kmol s), a total concentration of 5 kmol/m³.

8.8[1] Ammonia undergoes a reaction with oxygen in the presence of a platinum catalyst to form nitric oxide and water. The reaction is carried out in a microreactor (see Chapter 11), which results in a constant temperature. The reaction is

$$NH_3 + \tfrac{5}{2}O_2 \rightarrow NO + \tfrac{3}{2}H_2O \tag{8.70}$$

The reaction occurs on the surface of a tube in the microreactor as governed by the following rate of reaction (in $g\,mol/cm^2$) (Pignet and Schmidt, 1975). The partial pressures are in torr and the total pressure is maintained at 0.115 torr (as an approximation). The rate is in $g\,mol/cm^2$ s:

$$r_{NH_3} = \frac{A}{B \times C}, \quad A \equiv 3.4 \times 10^{-8}\exp(21{,}700/RT)p_{NH_3}p_{O_2}^{1/2}$$

$$B \equiv 1 + 8 \times 10^{-2}\exp(4400/RT)p_{O_2}^{1/2} \tag{8.71}$$

$$C \equiv 1 + 1.6 \times 10^{-3}\exp(25{,}500/RT)p_{NH_3}$$

The inlet stream contains ammonia and oxygen with partial pressures of 0.046 and 0.068 torr, respectively, and the flow rate is 100 L/h. Solve the problem for a temperature of (a) 473 K; (b) 673 K and compare the results. The microreactor channel is 40 × 300 µm, and its length is 1 cm. The governing equations are

$$\frac{dF_i}{dz} = Pv_i r_{NH_3}, \quad F_i(0) = F_{i0} \tag{8.72}$$

where v_i is the stoichiometric coefficient of ith species; P is the perimeter in cm; F_i is the flow rate of ith species in $g\,mol/s$; F_{i0} is the flow rate of ith species at inlet in $g\,mol/s$.

In addition,

$$F_T = \sum_{i=1}^{N} F_i, \quad y_i = \frac{F_i}{F_T}, \quad p_i = y_i p_T \tag{8.73}$$

8.9[1] When an enzyme is immobilized on a surface, there may be a mass transfer resistance that limits the concentration of the reacting species on the surface. The following problem (from Shuler, 1988) finds the concentration of substrate on the surface when there is a reaction accompanied by mass transfer resistance. The Michaelis–Menten reaction on a surface is governed by

$$\text{rate of reaction} = \frac{k_0 S}{S + K_m} \tag{8.74}$$

The equation for reaction at a surface with mass transfer limitations approaching the surface is

$$k_L(S_f - S) = \frac{\varepsilon k_0 S}{S + K_m} \tag{8.75}$$

where k_L is the mass transfer coefficient $= 10^{-3}$ cm/s; K_m is the kinetic constant $= 10^{-4}$ mol/L; k_0 is the kinetic constant $= 5 \times 10^{-4}$ mol/s mg enzyme; ε is the enzyme loading $= 1.6 \times 10^{-4}$ mg enzyme/cm^2; S_f is the external concentration $= 10^{-4}$–10^{-2} mol/L; S is the concentration at surface in mol/L.

Solve for S when the external concentration is (a) 10^{-4} mol/L and (b) 10^{-2} mol/L

8.10$_1$ The steady-state model for a nonisothermal CSTR representing a catalytic converter is

$$QC_{tot}(y_{in} - y) = \alpha V r(y, T)$$

$$QC_{tot}\hat{M}C_{pg}(T - T_{in}) = \alpha V(-\Delta H_{rxn})r(y, T)$$

$$r = \frac{0.05 k_1 y}{T(1 + K_1 y)^2}, \quad k_1 = 6.70 \ 10^{10} \exp(-12556/T), \ K_1 = 65.5 \ \exp(961/T)$$

$$(8.76)$$

The reaction rate expression is given for oxidizing carbon monoxide on a catalytic surface. When the inlet mole fraction CO is $y_{in} = 0.02$, and the inlet temperature is $T_{in} = 600$ K, find the solution to these equations. The parameters are $Q = 0.06555$ m^3/s, $C_{tot} = 0.0203$ kmol/m^3, $\alpha = 26{,}900$ m^2/m^3, $V = 6 \ 10^{-4}$ m^3, $\hat{M} = 30$ kg/kmol, r in kmol/m^{2}s, $C_{pg} = 1070$ J/kg K, $-\Delta H_{rxn} = 2.84 \ 10^8$ J/kmol.

8.11$_1$ Solve the problem in Eq. (8.53) but for a plug flow reactor instead of the CSTR.

$$Q\frac{dc}{dV_R} = -c \ \exp([\gamma(1 - 1/T)]), \quad c(0) = 1$$

$$Q\frac{dT}{dV_R} = \beta c \ \exp([\gamma(1 - 1/T)]), \quad T(0) = 1$$

$$(8.77)$$

Integrate from $V_R/Q = 0$ to 0.115 with $\beta = 0.15$, $\gamma = 30$. Compare the conversion to that achieved with a CSTR.

8.12$_1$ Repeat Problem 11 but with $V_R/Q = 0$ to 0.3 with $\beta = 0.25$, $\gamma = 30$.

8.13$_1$ Growth of cells in a transient CSTR is governed by

$$\frac{d\sigma}{dt} = 1 - \sigma - Da\frac{\sigma}{\omega + \sigma + \varepsilon\sigma^2}, \quad \sigma(0) = \sigma_0 \qquad (8.78)$$

For the parameters $\omega = 3.5625 \times 10^{-3}$, $\varepsilon = 2.5216$, $Da = 1.1905$, integrate to $t = 24$ for the following initial conditions for c_0: (a) 1; (b) 0; (c) 0.146019; (d) 0.142; (e) 0.150. Discuss.

8.14$_1$ During the cure of an epoxy in a prepreg operation to make polymeric composites, the extent of reaction, α, is sometimes modeled with the equation

$$\frac{d\alpha}{dt} = (1 + a\alpha^m)(1 - \alpha)^n, \quad \alpha(0) = 0 \qquad (8.79)$$

where $a = 3$, $m = 0.5$, $n = 0.7$. Solve this equation as a function of time.

8.15₁ Problem 8.10 is considered as a transient CSTR. The equations governing it are

$$\varepsilon V C_{tot} \frac{dy}{dt} = Q C_{tot}(y_{in} - y) - \alpha V r(y, T)$$

$$V \left[\varepsilon \rho_g C_{pg} + (1 - \varepsilon) \rho_s C_{ps} \right] \frac{dT}{dt} = Q C_{tot} \hat{M} C_{pg}(T_{in} - T) + \alpha V (-\Delta H_{rxn}) r(y, T)$$

$$(8.80)$$

The additional parameters are

$$C_{pg} = 1070 \text{ J/kg K}, \quad C_{ps} = 1000 \text{ J/kg K}, \quad \varepsilon = 0.68$$
$$\rho_g = C_{tot} \hat{M} = 0.609 \text{ kg/m}^3, \quad \rho_s = 2500 \text{ kg/m}^3$$

Initially the inlet mole fraction (of carbon monoxide) is 0.01 and the inlet temperature is 600 K. Integrate these equations under the following conditions:

$$t < 1 \text{ s} : y_{in} = 0.01, \ T_{in} = 600 \text{ K}$$
$$1 < t < 2 \text{ s} : \text{ inlet conditions increase linearly to}$$
$$y_{in} = 0.03, \ T_{in} = 700 \text{ K}$$
$$2 < t \text{ s} : y_{in} = 0.03, \ T_{in} = 700 \text{ K}$$

$$(8.81)$$

This model can be used to study the entire Federal Test Cycle covering many hours of operation of the catalytic converter.

8.16 **Project.** The water–gas shift reaction was modeled in the text using Aspen. Model it using (a) MATLAB and (b) Comsol.

8.17₂ The catalytic converter on your car is made from a honeycomb of ceramic, in which the gases flow down a tiny tube (about 1 mm² cross section). The catalyst is deposited on the walls of each tube, and that is where the reaction takes place. Consider here a very much simplified model; in reality one must solve for the flow of mass and energy in the tube (without reaction), transfer between the fluid and the wall (using mass and heat transfer coefficients) with reaction on the wall. The solid temperature and fluid temperature are not generally the same, but for this simple example we assume they are. This essentially assumes that the interchange of mass and heat between the fluid and solid is instantaneous. More information and examples are available (Young and Finlayson, 1976a, 1976b; Finlayson and Young, 1979; see also www.faculty.washington.edu/finlayso). The equations are a plug flow version of problem 8.10:

$$Q C_{tot} \frac{dy}{dV} = -\alpha r, \quad Q C_{tot} \hat{M} C_{pg} \frac{dT}{dV} = \alpha (-\Delta H_{rxn}) r, \quad y(0) = y_0, \ T(0) = T_0$$

$$(8.82)$$

The parameters are the same as in problem 8.10. Integrate these equations using MATLAB when the inlet mole fraction is 0.02 and the inlet temperature is 600, 550, and 500 K. If the temperature is low, does the converter work? This is the essence of the "first two minute" problem for reducing automobile pollution. Once the converter is warm, the pollution is very slight. It has been estimated that over 95% of the total pollution from an automobile comes out during this time when the converter is warming up.

8.18 Project. Beltrami (2002) and Gray (2006) post the following problem representing the dynamics of two competing populations of bacteria. The variables are the population densities of each species. The term in parentheses cap the growth due to limitations in the environment, and the last term represents the negative effects of competition between the species. The first term establishes the population density if only one species were present. The second term is proportional to the number of interactions between the species and represents the decline in population density due to competition:

$$\frac{dx_1}{dt} = 9x_1\left(1 - \frac{x_1}{9}\right) - 2x_1x_2, \quad \frac{dx_2}{dt} = 6x_2\left(1 - \frac{x_2}{12}\right) - x_1x_2 \quad (8.83)$$

(a) Determine all four steady-state solutions. (b) Integrate from the following points to steady state: $(x_1, x_2) = (3,1)$, $(7,4)$, $(0,0)$, $(0.01,0)$, $(5,2)$, $(5.01,2)$. (c) Write the solution as a steady solution plus a perturbation and linearize the equations for small perturbations. The right-hand side can then be represented by a matrix. Determine the eigenvalues for each of the steady states.

Numerical Problems (See Appendix E)

8.19₂ Solve the problem on page 144 for CA and CB using an Euler method programmed in MATLAB. Compare the exit concentrations as a function of mesh size, using at least three mesh sizes differing by a factor of 2.

8.20₂ Solve the problem on page 146 for SO_2 and temperature using an Euler method. Compare the exit concentration and temperature and peak temperature as a function of mesh size for 50, 100, and 200 steps.

8.21₁ Solve the problem in Figure 8.14 using the Newton–Raphson method.

8.22 Project. Solve the Problem 8.19 by using the second-order Runge–Kutta method programmed in Excel. Solve three times, each with a different step size in Δz: 0.2, 0.1, 0.05. Compare the accuracy as the step size decreases: does the error decrease with $\Delta z2$?

9

TRANSPORT PROCESSES
IN ONE DIMENSION

Chemical engineering processes involve the transport and transfer of momentum, energy, and mass. Momentum transfer is another word for fluid flow, and most chemical processes involve pumps and compressors, and perhaps centrifuges and cyclone separators. Energy transfer is used to heat reacting streams, cool products, and run distillation columns. Mass transfer involves the separation of a mixture of chemicals into separate streams, possibly nearly pure streams of one component. These subjects were unified in 1960 in the first edition of the classic book, *Transport Phenomena* (Bird et al., 2002). This chapter shows how to solve transport problems that are one dimensional, that is, the solution is a function of one spatial dimension; Chapters 10 and 11 treat two-dimensional (2D) and three-dimensional (3D) problems. The one-dimensional (1D) problems lead to differential equations, which are solved using the computer.

Diffusion problems in one dimension lead to boundary value problems. The boundary conditions are applied at two different spatial locations: at one side, the concentration may be fixed and at the other side the flux may be fixed. Because the conditions are specified at two different locations the problems are not initial value in character. It is not possible to begin at one position and integrate directly because at least one of the conditions is specified somewhere else and there are not enough conditions to begin the calculation. Thus, methods have been developed especially for boundary value problems. There is a baseball analogy that illustrates the difference. If you are at bat and there is no one on base, you can hit and run as fast as you can. This is like an initial value problem. If you are at bat and someone is on the first base, you can hit, but the way you run is influenced by what the runner ahead of you does. This is like a boundary value problem.

This chapter illustrates heat transfer, diffusion, diffusion with reaction, and flow in pipes, considering steady and unsteady processes. Most of the problems are solved using

Introduction to Chemical Engineering Computing, Updated Second Edition. Bruce A. Finlayson.
© 2014 John Wiley & Sons, Inc. Published 2014 by John Wiley & Sons, Inc.

Comsol Multiphysics, which is easy to use for these applications, but some are solved with MATLAB and Excel. Details of the finite element and finite difference methods are not given here, but you will find a brief description in Appendix E, along with references. In essence, you represent the solution at a set of grid points (finite difference) or nodal points (finite element). This is an approximation, so the answer is only as good as the mesh allows it to be, and the accuracy is not guaranteed. Thus, you have to solve the problem on more than one mesh to ensure that the results do not depend appreciably on the mesh. The solution is exact only if you use an infinite number of points, which no one has ever done. However, you can usually make it as accurate as needed, and the Comsol Multiphysics and MATLAB software makes it very easy to resolve the problem on a better mesh so that you can test the accuracy. A few numerical problems are provided that use Excel and require digging into the complexities of the numerical methods.

Instructional Objectives: After working through this chapter, you will be able to

1. Model heat transfer, diffusion and reaction, and fluid flow in one-dimensional situations like those occurring in your textbooks.
2. Solve boundary value problems by using MATLAB, Comsol Multiphysics, and Excel, for comparison.
3. Solve linear problems that you can compare with solutions in your textbooks.
4. Solve nonlinear problems that are not solvable with analytical methods and are probably not in your textbooks.
5. Practice using Comsol Multiphysics in preparation for Chapters 10 and 11.
6. solve problems in which convection dominates over diffusion to illustrate the oscillations that can occur and how to minimize them.

APPLICATIONS IN CHEMICAL ENGINEERING—MATHEMATICAL FORMULATIONS

Heat Transfer

Consider steady heat transfer across a slab of material that extends infinitely in two directions and has thickness L. In the thin direction, x, the equation for steady heat conduction is

$$\frac{d}{dx}\left(k(T)\frac{dT}{dx}\right) = Q$$

$$T(0) = T_0, \quad T(L) = T_L$$

(9.1)

The temperature, T, is a function of position, x. Notice that one condition is at $x = 0$ and the other at $x = L$. This makes it a two-point boundary value problem. The thermal conductivity k can depend on temperature, and when it does, the problem is nonlinear. The rate of energy generation is Q.

Diffusion and Reaction

Diffusion across a flat slab is very similar to Eq. (9.1):

$$\frac{d}{dx}\left(D\frac{dc}{dx}\right) = 0$$
$$c(0) = c_0, \quad c(L) = c_L \tag{9.2}$$

where the concentration, c, is a function of position, x. The diffusivity, D, can depend upon concentration. This is also a two-point boundary value problem because the concentration is specified at both $x = 0$ and $x = L$.

Many chemical reactions take place inside a catalyst pellet, which is a porous material (such as Al_2O_3 impregnated with the catalyst, usually metals of various kinds). The equation for a spherical catalyst pellet with radius R_p is

$$\frac{1}{r^2}\frac{d}{dr}\left(D_e r^2 \frac{dc}{dr}\right) = R(c)$$
$$\frac{dc}{dr}(0) = 0, \quad c(R_p) = c_R \tag{9.3}$$

where D_e is an effective diffusivity for the diffusion of mass inside the porous catalyst. The concentration, c, is a function of radial position, r. The reaction rate can have a variety of forms, all depending on concentration:

$$R = c, \quad R = c^2, \quad R = \frac{\alpha c}{1 + Kc} \tag{9.4}$$

In most cases, this problem is nonlinear. Again, this problem is a two-point boundary value problem because the boundary conditions are at two different spatial positions. At $r = 0$ there is no flux (center of the pellet), and at $r = R_p$ the concentration takes a specified value equal to the external concentration.

When there is a heat of reaction, the temperature increases due to the reaction. In that case, you have to also solve an energy balance on the catalyst pellet. A typical energy balance is

$$\frac{1}{r^2}\frac{d}{dr}\left(k_e r^2 \frac{dT}{dr}\right) = (-\Delta H_{rxn})R(c, T)$$
$$\frac{dT}{dr}(0) = 0, \quad T(R_p) = T_R \tag{9.5}$$

The temperature, T, is a function of radial position, r, and k_e is an effective thermal conductivity and $-\Delta H_{rxn}$ is the heat of reaction. Again, there is no flux at the center, and the temperature at the outer boundary is set at T_R. Now the reaction rate depends on both concentration and temperature [and Eq. (9.3) has to be written with $R(c,T)$]. Such problems can be very difficult due to the rapid change in the kinetic constants with temperature, and they provide a severe test of any numerical method.

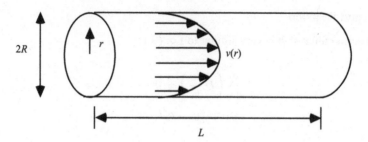

FIGURE 9.1 Fully developed pipe flow.

Fluid Flow

Consider the flow of a Newtonian fluid in a pipe, as illustrated in Figure 9.1. The governing differential equation is (Bird et al., 2002)

$$\frac{1}{r}\frac{d}{dr}\left(r\mu\frac{dv}{dr}\right) = -\frac{\Delta p}{L}$$

$$\frac{dv}{dr}(0) = 0, \quad v(R) = 0$$

(9.6)

where r is the radial position from the center of the pipe of radius R, v is velocity in the axial direction, μ is the viscosity, Δp is the pressure drop along the pipe, and L is the length over which the pressure drop occurs. The boundary conditions are at different locations, making it a two-point boundary value problem. The solution to this problem is Eq. (9.7):

$$v = \frac{1}{\mu}\frac{\Delta p}{L}\frac{R^2}{4}\left(1 - \frac{r^2}{R^2}\right)$$

(9.7)

When solving the problem analytically, it is also possible to use a boundary condition at $r = 0$ that says the velocity is finite. Both conditions give the same result, and in fact one can be derived from the other. Note that Δp is a positive number.

In pipe flow, the average velocity is defined by Eq. (9.8):

$$<v> = \frac{\int_0^{R_p} v(r)r\,dr}{\int_0^{R_p} r\,dr} = \frac{2}{R_p^2}\int_0^{R_p} v(r)r\,dr = 2\int_0^1 v(r')r'\,dr', \quad \text{where } r' = \frac{r}{R_p}$$

(9.8)

Note the $r\,dr$ in the integrand, which is necessary because cylindrical polar coordinates are used. The relationship between the average velocity and the pressure gradient can be obtained by integrating Eq. (9.7) using Eq. (9.8); it is the Hagen–Poiseuille law. The peak velocity is twice the average velocity in pipe flow:

$$<v> = \frac{1}{\mu}\frac{\Delta p}{L}\frac{R^2}{8} = \left(\frac{\Delta p}{L}\right)\frac{D^2}{32\mu}$$

(9.9)

When the flow is between two wide, flat plates instead of inside a cylinder, Eq. (9.6) becomes

$$\frac{d}{dx}\left(\mu\frac{dv}{dx}\right) = -\frac{\Delta p}{L}$$

$$\frac{dv}{dx}(0) = 0, \quad v(H) = 0$$

(9.10)

The plates are taken at a distance $2H$ apart. The solution is

$$v = \frac{1}{\mu}\frac{\Delta p}{L}\frac{H^2}{2}\left(1 - \frac{x^2}{H^2}\right)$$

(9.11)

Now the average velocity is

$$< v > = \frac{\int_0^H v(x)dx}{\int_0^H dx} = \frac{1}{H}\int_0^H v(x)dx = \int_0^1 v(x')dx', \quad \text{where} \quad x' = \frac{x}{H}$$

(9.12)

and the relationship between average velocity and pressure drop is

$$< v > = \frac{1}{\mu}\frac{\Delta p}{L}\frac{H^2}{3}$$

(9.13)

The peak velocity is 1.5 times the average velocity for flow between parallel plates. These formulas are provided here because they provide a good benchmark against which to check any numerical solution by integrating over boundaries.

When the fluid is non-Newtonian, it may not be possible to solve the problem analytically. For example, for the Bird–Carreau fluid (Bird et al., 1987, p. 171) the viscosity is

$$\eta = \frac{\eta_0}{\left[1 + \left(\lambda\frac{dv}{dr}\right)^2\right]^{(1-n)/2}}$$

(9.14)

The viscosity depends on the shear rate, dv/dr. New parameters are introduced that must be determined from experiments: η_0, λ, and n. Now the boundary value problem is

$$\frac{1}{r}\frac{d}{dr}\left(r\eta\frac{dv}{dr}\right) = -\frac{\Delta p}{L}$$

$$\frac{dv}{dr}(0) = 0, \quad v(R) = 0$$

(9.15)

It is not possible to solve this equation analytically for v, except for special values of η. For problems such as this, numerical methods must be used. If you set $\eta = \mu$ you get the same problem defined earlier for a Newtonian fluid.

Unsteady Heat Transfer

Consider the following heat transfer problem corresponding to heat transfer in a slab,

$$\frac{\partial T}{\partial t} = \alpha \frac{\partial^2 T}{\partial x^2} \tag{9.16}$$

The temperature depends both on time, t, and on position, x. The coefficient of thermal diffusion is

$$\alpha = \frac{k}{\rho C_p} \tag{9.17}$$

The slab is insulated at the left ($x = 0$) and has the temperature value of zero at the other side ($x = L$)

$$\frac{\partial T}{\partial x}(x = 0) = 0, \quad T = 0 \quad \text{at} \quad x = L \tag{9.18}$$

while the initial temperature distribution is

$$T(x, 0) = f(x) \tag{9.19}$$

You thus start with $T(x, 0)$ and find $T(x, t)$ as a function of x from 0 to L and t from 0 to infinity.

INTRODUCTION TO COMSOL MULTIPHYSICS

In this chapter, step-by-step instructions are provided for these 1D problems. Many features of Comsol Multiphysics are not used or needed until you solve 2D and 3D problems, and these features are explained in Chapters 10 and 11 and Appendix D. For the boundary value problems treated here, we generally follow these steps:

1. Open Comsol Multiphysics and choose the dimensions of the problem (here 1) and the physics modules we wish to solve.
2. Establish our preferences.
3. Create the geometry.
4. Insert the variables that define the problem and set the boundary conditions.
5. Create a mesh.
6. Solve the problem.
7. Examine the solution; here we limit ourselves to plots and integrals of the solution.

The icons shown in Figure 9.2 are ones that are used frequently in these examples. They are worth remembering.

next   finish   show   geometry/mesh build   plot   expressions   range of parameters

FIGURE 9.2 Icons used for setup: next, finish, show; geometry/mesh: build; plot: expression, plot; range of parameters.

EXAMPLE: HEAT TRANSFER IN A SLAB

Solution Using Comsol Multiphysics

Consider the problem of heat transfer in a slab with no internal heat generation but with a thermal conductivity that varies linearly with temperature—Eq. (9.1) with $k = 1 + T$.

$$\frac{d}{dx}\left[(1+T)\frac{dT}{dx} \right] = 0 \tag{9.20}$$
$$T(0) = 0, \quad T(1) = 1$$

The boundary conditions are taken as $T = 0$ at $x = 0$ and $T = 1$ at $x = 1$.

Step 1 (Open and Choose Physics) Open Comsol Multiphysics and choose 1D in the Model Wizard and click the next arrow. Then click on the arrow before "Heat Transfer" to open that folder. Choose "Heat Transfer in Solids ht" and click the next arrow. The Settings window will have a list of Preset options; select Stationary and then click the Finish flag. The nodes at the left of the screen are called the Model Builder and they provide a guide to the necessary steps (Figure 9.3).

Step 2 (Set Preferences) We will in turn use the Model 1, Study 1, and Results nodes. Note the icons at the top, right. The second one from the left is to "show" additional things, and we choose "Discretization" and "Equation View." In the root node we can choose a system of dimensions in the settings window; choose "none."

Step 3 (Geometry) Open the Model 1 node by clicking on the arrow to the left to expand this node (Figure 9.4). Then right click on the Geometry subnode and choose Interval. The default interval is zero to one, which is what we want, so we accept it by clicking on the "build" icon at the top of the settings window.

Step 4 (Define the Problem) With the Model 1 node open, click on the arrow to the left of the Heat Transfer node and choose Heat Transfer in Solids. Then click on the arrow

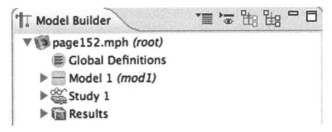

FIGURE 9.3 Outline of model builder nodes.

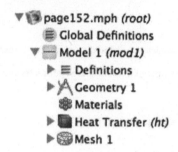

FIGURE 9.4   Outline of model 1 nodes.

before it and in the Settings window set the thermal conductivity to 1.0 (for testing first). Right click on Heat Transfer and choose Temperature. Choose the left node in the graphics window, use $+$ in the settings window to add it to the box in the settings window. Leave the value at 0. Do this again for the right end and set the value to 1.

Step 5 (Create a Mesh) Choose Mesh, and the Settings window for Size will appear. Accept them by clicking the Build icon. Sixteen nodes will be identified in the Graphics window. If you forget this step, it will be performed automatically for you when you solve the problem.

Step 6 (Solve the Problem) Choose Study and right click =. A plot of $T(x)$ should appear in the Graphics Window.

Step 7 (Examine the Solution) If it does not, or to create another graph, right click Results and choose 1D Plot Group. Right click 1D Plot Group and choose Line Graph. A plot of $T(x)$ should appear in the Graphics Window. The solution is a straight line from (0,0) to (1,1), as it should be. The heat flux is plotted by creating another Line Graph, but this time changing the expression for the y-Axis Data. Choose the expression button, select Heat Transfer, then Conductive Heat Flux, then the xx-component of heat flux (there is only an x-component in this problem though). The heat flux is a constant -1. If a plot does not appear, click the plot icon.

Step 8 (Solve the Real Problem) Go back to the Heat Transfer in Solids subnode and change the thermal conductivity to $1 + T$. Choose Study and right click =. Examine the solution in the same way. The temperature is shown in Figure 9.5 and the heat flux is shown in Figure 9.6. The temperature departs slightly from a straight line. The heat flux has some weird-looking oscillations between -1.5015 and -1.497, but these are an artifact of the numerical method. This is because the temperature is approximated using quadratic functions on finite elements, but the derivative (proportional to heat flux) is not continuous across element boundaries. If you click on 1D Plot Group, then Axis in the Settings window, you can check the Manual box and set the axis limits for the y to $-2 \rightarrow 0$; on that scale the oscillations are scarcely noticeable.

Step 9 (Calculate Averages) You can calculate the average temperature. Open the arrow the left of Results right click Derived Values; right click Derived Values and choose Integration and then Line Integration. In the Settings window, note that the variable being integrated

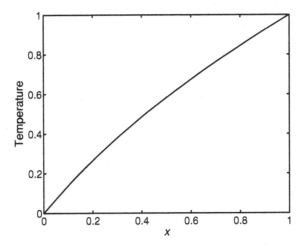

FIGURE 9.5   Solution to Eq. (9.20).

is T. Click $=$ and look at the Results icon (Figure 9.7, below the graph) to see the value: 0.55556. This is the value of the average temperature. You can check that this is the average by using Integration instead of Average and integrate T and 1 separately:

$$\int_0^1 T\,dx \,\bigg/ \int_0^1 dx = 0.55556 \tag{9.21}$$

Step 10 (Plot the Thermal Conductivity)   You can plot the thermal conductivity by right clicking on Results; choose 1D Plot Group. Then right click on it and choose Line Graph. In the Settings window, click on the expression button and choose heat transfer/thermal

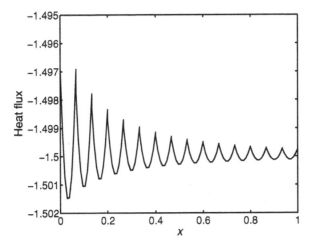

FIGURE 9.6   Heat flux in problem, Eq. (9.20).

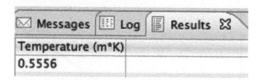

FIGURE 9.7   Messages/log/results.

conductivity to plot. Remove the units from the Settings window if they appear. The plot looks like Figure 9.5 except with the y-axis going from 1 to 2 $(1+T)$.

Solution Using MATLAB

The same problem can be solved using MATLAB. First, we write the problem in the form

$$\frac{dq}{dx} = 0, q = -k\frac{dT}{dx}, k = 1+T, T(0) = 1, T(1) = 0 \qquad (9.22)$$

The variable $y(1)$ will represent T and the variable $y(2)$ will represent the heat flux q. Thus, the equation becomes

$$\frac{dy(1)}{dx} = -\frac{y(2)}{k}, \frac{dy(2)}{dx} = 0, y(1) = 0 \text{ at left}, y(1) - 1 = 0 \text{ at right} \qquad (9.23)$$

We define the following functions (m-files) for the differential equation

```
function dydx=heattr(x,y,alpha)
alpha = 1.;
k = 1+alpha*y(1);
dydx(1) = -y(2)/k;
dydx(2) = 0;
dydx = dydx';
```

and boundary conditions (defined in terms of the residual, to be made zero.

```
function res = heatbc(ya,yb)
res(1) = ya(1);
res(2) = yb(1)-1;
res = res';
```

The initial guess is found using

```
solinit=bvpinit(linspace(0,1,5),[0 1])
```

The command linspace(0,1,5) sets a space from $x = 0$–1 with five intervals. The term [0 1] is a linear definition of an initial guess. Then the solution is found and plotted using

```
sol = bvp4c(@heattr,@heatbc,solinit)
x=linspace(0,1);
y = deval(sol,x);
```

```
plot(x,y(1,:));
xlabel('x')
ylabel('Temperature')
title('BVP using MATLAB')
```

The temperature plot is of course similar to Figure 9.5. The heat flux, $y(2,:)$ is constant at -1.5; it is determined better since the heat flux itself is a variable, which is smooth. In Comsol Multiphysics (in the form given), the heat flux is a derivative of the temperature and any discontinuities in slope are prominent.

EXAMPLE: REACTION AND DIFFUSION

The next example is a reaction-diffusion problem in a spherical domain. The reaction rate expression is a nonlinear function of concentration, of a type that is appropriate for the Michaelis–Menten reaction in biological systems. The nondimensional form of the problem is in Eq. (9.24), and it is solved for $\alpha = 5$, $K = 2$:

$$\frac{1}{r^2}\frac{d}{dr}\left(r^2\frac{dc}{dr}\right) = \frac{\alpha c}{1 + Kc}, \frac{dc}{dr}(0) = 0, c(1) = 1 \qquad (9.24)$$

Follow these steps to solve this problem in Comsol Multiphysics. Since the problem is in spherical geometry, we expand the terms to get

$$\frac{1}{r^2}\frac{d}{dr}\left(r^2\frac{dc}{dr}\right) \equiv \frac{d^2c}{dr^2} + \frac{2}{r}\frac{dc}{dr} = \frac{\alpha c}{1 + Kc}$$

$$\text{or} - \frac{d^2c}{dr^2} = \frac{2}{r}\frac{dc}{dr} - \frac{\alpha c}{1 + Kc} \qquad (9.25)$$

The descriptions of the Comsol Multiphysics steps are abbreviated now since they follow the first example closely. You need to remember how to right click and click the arrow to expand the menu. For convenience, a pdf version of the first example is available on the book website.

Step 1 (Choose the Problem) In Comsol, the 1D option will be appropriate for planar geometry; the 1D, axisymmetric option is appropriate for cylindrical geometry. Since we want spherical geometry, we will choose the 1D option and add the terms needed to make the equation agree with Eq. (9.24). Select from the first menu 1D, next; Transport of Diluted Species, next; and Stationary, finish. Set the preferences; be sure to choose "Discretization."

Step 2 (Define the Geometry) Open Model 1 and right click on Geometry; choose Intervals. The default interval is (0,1), which we keep; click Build. The independent variable is called x rather than r, but we could change it if desired.

Step 3 (Define the Problem) Open Model 1 and click on Transport of Diluted Species. If the domain is not listed in the Selection box, click on the domain and the $+$, or change Selection from "Manual" to "All Domains." You can also take convection out of the problem by unselecting the Convection box, or keep the velocity as zero, since there is no velocity in this problem. Near the bottom of the settings panel is the panel for Dependent Variables. Click it, and note that the concentration is called c. Under Transport, select Diffusion and change the diffusivity to 1 in the Settings panel (the problem is dimensionless). Go back and right click Transport and choose Reactions; select the domain and + to add it to the Domain box. Set R_C to $-$rate. [Why the minus sign? Check the equation in Transport and compare it with Eq. (9.25).] Right click on Definitions and choose variables. Define alpha as 5 and K as 2. Then insert the name "rate" and define it as alpha*$c/(1+K*c)$. Also click on Transport of Diluted Species and the Discretization tab (remember setting Discretiztion in the Preferences?). Change Linear to Quadratic.

Step 4 (Add the Spherical Terms) This is where we also have to add in the added term arising from the spherical coordinates: $(2/r)(dc/dr)$. When the independent variable is x and the dependent variable is c, the first derivative in the program is cx. Thus, we need to add $(2/x)*cx$ to the right-hand side of the equation, which is the R_C term. Add that so that the rate term is $-$rate $+ (2/x)*cx$.

Step 5 (Set the Boundary Conditions) To set the boundary conditions, right click on Transport and choose Concentration. Then click on point 2 (the right-most point), click + to add it to the box, click Species c, and set the value to 1. Select the No Flux button and notice that the boundary 1 is still using that boundary condition but boundary 2 has been overridden. Thus, the boundary conditions are correct.

Step 6 (Set the Mesh) In the Model Builder, right click on Mesh and choose Build All to create the mesh. Alternatively, select Mesh and then in the Settings window choose Build. The default mesh uses 16 points and the points are added to the plot of the geometry (or sometimes not).

Step 7 (Solve the Problem) Right click Study and choose = to solve the problem. If an error occurs, click on the Error node in the Model Builder to see what the problem is and trace it down. The plot appears automatically; see Study 1/Solver Configurations/Solver 1/Stationary Solver 1/Problems 1/Error.

Step 8 (Examine the Solution) Since the problem is dimensionless, and the default applies dimensions to the plot, select the 1D Plot Group Node and click the x-axis and y-axis labels to take out the dimensions or change them to other words. Click on the Title to change it to something more meaningful [solution to Problem (9.25)] and click the plot icon to redo the plot. To save the plot, choose the camera and follow the instructions. The solution is shown in Figure 9.8.

Parametric Solution

It is instructive to solve this problem for multiple parameters, α, and you can do that using the parametric solver. If you had not already used a variable name, alpha, in Physics/Subdomain Selection, you would have to use a variable name now.

Step 1 (Define the Problem) Go back and right click on Global Definitions and choose Parameters. Insert alpha, use 5 for now. Then go to Definitions under the Model and remove

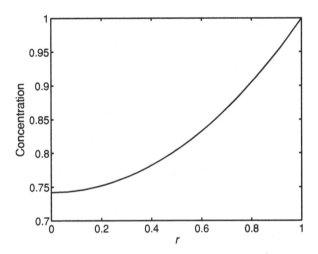

FIGURE 9.8   Concentration in spherical domain with Michaelis–Menten reaction.

alpha. Note: If you are using the parametric solver, the parameter *must* be defined in Global Definitions.

Step 2 (Define the Parametric Sweep) Right click on Study, choose Parametric Sweep. Choose the + sign to add a parameter; the Add window comes up. Since alpha is defined in Global Definitions, it is already there as a Parameter name; select it and click OK. In the Settings panel, alpha is listed as a parameter name; set the Parameter values to go from 0 to 20 in steps of 2. This can be done using the range symbol or by setting Parameter values to range(0,2,20).

Step 3 (Solve the Problem) Right click on Study = and a plot appears (Figure 9.11). To see the legend, select Line Graph 1 under the 1D Plot Group in the Model Builder/Results; open Legends and click on the Show legends box. Back up to 1D Plot Group and select where you would like the legend; here lower right. Click on the Plot icon to get the revised plot. The title has been changed to identify it as a plot of concentration for different reaction rate constants.

Step 4 (Examine the Solution) To see the rate, right click on Results and choose 1D Plot Group. A new node, 1D Plot Group 2 will be created. Set the title, x-axis, and y-axis to what you would like. Then right click on the 1D Plot Group 2 and select Line Graph 1; add the domain to the Selection box by clicking it in the Graphics window and using +, and change the expression to rate. Click on Show legends, and click the plot button to see the plot. Click on the camera to save the plot.

Figure 9.9 indicates that the concentration decreases from the outer part of the sphere to the inner part of the sphere and that this decrease is more dramatic with larger values of α. This is because the rate of reaction is faster and diffusion cannot keep up, thus decreasing the concentration. Figure 9.10 shows the rate of reaction as a function of position and α. For small values of α, the rate is the same at all positions, but as α increases the rate of reaction varies by a factor of two or more from the center to the outer boundary.

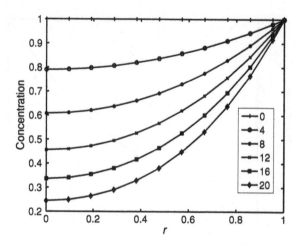

FIGURE 9.9   Solutions for several values of α.

EXAMPLE: FLOW OF A NEWTONIAN FLUID IN A PIPE

This example considers flow in a pipe for a Newtonian fluid, Eq. (9.6). You will solve this problem by using Comsol Multiphysics with $\Delta p = 2.8 \times 10^5$ Pa, $\mu = 0.492$ Pa s, $L = 4.88$, $R = 0.0025$ m. You must expand the differential equation because the 1D, axisymmetric option for Poisson's Equation does not include all the terms for cylindrical geometry. (This limitation is removed in 2D and 3D problems, but here a 1D solution suffices.) So, rewrite Eq. (9.6) as

$$\frac{d^2 v}{dr^2} = -\frac{1}{\mu}\frac{\Delta p}{L} - \frac{1}{r}\frac{dv}{dr}$$

$$\frac{dv}{dr}(0) = 0, \quad v(R) = 0$$

(9.26)

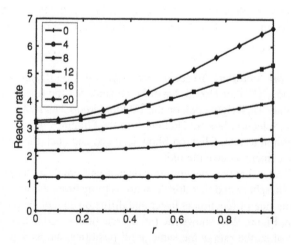

FIGURE 9.10   Rate of reaction as a function of $x(r)$ and α.

Step 1 (Choose the Problem) To solve Eq. (9.26), we must select the equation. In Comsol Multiphysics, the 1D option will be appropriate for planar geometry; the 1D, axisymmetric option is appropriate for cylindrical geometry, which is what we want. Select from the first menu: 1D, axisymmetric, next. The next screen indicates that there is no 1D option for Fluid Flow. Thus, we will use Mathematics/Classical PDEs/Poisson's Equation. Choose it, then the next arrow; then Stationary, and the finish flag. Leave the units as SI, but click on Equations View and Discretization.

Step 2 (Set the Variables) Working down the Model Builder, we right click on Definitions and choose variables. Define dpL as 2.8e5/(0.492*4.88) [alternatively you could insert the pressure drop, length, and viscosity here and define dpL as dp/(mu*L)].

Step 3 (Define the Geometry) Right click on Geometry and choose Intervals. The default interval is (0,1), so change it to (0,0.0025); click Build All.

Step 4 (Define the Problem) Click on Poisson's Equation. Select the domain if it is not highlighted in the Selection box. To see the equation being solved, choose Equation. Unfortunately, the cylindrical geometry is not reflected in the equation, but it is shown in the graphics window (with the axis). Near the bottom of the settings panel is the panel for Dependent Variables. Click it, and note that the dependent is called u; change it to v. Under Poisson's Equation, select Diffusion and leave c as 1 in the Settings panel. Make sure the domain is selected and appears in Domains. Set f to d$pL + vr/r$. Check the equation in Poisson's Equation and compare it with Eq. (9.26). The fact that you have to insert the vr/r is a mistake in the program that is only discovered by carefully checking the results. This happens because Poisson's equation usually refers to planar geometry only.

Step 5 (Set the Boundary Conditions) To set the boundary conditions, right click on Poisson's Equation and choose Dirichlet Boundary Condition. Then click on point 2 (the right-most point) in the Graphics window, click + to add it to the box, and keep the value of 0. Select the Zero Flux button and notice that the boundary 1 not applicable, since the program satisfies it without intervention by us due to the axisymmetric geometry. Thus, the boundary conditions are correct.

Step 6 (Set the Mesh) In the Model Builder, right click on Mesh and choose Build All to create the mesh. Alternatively, select Mesh and then in the Settings window choose Build All. The default mesh uses 16 points and the points are added to the plot of the geometry (or maybe not).

Step 7 (Solve the Problem) Right click Study and choose = to solve the problem. If an error occurs, click on the Error node in the Model Builder/Study 1 to see what the problem is and trace it down.

Step 8 (Examine the Solution) The plot appears automatically (Figure 9.11). To change the axis labels and title, select the 1D Plot Group Node and click the y-axis label and change it to velocity (m/s). Click on the Title to change it to Velocity for Newtonian Fluid and click the plot icon to redo the plot. To save the plot, choose the camera and follow the instructions.

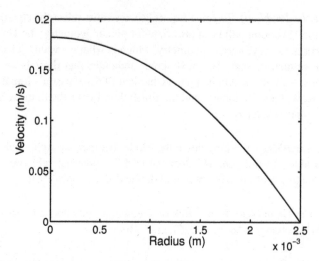

FIGURE 9.11   Velocity profile for fully developed Newtonian flow in a pipe.

Step 9 (Calculate the Average Velocity) To compute the average velocity, you need to be careful to get the integrals in Eq. (9.8). Right click on the node Derived Values under Results and choose Line Integration. Then in the settings window, open the panel Integration settings. A box there is clicked to have the integration be appropriate for cylindrical geometry. Click on the domain (line), then + to choose the line. First, insert the value 1.0 for the expression; click = . Look in the results for the tabular value, 1.9635e-5. This is appropriate for integrating $2\pi r\,dr$ from 0 to 0.0025. Next insert v for the expression. The integral this time is 1.779e-6. The ratio of these is the average velocity, 0.0906 m/s. Note that this is exactly half of the peak velocity, which is expected since that is true of the exact solution. Alternatively, you can choose Line Average and avoid the division. Be careful, though, because using Line Integration was how the error in Step 4 was discovered. It is also possible to look at the velocity gradient (a straight line from zero at the center to a maximum at the wall). The shear stress is the viscosity multiplied by the velocity gradient, and that is linear with distance, too, for a Newtonian fluid.

EXAMPLE: FLOW OF A NON-NEWTONIAN FLUID IN A PIPE

Next, you will consider flow in a pipe when the fluid is a non-Newtonian fluid, in particular a polymer melt. Take the viscosity as a Carreau function, Eq. (9.14). The parameters used here are $\eta_0 = 0.492$, $\lambda = 0.1$, and $n = 0.8$. The problem is defined by

$$\frac{1}{r}\frac{d}{dr}\left(r\eta\frac{dv}{dr}\right) = -\frac{\Delta p}{L} \text{ or } \frac{d}{dr}\left(\eta\frac{dv}{dr}\right) + \frac{\eta}{r}\frac{du}{dr} = -\frac{\Delta p}{L}$$

$$\frac{dv}{dr}(0) = 0, \quad v(R) = 0 \tag{9.27}$$

Step 1 (Choose the Problem) To solve problem (9.27), start with the solution for a Newtonian fluid.

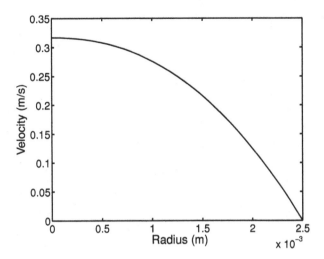

FIGURE 9.12   Velocity profile for fully developed non-Newtonian flow in a pipe.

Step 2 (Set the Variables) Change dpL to be just $\Delta p/L$, and include under Definitions the viscosity, called *eta*. First, the program will be changed with *eta* defined as 0.492 (the Newtonian value) to be sure we have set the problem up correctly. Then the Carreau function will be inserted for *eta*.

Step 3 (Define the Problem) The Geometry is the same, the boundary conditions are the same, and the mesh is the same. The only change in the equation is to use *eta* for c instead of 1.0 and eta^*dv/r in the expression for f.

Step 4 (Solve the Problem) Right click on Study = to solve the problem. We get the same result as before, so we are sure the program is correct, or at least as correct as our previous solution.

Step 5 (Use the Carreau Expression for Viscosity) Now go back to Definitions in the Model node, select the Variables node, and change the formula for *eta* to 0.492/ $(1+(lambda^*vr) \wedge 2) \wedge ((1-t)/2)$; add variables of *lambda* = 0.1 and $n = 0.8$. Right click on Study =. Note that the independent variable is r, not x, and the first derivative of the velocity v is vr.

Step 6 (Solve the Problem) Right click on Study = to solve the problem. It works best if you set the initial value of velocity (under Laminar Flow) to something other than zero.

Step 7 (Examine the Solution) Change the title to Velocity for non-Newtonian Fluid and click on the plot icon to change the plot. Click on the camera to save the plot, which is shown in Figure 9.12.

Step 8 (Plot Solution Features) To plot the viscosity across the domain, right click on Results and choose 1D Plot Group. Then change (in the Settings panel) the titles and labels. Right click on 1D Plot Group2 and choose Line Graph. Change the y-Axis Data Expression

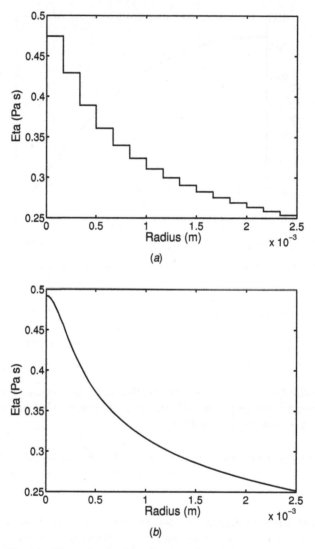

FIGURE 9.13   Viscosity, Eq. (9.14) for fully developed non-Newtonian flow in a pipe: (a) using linear elements for velocity; (b) using quadratic elements for velocity.

from v to *eta* and the Unit to blank. Click on the Plot icon to see the plot (Figure 9.13a). This is an unusual plot, but it emphasizes things both mathematical and physical. First, the viscosity seems to be constant in each region, which corresponds to each element. That is because the velocity is represented as a linear function in the element, so that the derivative is constant in the element. The physical lesson is that the viscosity approaches the rest viscosity near the centerline and decreases to its minimum at the wall, which is where the shear rate is the largest.

Step 9 (Improve the Solution)   To improve the plot, we need to solve it again. To change the trial functions [the functions representing the solution $v(r)$], we need to go back to the

Model Builder at the top, choose the "Show" icon (see Step 1 in the reaction-diffusion example). Select "Equation View" and "Discretization." Now when we go back to the Poisson's Equation node, one of the options (near the bottom of the Settings window) is "Discretization." Currently it says "Linear"; change it to "Quadratic." Resolve the problem and choose the plot of viscosity. Now we get a smooth curve (Figure 9.13b) since the velocity gradient can vary within an element. Another option would be to use linear functions for velocity and just choose lots of elements, so that the curve looked smooth.

Step 10 (Calculate Average Velocity) The average velocity is given by Eq. (9.8); calculate it the same way as for a Newtonian fluid. The average velocity increases from 0.091 to 0.168. The reason for this is shear thinning. The shear rate is zero at the centerline and is a maximum (in absolute value) at the solid boundary. Because the viscosity decreases as the shear rate increases [see Eq. (9.14)], the viscosity decreases as the radial position increases, as seen in Figure 9.15. One handy way to get all the solutions for several pressure drops (and avoid convergence problems) is to use the Parametric Sweep option (by right clicking the Study node); see Section Parametric Solution. The velocities vary over several orders of magnitude for the different pressure drops. You could plot the velocity divided by the pressure drop to improve your graph. An even better option would be to compute the average velocity of each solution, and plot the velocity divided by the average velocity, all on the same graph.

EXAMPLE: TRANSIENT HEAT TRANSFER

The next example is a transient heat transfer problem:

$$\frac{\partial T}{\partial t} = \alpha \frac{\partial^2 T}{\partial x^2}, \ \alpha = 2$$

$$\left.\frac{\partial T}{\partial x}\right|_{x=0} = 0, \ T(1,t) = 0, \ T(x,0) = 1 \tag{9.28}$$

Solution Using Comsol Multiphysics

Step 1 (Choose the Problem) To solve Eq. (9.28), open Comsol Multiphysics, choose 1D, next; Heat Transfer, then Heat Transfer in Solids, next; Time Dependent, finish.

Step 2 (Define the Geometry) Right click on Geometry and choose Interval. The default is (0,1), which we keep. Choose Build All.

Step 3 (Define the Problem) In the Model 1 node, choose Heat Transfer; in the Settings panel, look at the equation that Comsol Multiphysics will solve. Several of the terms need to be set to zero: U_{trans} and Q. Click on Heat Transfer in Solids, and change the parameters to User defined and $k = 2$, $\rho = 1$, $C_p = 1$. Then the equation is the same as Eq. (9.28).

Step 4 (Set the Boundary Conditions and Initial Conditions) To set the boundary conditions, right click on Heat Transfer and choose Temperature. Select the boundary point 2 at the right-hand side of the diagram and the $+$ symbol to add it to the Selection. Change the value from 293.15 to 0. Click on the Thermal insulation subnode and see that boundary point 1 has thermal insulation (which we want) and boundary point 2 has been overridden. To set the initial condition, click on the Initial Values node and set T to 1.

Step 5 (Set the Mesh)   Select the mesh and choose Build All; note that the nodal points are displayed in the graph.

Step 6 (Solve the Problem)   Right click on Study = to get the solution.

Step 7 (Examine the Solution)   Right click Data Sets and choose Parametric Extrusion 1D. The Settings window will list the times at which the solution will be plotted. Select the 1D Plot Group and change the Plot Settings to reflect what we have actually done, x, T, "Solution to Problem 9.28" and click on the plot icon to get the new plot. To get the legends for the graph, click on the Line Graph node, choose Legends (near the bottom) and click on the Show legends box. Then replot to obtain Figure 9.14. Note that it is not possible to have the temperature be both 0 (the boundary condition) and 1 (the initial condition) at the point $x = 1$, $t = 0$, and this introduces a small perturbation near that point. This perturbation quickly dampens out in time, though. It is really a manifestation of an improperly posed mathematical problem rather than an incorrect method, because such a discontinuity is impossible to create physically. This graph can be compared with graphs in your textbook to validate the computer solution.

Step 8 (Create a 3D Plot)   To see a colorful 3D plot, right click on Results, choose 2D Plot Group. Right click 2D Plot Group, choose Surface and make sure the data set chosen is Parametric Extrusion 1D. Right click again on Surface and choose Height Expression. A black and white version of the colorful 3D plot is shown in Figure 9.15.

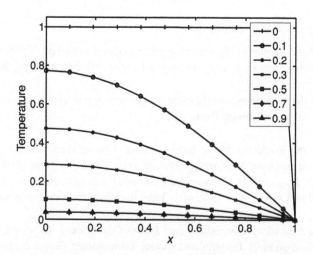

FIGURE 9.14   Solution of heat transfer problem, Eq. (9.28).

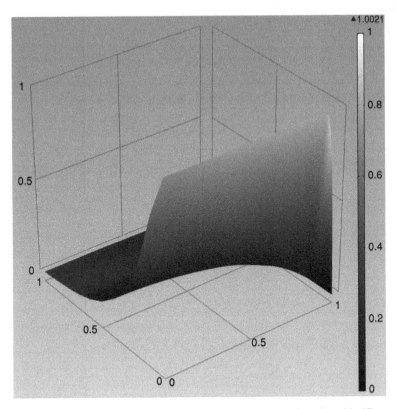

FIGURE 9.15   Solution of heat transfer problem, Eq. (9.28) plotted in 3D.

Solution Using MATLAB

MATLAB can solve partial differential equations (in one spatial dimension) of the form

$$c\left(x, t, u, \frac{\partial u}{\partial x}\right)\frac{\partial u}{\partial t} = x^{-m}\frac{\partial}{\partial x}\left(x^m f\left(x, t, u, \frac{\partial u}{\partial x}\right)\right) + s\left(x, t, u, \frac{\partial u}{\partial x}\right) \qquad (9.29)$$

The parameter m is 0, 1, or 2, for planar, cylindrical, or spherical geometry. For this problem we use $c = 1$ and $m = 0$. Here, the value of f is $2^*dT/dx$. The MATLAB function transtemp defines the differential equation.

```
function [c,f,s] = transtemp(x,t,u,dTdx)
c = 1;
s = 0;  %no source term
f = 2*dTdx;
```

The initial conditions are set with the function ictemp.

```
function T0=ictemp(x)
T0 = 1;
```

The boundary conditions can be

$$p(x, t, u) + q(x, t) f\left(x, t, u, \frac{\partial u}{\partial x}\right) = 0 \tag{9.30}$$

The function f is already defined in transtemp. Here, at the left we want

$$p = 0, \quad q = 1 \tag{9.31}$$

And at the right we want

$$p = T, \quad q = 0 \tag{9.32}$$

Thus, define the function bctemp.

```
function [pl,ql,pr,qr]=bctemp(xl,Tl,xr,Tr,t)
pl = 0;    % l for left boundary
ql = 1;
pr = Tr;   % r for right boundary
qr = 0;
```

The solution is obtained with the following MATLAB commands.

```
xmesh = linspace(0,1,11)
tspan = linspace(0,1,11)
sol = pdepe(0,@transtemp,@ictemp,@bctemp,xmesh,tspan)
```

Two types of plots are possible: a 3D plot and a line plot at different t; see Figure 9.16.

```
T = sol(:,:,1)
plot(xmesh,T(1,:,1),xmesh,T(2,:,1),xmesh,T(3,:,1),...
    xmesh,T(4,:,1),xmesh,...
T(5,:,1),xmesh,T(6,:,1),xmesh,T(7,:,1),xmesh,T(8,:,1),...
xmesh,T(9,:,1),xmesh,T(10,:,1),xmesh,T(11,:,1))
xlabel('x')
ylabel('Temperature')
title('Transient Heat Transfer with MATLAB)')
pause
surf(xmesh,tspan,T)
```

In Eq. (9.29) the variable u can be a vector, which makes possible solution of coupled partial differential equations.

EXAMPLE: LINEAR ADSORPTION

Adsorption is one way to remove a chemical species from a flowing stream. A cylindrical tube is packed with adsorbent material, oftentimes in the form of small spheres. The adsorbent has the property that some materials adsorb while others do not. The flowing stream goes in and out the interstices between the spheres, and contacts the adsorbent. The

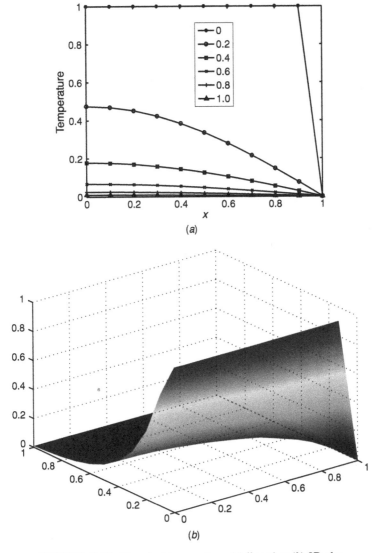

FIGURE 9.16   Transient temperature: (*a*) line plot; (*b*) 3D plot.

chemical that is strongly adsorbed is removed from the flowing stream and appears on the solid adsorbent. The next example illustrates that process.

The differential equations governing linear adsorption are given by two equations, the first a mass balance on the fluid phase and the second a mass balance on the stationary phase:

$$\phi \frac{\partial c}{\partial t'} + \phi V \frac{\partial c}{\partial x'} + (1 - \phi) \frac{\partial n}{\partial t'} = 0, \quad \frac{\partial n}{\partial t'} = k(\gamma c - n) \tag{9.33}$$

where c is the fluid concentration (moles per fluid volume), n is the concentration on the solid adsorbent (moles per solid volume), ϕ is the void fraction in the bed, V is the fluid velocity, t' is time, x' is distance down the bed, k is a mass transfer coefficient, and γ is the slope of the equilibrium line, n versus c. The initial conditions give the concentration in fluid phases and adsorbed phases at time zero, which in this example are no concentration at all:

$$c(x', 0) = 0, \quad n(x', 0) = 0 \tag{9.34}$$

The only boundary condition is the inlet concentration, here taken as 1.0:

$$c(0, t') = 1 \tag{9.35}$$

The equations are made nondimensional using the following variables:

$$x = \frac{x'k}{V}, \ t = kt' \tag{9.36}$$

Then Eq. (9.33) is

$$\frac{\partial c}{\partial t} + \frac{\partial c}{\partial x} + \frac{(1-\phi)}{\phi}(\gamma c - n) = 0, \quad \frac{\partial n}{\partial t} = \gamma c - n \tag{9.37}$$

This problem is linear, and an analytical solution exists, but it is very complicated to evaluate because it depends on integrals of Bessel functions (Rhee et al., 1986). Numerical solutions are also available for comparison (Finlayson, 1992). When you use Comsol Multiphysics, it is necessary to add some diffusion to the problem, and you solve an augmented equation for concentration. The added term involves a Peclet number, defined in terms of an effective diffusion coefficient:

$$\frac{\partial c}{\partial t} + \frac{\partial c}{\partial x} + \frac{(1-\phi)}{\phi}(\gamma c - n) = \frac{1}{Pe}\frac{\partial^2 c}{\partial x^2}, \ \frac{1}{Pe} = \frac{\phi V^2}{Dk}, \ \text{or } Pe = \frac{Dk}{\phi V} \tag{9.38}$$

Step 1 (Choose the Problem) To solve Eq. (9.38), open Comsol, choose 1D, next; Transport of Diluted Species, next; Time Dependent, finish.

Step 2 (Set the Variables) Right click on Global Definitions and define the following parameters: *phi* = 0.4, *gamma* = 2, *Pe* = 1000. Under the Model, right click Definitions and choose Variables. Set *raten* = *gamma*c* − *n* and *rate* = (1−*phi*)* *raten/phi*.

Step 3 (Define the Geometry) Under the Model, right click on Geometry and choose Interval. The default is (0,1), which we keep. Choose Build All.

Step 4 (Define the Problem for c) It can take the form of Eq. (9.38). Select the down arrow on Transport, and click Convection and Diffusion. Set the velocity to 1.0 and the D_C to 1/Pe. Right click on Transport and choose Reactions. Set R_C to −*rate*. Compare the equation format in Comsol with that in Eq. (9.38) to see why the minus is necessary.

Step 5 (Set the Boundary and Initial Conditions for c) Right click on Transport and choose Inflow. Select boundary point 1, then the + so that 1 appears in the Boundary Selection. Set the inflow concentration to 1.0. Right click on Transport again and choose boundary point 2 and then click +. Accept the boundary condition, which will allow convection out of the domain. Click on Initial Values and make sure the concentration is zero.

Step 6 (Define the Problem for n) Next, we add the equation for n, the amount of material adsorbed. Right click on Model 1, then choose "Transport of Diluted Species" again, next; then choose "Transient," finish. Select "Transport of Diluted Species 2" and check the equation. Change the name of the dependent variable to n. Click on "Convection and Diffusion" and set $D_n = 0$. Right click on "Transport of Diluted Species 2" and choose "Reactions." Set R_n to *raten*.

Step 7 (Set the Boundary and Initial Conditions for n) Click on Initial Values and verify that n is zero. We do not change the boundary conditions from no flux, since the adsorbed material cannot get out except by exchanging with the moving fluid.

Step 8 (Set the Mesh) Select the mesh and change the element size to "Extrafine" (in anticipation that many nodes will be needed). Choose "Build All"; notice that the nodal points are displayed in the graph.

Step 9 (Solve the Problem) Right click on Study = to get the solution.

Step 10 (Examine the Solution) Select the 1D Plot Group and change the Plot Settings to reflect what we have actually done, x, c, "Solution to Problem 9.37" and click on the plot icon to get the new plot. To get the legends for the graph, click on the Line Graph subnode, choose Legends (near the bottom) and click on the Show legends box. Then replot, choose the camera and save the figure. To see the plot of n, perform the same steps but plot n rather than c. The concentration in the fluid is shown in Figure 9.17*a* and the concentration adsorbed is shown in Figure 9.17*b*.

The solution is a good representation of the solution, but the front, where the concentration drops quickly, is not as steep as it should be (Finlayson, 1992). Some of the oscillations apparent in the figure are because the finite element method is not well suited to problems such as this, with convection but no diffusion. Comsol Multiphysics adds some stabilization features to smooth the oscillations without obscuring the essential details. A variety of specialized methods are available to do that, as described in Appendix E and by Finlayson (1992).

This example used a linear isotherm. It is easy to change to a nonlinear isotherm, since the only change is in Step 2; choose a different formula that represents another isotherm.

EXAMPLE: CHROMATOGRAPHY

In some columns, mass transfer is so fast that the concentration in the gas phase is in equilibrium with the concentration on the solid adsorbent. In that case the problem is slightly different. This case is a prototype of a chromatography column, which can be used to separate chemicals.

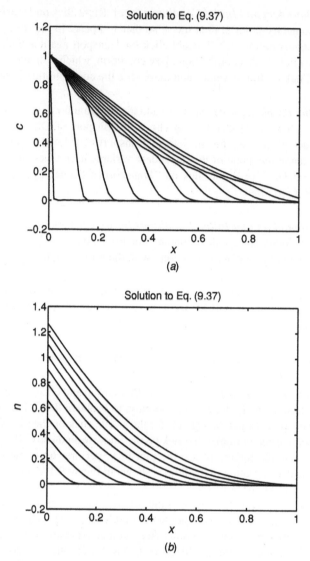

FIGURE 9.17   Solution of linear adsorption problem, Eq. (9.37), at times $0:0.1:1$: (*a*) fluid concentration; (*b*) solid concentration.

The equation is the first one of Eq. (9.33). Now, however, the solid concentration is an algebraic function of the fluid concentration, since they are in equilibrium. In this example, you will use a Langmuir isotherm with the algebraic expression:

$$n = \frac{\gamma c}{1 + Kc} \tag{9.39}$$

When this equation is substituted into Eq. (9.33) you obtain

$$\phi \frac{\partial c}{\partial t'} + \phi V \frac{\partial c}{\partial x'} + (1 - \phi)\frac{dn}{dc}\frac{\partial c}{\partial t'} = 0 \tag{9.40}$$

For the Langmuir isotherm, the equation is

$$\left[\phi + (1-\phi)\frac{dn}{dc}\right]\frac{\partial c}{\partial t'} + \phi V\frac{\partial c}{\partial x'} = 0, \quad \frac{dn}{dc} = \frac{\gamma}{(1+Kc)^2} \tag{9.41}$$

The nondimensional form of the problem is obtained using $t' = t$ and $x = x'/V$:

$$\left[1 + \frac{(1-\phi)}{\phi}\frac{dn}{dc}\right]\frac{\partial c}{\partial t} + \frac{\partial c}{\partial x} = 0, \quad \frac{dn}{dc} = \frac{\gamma}{(1+Kc)^2} \tag{9.42}$$

In this example, you will solve the problem with an initial concentration of one between $x = 0.25$ and 0.75. The inlet concentration is zero always:

$$c(x, 0) = \begin{array}{cc} 0, & x < 0.25 \\ 1, & 0.25 \le x \le 0.75 \\ 0, & 0.75 < x \end{array} \tag{9.43}$$

Step 1 (Choose the Problem) To solve Eq. (9.38), open Comsol Multiphysics, choose 1D, next; Transport of Diluted Species, next; Time Dependent, finish.

Step 2 (Set the Variables) Right click on Global Definitions and define the following parameters: $phi = 0.4$, $K = 2$, $gamma = 0.1$. Under the Model, right click "Definitions" and choose "Variables." Set $der = gamma/(1+K^*c) ^\wedge 2$ and $coeff = 1+(1-phi)^*der/phi$ to match Eq. (9.42).

Step 3 (Define the Geometry) Under the Model 1, right click on Geometry and choose "Interval." The default is $(0,1)$, so change it to $(0,2)$. Choose "Build All."

Step 4 (Define the Problem) First, set up the equation for c. Choose "Transport of Diluted Species" and be sure the "Convection" box is checked. Click on "Convection and Diffusion" and set u to 1/coeff and D_C to 0.001. There is no diffusion in this problem, but the finite element method requires some to get a solution. As the mesh is refined, the amount of diffusion decreases. The signal to you is oscillations in the figures, which indicates that either more elements or more artificial diffusion is needed.

Step 5 (Set the Boundary and Initial Conditions) Right click on "Transport of Diluted Species" and choose "Inflow." Select boundary point 1, then the $+$ so that 1 appears in the "Boundary Selection." Set the inflow concentration to zero. Right click on "Transport of Diluted Species" again and choose "Outflow." Select boundary point 2 and then click $+$. Accept the boundary condition, which will allow convection out of the domain. Click on "Initial Values" make sure that the domain is included in the "Domain Selection." We want the concentration to start as 0 from $x = 0$ to 0.25, then change to 1 until $x = 0.75$, then change back to zero. Thus, we use $1^*(x > 0.25)-1^*(x > 0.75)$. The $(x > 0.25)$ means the value multiplying it will be used when $x > 0.25$ but not otherwise. Thus, the initial value will be zero until $x = 0.25$, then 1, and from $x = 0.75$ on it will be zero again since both functions are used and are identical and subtracted from each other.

Step 6 (Set the Mesh) Select the mesh and change the element size to extremely fine (in anticipation that many nodes will be needed). Choose Build All; note that the nodal points are displayed in the graph.

Step 7 (Solve the Problem) Before solving the problem click on Step 1: Time Dependent under Study 1. Change the Times to see the solution to range(0,1) so that there will be two curves. Right click on "Study 1" = to get the solution. [You can use range(0,0.2,1) to get more curves, but only two are used here to allow easier comparison with Figure 9.18*b*.]

Step 8 (Examine the Solution) Select the 1D Plot Group and change the Plot Settings to reflect what we have actually done, *x*, *c*, "Solution to Problem 9.42" and click on the plot icon

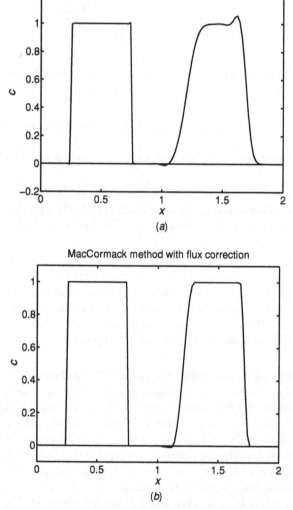

FIGURE 9.18   Solution of chromatography problem, Eq. (9.42): (*a*) Comsol Multiphysics; (*b*) MacCormack flux-corrected method (Appendix E).

to get the new plot. To get the legends for the graph, click on the Line Graph node, choose "Legends" (near the bottom) and click on the "Show legends" box. Then replot, choose the camera and save the figure. The concentration in the fluid is shown in Figure 9.18a.

Now the question arises as to the effect of the added diffusion. If we change the diffusion coefficient to 0.01, right click Study = , and look at the plot, we find that it is smoothed out considerably. If we change the diffusion coefficient to 0.0001, we find the oscillations increase dramatically. Thus, the value of 0.001 seems to be enough to get a reasonable solution, but not so much that it makes the solution incorrect. This is the meaning of the term 'streamline diffusion', and here we are just introducing it ourselves. The solution obtained in Comsol Multiphysics can be compared with Figure 9.18b, which is generated with a specialized method called MacCormack method with Flux-correction described in Appendix E. The Comsol Multiphysics solution is a reasonable solution if we ignore the small oscillations.

PRESSURE SWING ADSORPTION

Hydrogen is made everyday in refineries, and it is used to make ammonia and methanol. In the future, more may be needed for automobile fuel cells. Most of the hydrogen is obtained by purifying a gas mixture of 60–90 mol percent hydrogen (Sircar and Golden, 2010). The primary method is pressure swing adsorption. The gaseous stream flows at high pressure through a packed bed of adsorbent. The carbon dioxide, carbon monoxide, methane, and nitrogen are adsorbed on the packing, and quite pure hydrogen exits. This requires a good adsorbent, such as activated carbon or a xeolite. Then the pressure is reduced and the output, still pure hydrogen, is directed to another vessel. Finally a stream of hydrogen flows in the reverse direction at the lowest pressure to remove the adsorbed gases. For this to be efficient, processes lower the pressure several times so that there are several co-current depressurizations steps and several countercurrent decompression steps (Sircar and Golden, 2010). In this problem, an extremely simplified version is considered.

It is assumed that the process is isothermal and dilute so that the velocity does not change as the material adsorbs; these limitations are made strictly for simplification here, since they are not usually true (Ruthven et al., 1994). The equations are an expanded version of Eq. (9.37) so that the actual nonlinear adsorption process can be modeled. For each of the four species the equations are

$$\frac{\partial y_i}{\partial t} + v\frac{\partial y_i}{\partial x} = D\frac{\partial^2 y_i}{\partial x^2} - \frac{1}{\phi}\frac{RT}{p}\rho_B\frac{\partial \bar{q}_i}{\partial t} \tag{9.44}$$

$$\frac{\partial \bar{q}_i}{\partial t} = k_i(q_i^* - \bar{q}_i), \quad q_i^* = \frac{a_i p_i}{1 + \sum_{j=1}^n b_j p_j} \tag{9.45}$$

The y_i is the mole fraction, v is the interstitial velocity, D is an effective diffusion coefficient (needed for stability), ϕ is the void fraction, R is the gas constant, T is the temperature, p is the pressure, t is the time, x is the distance from the inlet, and k_i is the mass transfer coefficient. The concentration of the ith species on the adsorbant is $\bar{q}_i$ and the concentration in equilibrium with the partial pressure p_i is q_i^*. The constants a_i and b_i are parameters in the adsorption relation which is written as a multicomponent Langmuir formula.

When nondimensionalized, the equations are

$$\frac{\partial y_i}{\partial t'} + \frac{\partial y_i}{\partial x'} = \frac{1}{Pe} \frac{\partial^2 y_i}{\partial x'^2} - \text{coeff}\frac{\partial \bar{q}_i}{\partial t'}$$

$$\frac{\partial \bar{q}_i}{\partial t'} = \frac{k_i L}{v}(q_i^* - \bar{q}_i), \quad q_i^* = \frac{a_i \, p_T \, y_i}{1 + p_T \sum_{j=1}^{n} b_j y_j} \tag{9.46}$$

There will be four Transport equations for the four components and four Transport equations (without velocity) for the concentration on the adsorbent.

In the real process, this step takes place, then the pressure is reduced and there is back flow, then the pressure is reduced some more. One would need the adsorbent concentration at the end of one step as the initial condition for the next step. Then the steps would have to be run over and over until a cyclic condition was reached in which each step was the same. They will not be the same at the start, however, because we start with nothing adsorbed, whereas in the cyclic operation, there will be something, and it will be the same each cycle. Thus, this process can become very complicated; Problem 9.29 is only a start.

CHAPTER SUMMARY

You saw how the equations governing energy transfer, mass transfer, and fluid flow were similar, and examples were given for 1D problems. Examples included heat conduction, both steady and transient, reaction and diffusion in a catalyst pellet, flow in pipes and between flat plates of Newtonian or non-Newtonian fluids. The last two examples illustrated an adsorption column, in one case with a linear isotherm and slow mass transfer and in the other case with a nonlinear isotherm and fast mass transfer. Specific techniques you demonstrated include parametric solutions when the solution is desired for several values of one parameter and the use of artificial diffusion to smooth time-dependent solutions which have steep fronts are large gradients.

PROBLEMS

Note: When solving boundary value problems in MATLAB, the differential equation is evaluated at each boundary, as well as points between. Thus, for cylindrical and spherical geometry, you need to use l'Hopital's rule to evaluate the differential equation:

$$\lim_{r \to 0} \frac{1}{r} \frac{dc}{dr} = \frac{d^2c/dr^2(0)}{dr/dr(0)} = \frac{d^2c}{dr^2} \tag{9.47}$$

Chemical Reaction

9.1₁ A chemical reactor with axial dispersion is governed by the following equations:

$$\frac{1}{Pe} \frac{d^2c}{dx^2} - \frac{dc}{dx} = Da\frac{c}{c+v}, \quad -\frac{1}{Pe}\frac{dc}{dx}(0) = 1 - c(0), \quad \frac{dc}{dx}(1) = 0 \tag{9.48}$$

Solve for $Da = 8$, $v = 3$, and $Pe = 15$, 150, and 1500. (Hint: include streamwise diffusion as needed.)

9.2₂ The following equation governs diffusion and reaction of carbon monoxide in an isothermal catalyst. Find the average reaction rate and compare the average with the reaction rate if the concentration were everywhere the same as the external concentration (here $c = 1$). The parameters are $\phi = 32$, $Bi_m = 10$, $\alpha = 20$. Solve using (a) Comsol and (b) MATLAB:

$$\frac{1}{r^2}\frac{d}{dr}\left(r^2\frac{dc}{dr}\right) = \phi^2 R(c), \quad \frac{dc}{dr}(0) = 0,$$

$$-\frac{dc}{dr}(1) = Bi_m(c(1) - 1),$$

$$R(c) = \frac{c}{(1 + \alpha c)^2} \tag{9.49}$$

9.3₁ The transient reaction and diffusion in a packed bed with axial dispersion is governed by the following equations. Begin with an initial concentration in the bed of zero. At time zero start flowing with the inlet concentration $= 1.0$. Integrate to steady state. Use the parameters $Pe = 100$, $0 \le x \le 1$, $Da = 2$, $v = 2$:

$$\frac{\partial c}{\partial t} + \frac{\partial c}{\partial x} = \frac{1}{Pe}\frac{\partial^2 c}{\partial x^2} - Da\frac{c}{c + v}, \quad -\frac{1}{Pe}\frac{\partial c}{\partial x}(0, t) = 1 - c(0, t),$$

$$\frac{\partial c}{\partial x}(1, t) = 0,$$

$$c(x, 0) = 0 \tag{9.50}$$

9.4₁ Bratu's equation is an interesting boundary value problem that arises in a simplification of a combustion reaction within a cylindrical vessel (Frank-Kamenetskii, 1969; Lin et al., 2008). The problem has more than one solution for the same conditions; thus, one must start from different initial guesses to find it. In addition, time-dependent solutions can become unstable and jump from one steady solution to another as a parameter changes. Or, the steady-state solution may be unstable and not realizable. The equations are

$$\frac{d^2 y}{dx^2} + \lambda \exp(y) = 0, \quad y(0) = y(1) = 0 \tag{9.51}$$

Use Comsol Multiphysics to solve this problem for $\lambda = 1$. Try to find more than one solution.

Chemical Reaction and Heat Transfer

9.5₁ A chemical reactor with axial dispersion of both mass and energy is governed by the following equations:

$$\frac{1}{Pe}\frac{d^2 c}{dx^2} - \frac{dc}{dx} = R(c, T), \quad -\frac{1}{Pe}\frac{dc}{dx}(0) = 1 - c(0), \quad \frac{dc}{dx}(1) = 0 \quad (9.52)$$

$$\frac{1}{Pe_H}\frac{d^2 T}{dx^2} - \frac{dT}{dx} = \beta R(c, T), \quad -\frac{1}{Pe_H}\frac{dT}{dx}(0) = 1 - T(0), \quad \frac{dT}{dx}(1) = 0 \quad (9.53)$$

The reaction rate expression is

$$R(c, T) = 3.817c^2 \exp(\gamma - \gamma/T) \tag{9.54}$$

and the parameters are $Pe = Pe_H = 96$, $\gamma = 17.6$, $\beta = -0.056$.

9.6₁ The following equations govern diffusion and reaction in a catalyst with significant heat effects. Solve for $\phi = 1.1$, $\gamma = 30$ and $\beta = 0, 0.01, 0.03, 0.07, 0.10, 0.12, 0.15$:

$$\frac{1}{r^2}\frac{d}{dr}\left(r^2\frac{dc}{dr}\right) = \phi^2 R(c, T), \qquad \frac{dc}{dr}(0) = 0, \quad c(1) = 1 \tag{9.55}$$

$$\frac{1}{r^2}\frac{d}{dr}\left(r^2\frac{dT}{dr}\right) = -\beta\phi^2 R(c, T), \qquad \frac{dT}{dr}(0) = 0, \quad T(1) = 1,$$

$$R(c, T) = c\exp[\gamma(1 - 1/T)] \tag{9.56}$$

9.7₁ Solve the following problem for reaction and heat transfer in a porous catalyst pellet for the parameters: $\phi = 1.1$, $\gamma = 30$, $\beta = 0.15$, $Le = 1050$ (see Hellinckx et al., 1972), for a simplified form of the problem. Use initial conditions of $c = 0$, $T = 0.8$. Integrate from $t = 0$ to 1050:

$$\frac{\partial c}{\partial t} = \frac{1}{r^2}\frac{\partial}{\partial r}\left(r^2\frac{\partial c}{\partial r}\right) - \phi^2 R(c, T), \qquad \frac{\partial c}{\partial r}(0, t) = 0, \quad c(1, t) = 1 \tag{9.57}$$

$$Le\frac{\partial T}{\partial t} = \frac{1}{r^2}\frac{\partial}{\partial r}\left(r^2\frac{\partial T}{\partial r}\right) + \beta\phi^2 R(c, T), \qquad \frac{\partial T}{\partial r}(0, t) = 0,$$

$$T(1, t) = 1, \quad R(c, t) = c\exp[\gamma(1 - 1/T)] \tag{9.58}$$

9.8₁ Solve problem 9.6 with $Le = 0.1$. Integrate from $t = 0$ to 1.

9.9₁ Hlaváček et al. (1968) provide the following example of heat and mass transfer in a porous catalyst that is a flat particle. The parameters are chosen so that multiple solutions exist; thus one must start from different initial guesses to find them. In addition, time-dependent solutions can become unstable and jump from one steady solution to another as a parameter changes. Or, the steady-state solution may be unstable and not realizable. The equations are

$$\frac{d^2 y}{dx^2} = \lambda y \exp\left[\frac{\gamma\beta(1 - y)}{1 + \beta(1 - y)}\right], \qquad y'(0) = 0, y(1) = 1 \tag{9.59}$$

Use Comsol Multiphysics to solve for $\lambda = 0.05, 0.1$, and 0.15. Find as many solutions as you can for each parameter, λ. Lin et al. (2008) provide a method to be sure all the solutions have been found.

Mass Transfer

9.10$_2$ Solve the following diffusion problem with a variable diffusivity (appropriate for some polymers). Solve using (a) Comsol and (b) MATLAB:

$$\frac{\partial c}{\partial t} = \frac{\partial}{\partial x}\left(D\frac{\partial c}{\partial x}\right), \quad c(0,t)=1, \quad c(1,t)=0, \quad c(x,0)=0, \quad D=e^{0.5c}$$

$$(9.60)$$

Heat Transfer

9.11$_1$ Solve the following heat transfer problem with radiation in cylindrical geometry. The parameters are $G=2$, $Nu=10$:

$$\frac{d^2T}{dr^2} + \frac{1}{r}\frac{dT}{dr} + G, \quad \frac{dT}{dr}(0)=0, \quad -\frac{dT}{dr}(1)=Nu[T^4(1)-1] \qquad (9.61)$$

9.12$_1$ Solve the following transient problem and integrate to steady state. Do you get the same solution at steady state as found in Problem 9.9? Solve using (a) Comsol and (b) MATLAB:

$$\frac{\partial T}{\partial t} = \frac{\partial^2 T}{\partial r^2} + \frac{1}{r}\frac{\partial T}{\partial r} + G, \quad \frac{\partial T}{\partial r}(0,t)=0,$$

$$-\frac{\partial T}{\partial r}(1,t)=Nu[T^4(1,t)-1], \qquad T(r,0)=1 \qquad (9.62)$$

9.13$_1$ Solve the following heat transfer problem. The parameters are $\rho=491$ lb$_m$/ft^3, $C_p=0.11$ Btu/lb$_m$, $R=0.4$ ft, $h=30$ Btu/h ft^2°F, $k=3$ Btu/h ft^2°F, $T_0=120$°F, $T_{surr}=68$°F:

$$\rho C_p\frac{\partial T}{\partial t} = \frac{k}{r}\frac{\partial}{\partial r}\left(r\frac{\partial T}{\partial r}\right), \qquad \frac{\partial T}{\partial r}(0,t)=0,$$

$$-k\frac{\partial T}{\partial r}(R,t)=h[T(R,t)-T_{surr}], \qquad T(r,0)=T_o \qquad (9.63)$$

9.14$_1$ Solve the following heat transfer problem. The parameters are $\rho=491$ lb$_m$/ft^3, $C_p=0.11$ Btu/lb$_m$, $R=0.4$ ft, $U=12.9$ Btu/h ft^2°F, $T_0=120$°F, $T_{surr}=68$°F. Compare the average temperature versus time with the solution found in Problem 9.13:

$$\rho C_p\pi R^2 L\frac{dT}{dt} = -U2\pi RL(T-T_{surr}), \quad T(0)=T_0 \qquad (9.64)$$

Electrical Fields

9.15$_2$ The following (nondimensional) problem governs the electric potential outside a cylinder that is kept at potential Ψ_0 whereas the potential at a distance L away from the center of the cylinder is kept at zero; L is taken as 10. The parameter κ is 0.8, the applied potential is 2, and the radius of the cylinder is 1. Find the potential distribution around the cylinder. Solve using (a) Comsol and (b) MATLAB:

$$\frac{d^2\Psi}{dr^2} + \frac{1}{r}\frac{d\Psi}{dr} = \kappa^2\Psi, \quad \Psi(R) = \Psi_0, \quad \Psi(L) = 0 \tag{9.65}$$

Parts (a) and (b) are applicable when the dependent variable remains small (linearized Poisson–Boltzmann equation). Solve the problem with the right-hand side = $\kappa^2 \sinh(\Psi)$ (Kirby, 2010, pp. 206–208). Comment on the validity of the Debye approximation.

9.16₁ The transient charge distribution in a semiconductor is governed by the following partial differential equation. Integrate the equation to steady state with $\beta = 1.5$:

$$\frac{\partial c}{\partial t} = \frac{\partial^2 c}{\partial x^2} - \beta e^c, \quad c(x, 0) = 0.5, \quad c(0, t) = 1, \quad c(1, t) = 1 \tag{9.66}$$

9.17₁ Use Comsol Multiphysics to solve the problem

$$\frac{d^2\Psi}{dy^2} = \kappa^2 \sinh(\Psi), \quad \Psi(0) = 2, \quad \frac{d\Psi}{dy}(\infty) = 0 \tag{9.67}$$

using $\kappa = 0.8$. Use a finite, but large, domain. Show the difference when the Debye approximation is used and the right-hand side is = $\kappa^2\Psi$.

9.18₁ Use Comsol Multiphysics to solve the Boltzmann problem between two flat plates

$$\frac{d^2\Psi}{dy^2} = \kappa^2 \sinh(\Psi) \quad \text{and} \quad \Psi(0.5) = 2, \quad \frac{d\Psi}{dy}(0) = 0 \tag{9.68}$$

in a finite domain, 0 to 0.5, with zero slope at $y = 0$ (the center) and a boundary condition of 2 at $y = 0.5$. The $\kappa = 0.8$. Also do this for the Debye approximation, with the right-hand side replaced by Ψ. Is the Debye approximation valid?

Fluid Flow

9.19₂ Solve for the flow of a Newtonian fluid in a pipe, following the example. Plot the shear rate as a function of radial position. Calculate the average velocity. Plot the shear stress as a function of radial position. Solve using (a) Comsol and (b) MATLAB.

9.20₂ Solve the flow of a non-Newtonian fluid in a pipe, following the example, for pressure drops of 10, 10^3, 10^5, 10^7 Pa. The parameters are $\eta_0 = 0.492$ Pa s, $\lambda = 0.1$ s^{-1}, and $n = 0.8$. Plot the velocity, shear rate and shear stress as a function of radial position. How do these curves change as the pressure drop is increased?

9.21₁ A fiber spinning problem is governed by the following equations from Middleman (1977):

$$\frac{d^2u}{dx^2} = \frac{\rho}{3\mu}u\frac{du}{dx} + \frac{1}{u}\left(\frac{du}{dx}\right)^2, \quad u(0) = u_0, \quad u(L) = u_L \tag{9.69}$$

The ratio of the two velocities is the drawdown ratio,

$$D_R = \frac{u_L}{u_0} \tag{9.70}$$

When $\rho = 0$ the solution for velocity and radius of the fiber is (Middleman, 1977, p. 237)

$$u = u_0 \exp\left(\frac{x \ln D_R}{L}\right), \quad R(x) = R_0 \exp\left(-\frac{1}{2}\frac{x \ln D_R}{L}\right) \tag{9.71}$$

Solve the problem without inertia ($\rho = 0$) using $u_0 = 16.7$ cm/s, $u_L = 1666.7$ cm/s, and $L = 300$ cm.

9.22₁ Solve the fiber-spinning problem (9.21) with $\rho = 0.96$ g/cm³ and $\mu = 5000$ poise. Is inertia important? How does the fiber radius compare with the solution to Problem (9.21)?

9.23₁ Solve the fiber-spinning problem for a non-Newtonian fluid with viscosity that depends on shear rate. The problem without inertia is

$$\frac{d}{dx}\left(\eta\frac{du}{dx}\right) = \frac{\eta}{u}\left(\frac{du}{dx}\right)^2, \quad u(0) = u_0, \quad u(L) = u_L,$$

$$\eta = K(2\times 3^{(n-1)/2})\left|\frac{du}{dx}\right|, \quad K = 5000 \tag{9.72}$$

Solve it for $u_0 = 16.7$ cm/s, $u_L = 1666.7$ cm/s, $L = 300$ cm, and $n = 1$ to 0.5 using the parametric solver.

9.24₂ Berg (2010, p. 712) discusses steady thermocapillary flow in one dimension. Consider a shallow trough of fluid, with a temperature gradient imposed from left to right (hot on the left); see Figure 9.19. First, we assume that the solution depends only on the vertical distance, y, and that the trough is very long and very wide. Due to the imposed temperature gradient on the surface, and the temperature dependence of surface tension, a force is applied on the top that requires the fluid to move from left to right. To obey continuity, it must (somewhere) also move from right to left. Thus, the problem is

$$\mu\frac{d^2v}{dy^2} = \frac{dp}{dx}, \quad v(h) = 0, \quad \left[\mu\frac{dv}{dy} + \frac{d\sigma}{dT}\frac{dT}{dx}\right]_{y=0} = 0 \tag{9.73}$$

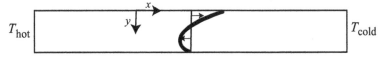

FIGURE 9.19 Thermocapillary flow.

The pressure drop is eliminated by the condition that the new flow rate (left to right) be zero:

$$\int_0^h v \, dy = 0 \tag{9.74}$$

Use Comsol Multiphysics to solve for the velocity profile when $L = 10$ cm, $h = 1$ mm, $\Delta T = 40°C$, $\mu = 5$ cP, and $d\sigma/dT = 0.1$ mN/mK. (Hint: you can program it with a specific dp/dx and vary it until you get the integral to be zero; or you can use the parametric solver and solve for a variety of values of dp/dx and interpolate and extrapolate; or you can use integration coupling to cause the dp/dx to be chosen such that the average velocity is zero; see Appendix D.)

9.25₁ When colloids flow under conditions of low volume fraction, the viscosity of the solution is constant. When the volume fraction is larger, though, the particles of a suspension of hard, noninteracting (i.e., nonaggregating) spherical particles, cause the viscosity to vary with the dimensionless shear rate, called Peclet Number, Pe, as follows (Berg, 2010, p. 626):

$$\frac{\eta(Pe) - \eta(\infty)}{\eta(0) - \eta(\infty)} = \frac{\sinh^{-1}(Pe)}{Pe} \tag{9.75}$$

This is often approximated by the Cross equation:

$$\frac{\eta(\dot\gamma) - \eta(\infty)}{\eta(0) - \eta(\infty)} = \frac{1}{1 + (C\dot\gamma)^p} \tag{9.76}$$

This is reminiscent of polymers with viscosity functions like Eq. (9.14). The C is the "Cross constant" and the p is the "Cross exponent." Use Comsol Multiphysics to solve for 1D flow of a colloid between two flat plates that are 1 cm apart, when the constants are $\eta(0) = 0.5$, $\eta(\infty) = 0.01$, $C = 2$, $p = 0.4$ under a variety of pressure drops from 1 to 1e6.

9.26₂ Kirby (2010, p. 259) suggests a change of variable for problems with high Peclet numbers. The problem

$$\frac{\partial c}{\partial t} + Pe\frac{\partial c}{\partial x} = \frac{\partial^2 c}{\partial x^2}, \quad c(0, t) = 1, \quad \left.\frac{\partial c}{\partial x}\right|_{x=1} = 0, \quad c(x, 0) = 0 \tag{9.77}$$

would be converted using $c = \exp(\gamma)$ to

$$e^\gamma \frac{\partial \gamma}{\partial t} + Pe\, e^\gamma \frac{\partial \gamma}{\partial x} = \frac{\partial}{\partial x}\left(e^\gamma \frac{\partial \gamma}{\partial x}\right), \quad \gamma(0, t) = 0, \quad \left.\frac{\partial \gamma}{\partial x}\right|_{x=1} = 0,$$
$$\gamma(x, 0) = -10. \tag{9.78}$$

The value -10 is simply taken as a large negative number. Use Comsol Multiphysics to solve this problem on the domain $x = 0,1$ for $Pe = 1000$ using 30 and 60 elements.

Compare the solution with that obtained for the original problem by using the same number of elements.

9.27₁ Thermal flow field fractionation involves mass transfer induced by temperature gradients. Consider two flat plates kept at different temperatures. Consider two flat plates separated by 5×10^{-5} m, kept at different temperatures. The fluid between is a two-component system. The theory is based on equations from deGroot and Mazur (1954) and Bird et al. (2002, p. 770). The mass flux of component one is given by Eq. (9.79):

$$j_1 = -\rho D \nabla c - \rho D_T c(1 - c)\nabla T \tag{9.79}$$

where j_1 is the mass flux, ρ is the density of the solution, D is the diffusion coefficient of the solute in the solvent, c is the mass fraction of the solute (component 1) in the solvent (component 2), D_T is the thermal diffusion coefficient, and T is the temperature. If the solute is dilute, this reduces to

$$j_1 = -\rho D \nabla c - \rho D_T c \nabla T \tag{9.80}$$

Dividing by the average molecular weight of the mixture (assumed constant in a dilute system) converts this to molar flux. The problem is to solve the following equation for the dilute system:

$$\rho C_p \frac{\partial T}{\partial t} = k\nabla^2 T, \quad \frac{\partial c}{\partial t} = \nabla \cdot [D\nabla c + D_T \nabla T] \tag{9.81}$$

The left surface is at 308 K, the right one is at 298 K. The initial temperature is 298 K and the initial relative mass fraction was 1.0. This is a dimensionless value, the actual mass fraction divided by itself.

9.28₂ Flow through a partially saturated soil is often described macroscopically using the following equation (for one dimension):

$$\frac{\partial}{\partial t}(\rho\phi S) = -\frac{\partial(\rho q)}{\partial x} \tag{9.82}$$

where ρ is the density (kg/m³), ϕ is the porosity of the rock/soil, and S is the saturation, defined as the fraction of free space that is occupied by water. The combination $\rho\phi S$ is the mass density per system volume. The mass flux is ρq in units of mass of water per unit time per unit total cross-sectional area (kg/m² s); q is the volumetric flux (a velocity, m/s). The mass flux is related to the pressure gradient by Darcy's law

$$\rho q = -\frac{\rho k}{\mu}\left(\frac{\partial p}{\partial x} - \rho g\right) \tag{9.83}$$

In this equation, k is the permeability (m²), μ is the viscosity of the liquid (kg/m s), p_w is the pressure (Pa), and g is the acceleration of gravity (m/s²).

The mass balance is

$$\phi \frac{\partial S}{\partial t} = \frac{\partial}{\partial x}\left[\frac{k}{\mu}\left(\frac{\partial p}{\partial x} - \rho g\right)\right] \tag{9.84}$$

The saturation and permeability depend on the capillary pressure. Here, we consider water in dry soil, in which the void volume not filled with water is filled with air at a constant pressure. The capillary pressure is $p_c = p_{air} - p$, which can be reduced to $p_c = -p$ since p_{air} is constant. The saturation time derivative is then

$$\frac{\partial S}{\partial t} = -\frac{dS}{dp_c}\frac{\partial p}{\partial t} \quad \text{so that} \quad -\phi\frac{dS}{dp_c}\frac{\partial p}{\partial t} = \frac{\partial}{\partial x}\left[\frac{k}{\mu}\left(\frac{\partial p}{\partial x} - \rho g\right)\right] \tag{9.85}$$

The equation is made nondimensional using

$$p' = \frac{p}{\rho g L}, \quad k_r = \frac{k}{k_0}, \quad x' = \frac{x}{L}, \quad t' = \frac{t}{t_c}, \quad t_c = \frac{\phi L \mu}{k_0 \rho g},$$

$$-\frac{dS}{dp_c'}\frac{\partial p'}{\partial t'} = \frac{\partial}{\partial x'}\left(k_r \frac{\partial p'}{\partial x'}\right) - \frac{\partial k_r}{\partial x'} \tag{9.86}$$

The dependence of S and k_r on p_c is empirical and must be determined from measurements of the soil in question. The following formulae were developed to model very dry soil of the type in Richland, Washington, where nuclear waste is stored:

$$k_r = \frac{1}{1 + (p_c'L/B)^\lambda}, \quad \frac{S - S_r}{1 - S_r} = \frac{1}{1 + (p_c'L/A)^\eta} \tag{9.87}$$

For this soil the parameters are $k_r = 0.50 \, \pi$ m, $\phi = 0.485$, $S_r = 0.32$, $A = 231$ cm, $B = 146$ cm, $\eta = 3.65$, $\lambda = 6.65$. Solve a problem for a soil of height 100 cm initially at $p/\rho g = -1000$ cm (the lower the value the drier the soil). In nondimensional terms, solve for $x = 0$ to 0.4 (the solution never gets past that point). Let the left hand side (top) be at $p_c' = -0.01$ (almost saturated), the initial condition be $p_c' = -10$ (very dry) and have zero flux out the right-hand side. First, solve the problem without gravity and then include gravity.

9.29₂ Solve the simplified pressure swing adsorption problem, Eq. (9.46). For the purposes of this problem, reasonable values will be taken for the parameters, and the Langmuir formula for adsorption of a pure component models that in Sircar and Golden (2010). First, plot the equilibrium concentration versus pressure for pressures from 0.1 to 100 atm. Next, solve the example problem, linear adsorption. Finally, change that problem to involve one nonlinear Langmuir isotherm. Run that to verify that the basic format of your model is correct. Then add several Transport equations so that there are four mole fractions and four adsorbent concentrations. For the concentration on the solid, use a diffusivity of zero and no-flux boundary conditions at each end. Begin the calculation with nothing adsorbed and integrate until $t' = 1$ when the column pressure is 20 atm. Look at the gas mole fraction, the concentration on the adsorbant, and the breakthrough curve (concentrations at the exit versus time). For CO_2, CH_4,

CO, and H_2, the values of a are (mmol/g atm): 4, 1.6, 0.4, 0.03. The values of b (atm^{-1}) are 0.188, 0.172, 0.068, 0. The inlet mole fractions are 0.2, 0.04, 0.02, 0.74. Solve with $Pe = 1000$, coeff $= 7.7$ and $k_i L/v = 0.05$.

Numerical Problems (See Appendix E)

9.30₁ Solve Problem 9.17 using a finite difference method in Excel.

9.31₁ Solve Eq. (9.26) using a finite difference method in Excel with the parameters used to prepare Figure 9.13.

9.32₂ Solve Eq. (9.28) using a finite difference method in space and the Euler method in time for $t = 0$ to 1. Program it in Excel.

9.33₂ Solve the problem of start-up flow in a pipe using the finite difference method in space and ode45 in MATLAB for time. Use the parameters used to create Figure 10.13.

9.34₂ Investigate the accuracy of the finite element method applied to the reaction-diffusion problem, Eq. (9.25) when $\alpha = 20$ and $K = 2$. To do this, start with linear discretization and solve the problem on increasingly dense meshes. Keep a record of the flux at the $r = 1$ and the number of degrees of freedom. Then do this again with quadratic discretization, then with cubic discretization.

What is the exact solution? Plot the flux versus mesh size and degrees of freedom. As an additional test, calculate the absolute value of the residual, too. This is the differential equation with the finite element solution put into it. The goal is to make it zero, but this will occur only for an infinite number of degrees of freedom. For this problem the residual is

$$\text{Residual} = crr + 2 * cr/r - alpha * c/(1 + K * c) \qquad (9.88)$$

When you use linear discretization the second derivatives will not be defined. Integrate the absolute value of the residual over the entire domain; the residual will be oscillating around zero so the absolute value is necessary.

10

FLUID FLOW IN TWO AND THREE DIMENSIONS

Most chemical processes involve fluids flowing in turbulent flow. However, polymer melts (the precursors to plastics) are so viscous that they flow in laminar flow. In modern chemical engineering, there are more and more applications in small devices in which laminar flow prevails. In microfludics, these devices may be used for detection of biological or chemical weapons (after the payload is dispersed into the air), for medical diagnostics for cheap and rapid measurements, and for analytical tools used in process control. There are even applications where whole chemical processes are done on a microscale. This may be advantageous because the temperature can be carefully controlled (making a certain reaction possible), or the conversion can be increased due to precise mixing (minimizing waste products and recycle streams), or it may be advantageous to have distributed manufacture for safety reasons. In microfluidics, the dimensions of the channels are 1 mm or less and the cross sections of the channels are square or rectangular (so they are no longer called pipes), and the flow is laminar even in complex geometries. Thus, it is worthwhile to learn how to solve laminar flow problems in complex geometries.

Two methods are used in commercial computational fluid dynamics (CFD) codes: the finite volume method and the finite element method. To a beginner, it probably makes little difference which method is used. The author has used finite element methods for fluid flow for over 35 years, and that is the method used in Comsol Multiphysics, which is the CFD program illustrated here.

What does matter, though, is the difference between laminar and turbulent flow. A common misconception is that if you see vortices (swirling motion), the flow is turbulent. That is not necessarily true. In turbulent flow, there are fine-scale oscillations on time

Introduction to Chemical Engineering Computing, Updated Second Edition. Bruce A. Finlayson.
© 2014 John Wiley & Sons, Inc. Published 2014 by John Wiley & Sons, Inc.

scales of 0.1 ms, and spatial fluctuations on the scale of 30 μm (30 × 10^{-6} m, or 3% of a millimeter). To model turbulent flow, one option is direct numerical simulation (DNS). In DNS, the Navier–Stokes equations are integrated in time on a very small spatial scale, with millions of nodes; this is fundamental (no assumptions are made), but it is also very time consuming. A DNS simulation for even simple problems, such as turbulent flow in a pipe), may take hours for a network of a few computers to simulate what happens in seconds in reality. The computer needs to solve for three fluctuating velocities on a very small spatial scale and as a function of time. To overcome this problem, commercial CFD codes use k-ε models, which introduce new variables representing the turbulent kinetic energy (k) and rate of dissipation (ε), and add equations to predict them. The commercial codes then solve for the time-averaged velocities and use k and ε to represent the effect of the fluctuating velocities without solving the fluctuating velocities. The equations have several empirical parameters that have been determined by comparison with experimental data, but these parameters change depending on the type of flow, and some uncertainty exists in realistic situations. The k-ε model also involves steep boundary layers near solid surfaces, and the CFD codes try to take that into account analytically. To summarize the k-ε model: it contains empiricism but is relatively fast to compute with. There is an intermediate method—large eddy simulation (LES)—which is more reliable than the k-ε models and uses parameters from DNS studies, but it, too, involves lengthy calculations, somewhere between DNS and k-ε.

This chapter focuses on fluid flow, leaving the combination of fluid flow, heat transfer, and diffusion to Chapter 11. Examples of fluid flow include entry flow into a pipe, flow in a microfluidic T-sensor, turbulent flow in a pipe, time-dependent start-up of pipe flow, flow in an orifice, and flow in a serpentine mixer. The examples demonstrate many of the techniques that are useful in the program Comsol Multiphysics.

Instructional Objectives: After working through this chapter, you will be able to

1. Model laminar fluid flow in two-dimensional (2D) and three-dimensional (3D) situations, including axisymmetric geometries.
2. See how to model turbulent flow, which depends on the specific modules of Comsol Multiphysics in your installation.
3. Set the options you need: units and discretization.
4. Apply various boundary conditions, including periodic boundary conditions.
5. Handle variable viscosity, depending upon shear rate (non-Newtonian fluid and colloids).
6. Make a variety of plots (velocity, streamlines, arrow plots, pressure plots) and calculate properties of the solution (average values at boundaries).
7. Use the parametric solver to compute a number of solutions with one parameter changing.
8. Check your solution for continuity of flow.
9. Learn to use coupling variables to solve for a fully developed velocity in two dimensions that is used for an inlet velocity in a 3D problem.

In addition, you will be introduced to the field of microfluidics.

MATHEMATICAL FOUNDATION OF FLUID FLOW

Navier–Stokes Equation

The flow equations are taken here as the Navier–Stokes equations, with some extensions to purely viscous non-Newtonian fluids. In vector form, the Navier–Stokes equation is

$$\rho\frac{\partial \mathbf{u}}{\partial t} + \rho\mathbf{u}\cdot\nabla\mathbf{u} = -\nabla p + \mu\nabla^2\mathbf{u} \tag{10.1}$$

This vector equation applies to the case when the gravitational term is absorbed into the pressure. It also takes the density as constant, leading to an incompressible fluid:

$$\nabla\cdot\mathbf{u} = 0 \tag{10.2}$$

For steady flows, the equation reduces to

$$\rho\mathbf{u}\cdot\nabla\mathbf{u} = -\nabla p + \mu\nabla^2\mathbf{u} \tag{10.3}$$

In component notation in Cartesian geometry and two dimensions, these equations are

$$\rho\left(\frac{\partial u}{\partial t} + u\frac{\partial u}{\partial x} + v\frac{\partial u}{\partial y}\right) = -\frac{\partial p}{\partial x} + \mu\left(\frac{\partial^2 u}{\partial x^2} + \frac{\partial^2 u}{\partial y^2}\right)$$
$$\rho\left(\frac{\partial v}{\partial t} + u\frac{\partial v}{\partial x} + v\frac{\partial v}{\partial y}\right) = -\frac{\partial p}{\partial y} + \mu\left(\frac{\partial^2 v}{\partial x^2} + \frac{\partial^2 v}{\partial y^2}\right) \tag{10.4}$$
$$\frac{\partial u}{\partial x} + \frac{\partial v}{\partial y} = 0$$

Because the program Comsol Multiphysics includes all these terms, you merely need to define the problem you want to solve.

When evaluating pressure, one complication in the finite element method is a mathematical condition saying that the approximating functions for pressure have more restricted choices than those for velocity. This is called the LBB condition [named after Ladyshenskaya, Brezzi, and Babuska (Gresho et al., 1998, p. 593)], and the designers of the computer program have taken it into account for you. A variety of consistent methods of discretization are included in Comsol Multiphysics. Another feature of the finite element method is that the computer problem can get very large. Often you will have to make compromises between the accuracy you would like and how long you are willing to wait for a solution. With modern computers, this is not as big a problem as it once was, but in complicated 3D cases it is still a concern. Thus, one aspect of your solution method will be mesh refinement to estimate the accuracy.

The finite element method in two and three dimensions is similar to the finite element in one dimension, except the bookkeeping is harder. But again, Comsol Multiphysics does that bookkeeping for us. For one type of element (there are several choices), the domain of the problem is covered by small triangles (in two dimensions) or small tetrahedrons (in three dimensions), and on each triangle (or tetrahedron) the dependent variable (*u* or *v* velocity or *w*) is approximated by a polynomial, even a straight line along the boundary. In two dimensions, the approximation is a triangular patch looking like a geodesic dome. Engineers and scientists have determined how to make the approximation satisfy the equations as best as possible, but that is beyond the scope of this book; see Appendix E for a brief description of the finite element method. When the solution is represented this way, you need to recognize that you have introduced an approximation, and the solution is no better than the mesh (or set of triangles or tetrahedrons) allows. Thus, you will usually solve the problem on more than one mesh, each one more refined than the last, to ensure that the results do not depend upon the mesh.

Non-Newtonian Fluid

For purely viscous non-Newtonian fluids, the viscosity is a function of shear rate, as illustrated in Chapter 9, and the steady-state equation must be written as

$$\rho \mathbf{u} \cdot \nabla \mathbf{u} = -\nabla p + \nabla \cdot [\eta(\nabla \mathbf{u} + \nabla \mathbf{u}^T)]$$

$$\eta = f\left(\frac{1}{2}\Delta : \Delta\right), \quad \Delta = \nabla \mathbf{u} + \nabla \mathbf{u}^T$$

(10.5)

$$\frac{1}{2}(\Delta : \Delta) = 2\left[\left(\frac{\partial u}{\partial x}\right)^2 + \left(\frac{\partial v}{\partial y}\right)^2\right] + \left[\frac{\partial u}{\partial y} + \frac{\partial v}{\partial x}\right]^2$$

(10.6)

The expression for the shear rate in 2D flow, Eq. (10.6) is obtained from Bird et al. (2002, p. 849). The expression for cylindrical geometry is Eq. (10.7):

$$\frac{1}{2}(\Delta : \Delta) = 2\left[\left(\frac{\partial v_r}{\partial r}\right) + \frac{v_r^2}{r^2} + \left(\frac{\partial v_z}{\partial z}\right)^2\right] + \left[\frac{\partial v_r}{\partial z} + \frac{\partial v_z}{\partial r}\right]^2$$

(10.7)

Note the notation is different from that used in Comsol Multiphysics. With the transformation

$$v_r \to u, \quad v_z \to v, \quad \frac{\partial v_r}{\partial r} \to ur, \quad \frac{\partial v_r}{\partial z} \to uz, \quad \frac{\partial v_z}{\partial r} \to vr, \quad \frac{\partial v_z}{\partial z} \to vz$$

(10.8)

this is

$$\tfrac{1}{2}(\Delta : \Delta) = 2\left[ur^2 + (u/r)^2 + (vz)^2\right] + [uz + vr]^2$$

(10.9)

The shear rate is also defined in Comsol Multiphysics. To find out what it is called, go to Results, choose a plot, choose expressions, and look for "shear rate." Here it is mod1.spf.sr. Thus, one can either define the shear rate as shown in Eq. (10.9) or use mod1.spf.sr.

This chapter gives several illustrations applying the finite element method using Comsol Multiphysics. One decision you must make is whether to solve the equation in dimensional form or nondimensional form. The dimensional form will solve one problem at a time, whereas the nondimensional form will be solving an infinite number of problems each simulation. The nondimensional form also allows you (forces you) to consider the relative importance of different phenomena for your problem. Three-fourths of the problems in the classic transport book, *Transport Phenomena* (Bird et al., 2002) are formulated in a nondimensional format, for example. On the other hand, if you are modeling a specific device, as you would when working for a company, the more general nondimensional form is not as necessary and you may use the dimensional form. Both are illustrated here, but most of the examples are in a nondimensional format. Shown below is the method to make the problem nondimensional and instructions are given for doing this with Comsol Multiphysics, too.

NONDIMENSIONALIZATION

This section takes the dimensional Navier–Stokes equation, Eq. (10.10) and derives two different dimensionless versions:

$$\rho \frac{\partial \mathbf{u}}{\partial t} + \rho \mathbf{u} \cdot \nabla \mathbf{u} = -\nabla p + \mu \nabla^2 \mathbf{u} \tag{10.10}$$

The velocity, pressure, and distance are made dimensionless by dividing them by a constant standard for velocity, pressure, and distance. The nondimensional variable is denoted by a prime:

$$\mathbf{u}' = \frac{\mathbf{u}}{u_s}, \quad p' = \frac{p}{p_s}, \quad \mathbf{x}' = \frac{\mathbf{x}}{x_s}, \quad \nabla' = x_s \nabla \text{ or } \mathbf{u} = u_s \mathbf{u}', \quad p = p_s p', \quad \mathbf{x} = x_s \mathbf{x}', \quad \nabla = \frac{1}{x_s} \nabla'$$

$$\tag{10.11}$$

The variables are substituted into the Navier–Stokes equation, Eq. (10.10); note that the dimensional standards are constants and can be taken through the differentials:

$$\frac{\rho u_s}{t_s} \frac{\partial \mathbf{u}'}{\partial t'} + \frac{\rho u_s^2}{x_s} \mathbf{u}' \cdot \nabla' \mathbf{u}' = -\frac{p_s}{x_s} \nabla' p' + \frac{\mu u_s}{x_s^2} \nabla'^2 \mathbf{u}' \tag{10.12}$$

Option One: Slow Flows

This option is especially useful for slow flows such as those that occur in microfluidics, since then the inertial terms are multiplied by the Reynolds number, which is small and the term can often be neglected. Multiply Eq. (10.12) by $x_s^2 / \mu u_s$ to give

$$\frac{\rho x_s^2}{\mu t_s} \frac{\partial \mathbf{u}'}{\partial t'} + \frac{\rho u_s x_s}{\mu} \mathbf{u}' \cdot \nabla' \mathbf{u}' = -\frac{p_s x_s}{\mu u_s} \nabla' p' + \nabla'^2 \mathbf{u}' \tag{10.13}$$

Choose the pressure standard so that the coefficient of the pressure gradient term is 1.0; make a similar choice for the time standard to get

$$\frac{p_s x_s}{\mu u_s} = 1, \quad \text{or } p_s = \frac{\mu u_s}{x_s}, \quad \frac{\rho x_s^2}{\mu t_s} = 1 \text{ or } t_s = \frac{\rho x_s^2}{\mu} \tag{10.14}$$

Then the nondimensional Navier–Stokes equation is

$$\frac{\partial \mathbf{u}'}{\partial t'} + Re\,\mathbf{u}'\cdot\nabla'\mathbf{u}' = -\nabla' p' + \nabla'^2\mathbf{u}', \quad Re = \frac{\rho u_s x_s}{\mu} \tag{10.15}$$

An alternative is to use a different standard for time,

$$t_s = \frac{x_s}{u_s}, \quad \text{giving} \quad \frac{\rho x_s^2}{\mu t_s} = \frac{\rho u_s x_s}{\mu} = Re \tag{10.16}$$

in which case the nondimensional Navier–Stokes equation is

$$Re\frac{\partial \mathbf{u}'}{\partial t'} + Re\,\mathbf{u}'\cdot\nabla'\mathbf{u}' = -\nabla' p' + \nabla'^2\mathbf{u}' \tag{10.17}$$

If you want the pressure in pascals, take the nondimensional pressure (from the computer) and multiply it by the pressure standard:

$$p = p_s p' = \frac{\mu u_s}{x_s} p' \tag{10.18}$$

Option Two: High-Speed Flows

This option is most applicable for a high Reynolds number. Multiply Eq. (10.12) by $x_s/\rho u_s^2$ to give

$$\frac{x_s}{t_s u_s}\frac{\partial \mathbf{u}'}{\partial t'} + \mathbf{u}'\cdot\nabla'\mathbf{u}' = -\frac{p_s}{\rho u_s^2}\nabla' p'' + \frac{\mu}{\rho u_s x_s}\nabla'^2\mathbf{u}' \tag{10.19}$$

The symbol p'' is used now for Option Two. Choose the pressure standard so that the coefficient of the pressure gradient term is 1.0 and make a similar choice for the time standard:

$$\frac{x_s}{t_s u_s} = 1, \quad t_s = \frac{x_s}{u_s} \quad \text{or} \quad \frac{p_s}{\rho u_s^2} = 1, \quad \text{or} \quad p_s = \rho u_s^2 \tag{10.20}$$

Then the nondimensional Navier–Stokes equation is

$$\frac{\partial \mathbf{u}'}{\partial t'} + \mathbf{u}'\cdot\nabla'\mathbf{u}' = -\nabla' p'' + \frac{1}{Re}\nabla'^2\mathbf{u}' \tag{10.21}$$

If you want the pressure in pascals, take the nondimensional pressure (from the computer) and multiply it by the pressure standard:

$$p = p_s p'' = \rho u_s^2 p'' \tag{10.22}$$

By comparing Eq. (10.17) with Eq. (10.21) you see that the two different dimensionless pressures are related by

$$p' = Re p'' \qquad (10.23)$$

It is easy to show that the pressure in pascals is the same in both options. The dimensional pressure drop in Option One is given by Eq. (10.18). This is also

$$\Delta p = p_s \Delta p' = \frac{\mu u_s}{x_s} \Delta p' = \frac{\rho u_s^2}{Re} \Delta p', \text{ since } \frac{\rho u_s^2}{Re} = \frac{\mu u_s}{x_s} \qquad (10.24)$$

The dimensional pressure drop in Option Two is given by Eq. (10.22). By using Eq. (10.23) you get

$$\Delta p = \rho u_s^2 \Delta p'' = \frac{\rho u_s^2}{Re} \Delta p' \qquad (10.25)$$

which is the same as Eq. (10.24).

The reason for having both of them is that for slow flows the dimensionless result often does not depend upon the Reynolds number and for high-speed flows the convergence is better in the form of Eq. (10.21). In this book, we mainly use Option One, but sometimes convert the results to Option Two for comparison with the literature. In Comsol Multiphysics, to use the nondimensional version we must change the units to "none" in Preferences. Once you have solved a nondimensional equation, to get the results in units for a specific case, you must identify for your case the standards used to make the equation dimension, such as distance (e.g., a radius and length), and a velocity (average velocity somewhere, where?).

EXAMPLE: ENTRY FLOW IN A PIPE

Next, consider entry flow into a pipe as illustrated in Figure 10.1. The flow is taken as uniform in radius at the entrance (taking the value 1.0 at each radial position), and the velocity has to develop into its fully developed profile. The pipe radius is taken as 0.5. How far downstream do you have to go to get a fully developed solution? (Another way to pose this question is how much error do you make if you assume fully developed flow when it is not?) As discussed in the introduction to Chapter 9, the major steps for applying Comsol Multiphysics are to open it, choose the problem we wish to solve, set our preferences, create a geometry, identify parameter values, choose and set the boundary conditions, create a mesh, solve the problem, and then examine the solution. These steps are described carefully in this first example, but be sure to review the icons in Figures 9.2–9.4.

Step 1 (Open and Set Preferences) Open Comsol Multiphysics and choose 2D axisymmetric in the Model Wizard and click the next arrow. Then open "Fluid Flow" and "Single Phase Flow" and choose "Laminar Flow," click the next arrow. The Settings window will have a list of Preset options; select Stationary and then click the Finish flag. The nodes at the left of the screen are called the Model Builder and they provide a guide to the necessary steps.

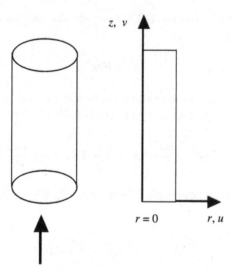

FIGURE 10.1   Entry flow into a pipe.

We will in turn use the Model 1, Study 1, and Results nodes. Note the icons at the top, right. The second one from the left is to "show" additional things, and we choose "Discretization" and "Equation View." In the root node we can choose a system of dimensions in the settings window; for the Unit System choose "none."

Step 2 (Define the Geometry)   Open the Model 1 node by clicking on the arrow to the left to expand this node. Then right click on the Geometry node and choose Rectangle. Set the width to 0.5 and the height to 2. Then click the "build" icon at the top of the settings window.

Step 3 (Define the Problem)   With the Model 1 node open, click on the "Laminar Flow' node and change the Compressibility option to Incompressible flow in the settings window. Open the "Laminar Flow" node and select "Fluid Properties." Click on the domain, the + to add the domain to the Selection Box if it is not selected. Change the density and viscosity to User defined, with $\rho = 10$, $\mu = 1$.

Step 4 (Set the Boundary Conditions)   Right click Laminar Flow and choose Inlet. Select the boundary 2 and + to add it to the Selection box; set the normal velocity to 1.0. Right click Laminar Flow and choose Outlet. Select the boundary 2, then + to add it to the box; and change the boundary condition from Pressure, no viscous stress (since we know there is viscous stress in fully developed flow) to Pressure, $p_0 = 0$. The centerline boundary condition is automatic in axisymmetric geometry (symmetry about $r = 0$), and allows slip, and the default boundary condition is velocity zero, which is appropriate for the wall at $r = 0.5$. A neutral boundary is one through which flow occurs but without our placing specific conditions on it (like a hole into another domain).

Step 5 (Create a Mesh)   Choose "Mesh 1" and the Settings window for Element Size will appear. Accept "Normal" and then click the Build icon. A mesh will be created in your

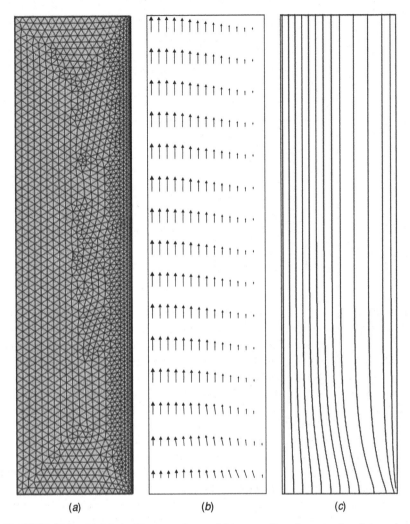

(a)   (b)   (c)

FIGURE 10.2 (a) Mesh, (b) arrow plot, and (c) streamlines for entry flow into a pipe.

domain (see Figure 10.2a). If you forget this step, it will be performed automatically for you when you solve the problem.

Step 6 (Solve the Problem) Choose "Study 1" and right click =. After a moment, a color plot of the velocity magnitude will appear in the Graphics Window. What is plotted is the value of $U = \text{sqrt}(u^2 + w^2)$.

Step 7 (Examine the Solution) To obtain more specific information, right click Results, choose "2D Plot Group," right click "2D Plot Group 2" and choose Arrow Surface. Click the plot icon to see the arrow plot (Figure 10.2b). Also, right click again "2D Plot Group 2" and choose streamline; choose a "Streamline Positioning" and click the plot icon to get the streamline plot (see Figure 10.2c).

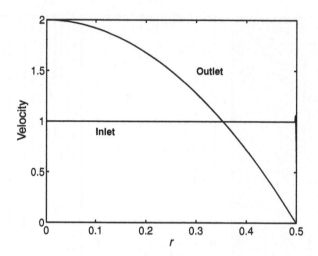

FIGURE 10.3 Velocity at inlet and outlet.

Step 8 (Make More Plots) The inlet velocity is found by right clicking Results, choosing "1D Plot Group." Right click that and choose "Line Graph." Then select boundary 2, change the "X-Axis Data Expression" to r, and click the plot icon. The result is a step function. Change the title and axis labels to "Inlet Velocity," r and w (see Figure 10.3). The velocity is not exactly 1.0 for all radii at the inlet, because there is a discontinuity in the boundary conditions. The boundary conditions use $v = 1$ on one boundary and $v = 0$ on an adjoining boundary, and the program has to choose which one to use at the corner. Comsol does a good job of satisfying these conflicting conditions. (It is not possible to achieve those boundary conditions physically.) Repeat this procedure under the same "Plot Group" but choose the outlet boundary (see Figure 10.3). To see the approach to fully developed flow, plot the axial velocity along the centerline. Right click Results, choose "1D Plot Group." Right click "1D Plot Group" and select "Line Graph," then select boundary 1, the centerline, then the + to add it to the Selection box. Change the "X-Axis Data Parameter" to "Expression," z. Change the titles and axis labels and click the plot icon to get Figure 10.4.

Step 9 (Improve the Solution) There is a small increase in the centerline velocity near the exit; this is not a valid solution. The reason is that the finite element functions are linear functions. For laminar flows, quadratic functions usually give better results. Click the Show button (see Figure 9.2) and click Discretization. Then go back to Laminar Flow; there is now an option for Discretization of fluids. The default value is "P1+P1," which means that both the velocity and the pressure are approximated with linear elements. Change that to "P2+P1"; then the velocity is approximated with quadratic elements. Right click Study = to get the new solution. Select the "Line Graph" that has the centerline velocity plotted and you see that the uptick at the end has disappeared. Figures 10.3 and 10.4 are made with this solution.

Step 10 (Calculate Average Properties) To calculate average quantities, right click on Derived Values (under Results) and select "Integration/Line Integration." Click on the inlet boundary 2, then + to add it to the Selection box and choose to integrate velocity. Click

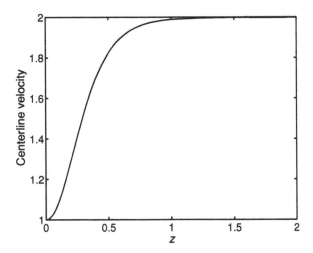

FIGURE 10.4 Centerline velocity in pipe flow.

the box "Include Surface Integral" in "Integration Settings" and then = (at the top) to get the value. It is listed in Results as 0.7846. If you wonder whether the integration has been done using $r dr$ or just dr, change the expression to 1 and click the = sign. The value that appears is 0.7846, which is the area of a circle with radius 0.5:

$$2\pi \int_0^R r dr = \pi R^2 = \frac{\pi}{4} = 0.7854 \tag{10.26}$$

Thus, the $2\pi r dr$ is included automatically and the average velocity is $0.7846/0.7854 = 0.9990$. This is not quite 1.0 due to the discontinuity of the inlet velocity at the wall.

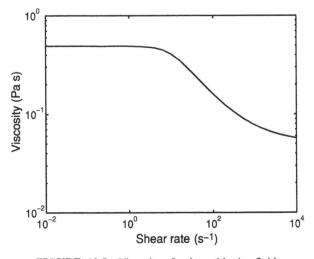

FIGURE 10.5 Viscosity of a shear-thinning fluid.

Step 11 (Report)   To document your work, save a picture of the domain and mesh; do this by clicking the camera, then Browse, give a name, then Save, then OK. Be sure to give the dimensions, identify the boundary conditions, give values of parameters you used, list how many elements and degrees of freedom were used, and indicate how solutions with different number of elements compare. Then give the results and indicate why they are reasonable. Next, change the problem slightly and take the inlet velocity as the fully developed velocity profile:

$$v(r, 0) = 2 \left[1 - \left(\frac{r^2}{0.5} \right) \right] \tag{10.27}$$

Step 12 (Solve Again)   Use a fully developed inlet velocity. This is inserted in Laminar Flow, as the v velocity:

$$2 * (1 - (r/0.5)\,^\wedge\, 2) \tag{10.28}$$

The streamlines are now straight.

EXAMPLE: ENTRY FLOW OF A NON-NEWTONIAN FLUID

Next, consider the entry flow of a non-Newtonian fluid, essentially the same problem as before, except with a different fluid. The only thing you need to change is the expression for viscosity, which is no longer constant. The shear viscosity is modeled as a Carreau function, which is a slight extension of Eq. (9.14):

$$\frac{\eta - \eta_\infty}{\eta_0 - \eta_\infty} = \left[1 + (\lambda \dot{\gamma})^2 \right]^{(n-1)/2} \tag{10.29}$$

The $\eta_\infty, \eta_0, \lambda$, and n are all parameters that are used to fit data, taken here as $\eta_\infty = 0.05$, $\eta_0 = 0.492$, $\lambda = 0.1$, and $n = 0.4$. A plot of the viscosity versus shear rate is given in Figure 10.5. For small shear rates, the viscosity is essentially constant, as it is for a Newtonian fluid. For extremely large shear rates, the same thing is true. For moderate shear rates, though, the viscosity changes with shear rate. In pipe flow, or channel flow, the shear rate is zero at the centerline and reaches a maximum at the wall. Thus, the viscosity varies greatly from the centerline to the wall. This complication is easily handled in Comsol Multiphysics. Once you know how to solve the problem for a Newtonian fluid (as in Figure 10.3), there are two major changes you must make for a non-Newtonian fluid. One of those is to represent the viscosity, Eq. (10.29), in Comsol Multiphysics. The other is to be smart about how you obtain a converged solution.

Step 1 (Open and Set Preferences)   Open the solution for entry flow of a Newtonian fluid and save it with a new name.

Step 2 (Define the Problem)   Open the "Laminar Flow" node and select "Fluid Properties." Change the viscosity to eta. We next need to define "eta." That in turn requires definition of the "shearrate." Right click on the "Global Definitions" and click Parameters. Enter the follow names and values: "etainf," 0.05; "eta0," 0.492; "lambda," 0.1; "*n*," 0.4. Go to the "Model node 1" and right click Definitions and open Variables. Enter the following formulas.

```
eta = etainf+(eta0-etainf)*(1+(lambda*gammadot)^2)^((n-1)/2)
gammadot2 = 2*(ur^2+(u/r)^2+wz^2)+(uz+wr)^2
gammadot = sqrt(gammadot2).
```

Alternatively, use the variable mod1.spf.sr for the shear rate in place of gammadot since Comsol Multiphysics calculates it automatically.

Step 3 (Solve the Problem) In this case, change click the "Initial Value" node under "Laminar Flow" and put in an axial velocity of $2*(1 - (r/0.5)^2)$. This gives the program something to work with. If one starts with an initial velocity of $v = 0$ (the default), the program may return a message saying it cannot converge or that it tried to divide by zero. Thus, you should always obtain a solution for a Newtonian fluid first, before invoking the non-Newtonian viscosity: simply replace the word "eta" with 1.0 or some number appropriate to your case, solve the problem, put back the "eta," and use the resolve option. This time the computer has a guess of a solution to work from, and a solution is found.

Step 4 (Check Reynolds Number) Next, check the Reynolds number, which is 390, ensuring that the flow is laminar. The peak average velocity is 0.078 m/s, and in laminar flow the peak velocity should be twice that, or 0.156 m/s, which it is.

EXAMPLE: FLOW IN MICROFLUIDIC DEVICES

Next, consider flow in what is called a T-sensor. Two flows come together, join, and traverse down one channel, as illustrated in Figure 10.6. This device is used in microfluidic medical devices, which are discussed further in Chapter 11. Here, you will consider only the flow (which has no special utility until the convective diffusion equation is added in Chapter 11).

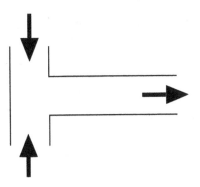

FIGURE 10.6   Flow in a T-sensor.

Step 1 (Open and Set Preferences)   Open Comsol Multiphysics and choose two dimensions; next, Laminar Flow; next, Stationary, finish.

Step 2 (Define the Geometry)   In "Model 1" right click Geometry and choose Rectangle; make one of width 2, height 0.5, base corner at $x = -0.5$, $y = -0.25$; click Build All. Do this again with the rectangle of width 0.5, height 1.5, base corner at $x = -1$, $y = -0.75$. You can either retain internal boundaries or not; if they are there, the mesh will be affected near them, so if the internal boundaries have no purpose it is best not to retain them.

Step 3 (Define the Problem)   Open the "Laminar Flow" node, then the "Fluid Properties" node. Under "Fluid Properties," change the density to User Defined, set the value to 1.0. Do the same for the viscosity. This corresponds to a nondimensional equation with a Reynolds number of 1.

Step 4 (Set the Boundary Conditions)   Right click the "Laminar Flow" node and choose Inlet. Select the top boundary, +, to add it to the "Boundary Selection" box. Set the Normal inflow velocity to $U_0 = 1$. Do this again and select the bottom boundary; set its Normal inflow velocity to $U_0 = 1$. Again right click the "Laminar Flow" node and choose Outlet. Select the right-most boundary, +, to add it to the "Boundary Selection" box. Change the boundary condition from "Pressure, no viscous stress" to "Pressure, with $p_0 = 0$". We know there are viscous stresses when there is flow between two walls.

Step 5 (Create a Mesh)   Click the Mesh node and Build All to get the mesh (see Figure 10.7).

Step 6 (Solve the Problem)   Then right click Study = to get the solution.

Step 7 (Examine the Solution)   To see the streamlines, right click results and choose 2D Plot Group. Then right click it and choose Streamlines; you may have to make a choice in "Streamline Positioning" (see Figure 10.8).

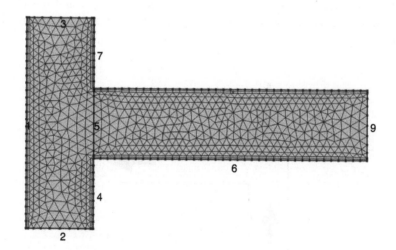

FIGURE 10.7   Mesh for flow in a T-sensor.

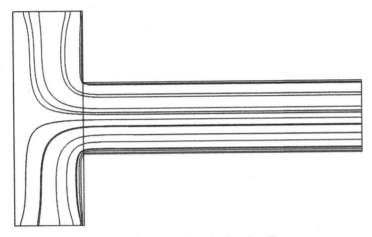

FIGURE 10.8 Streamlines for flow in a T-sensor.

Step 8 (Validate the Solution) To validate part of the solution, the overall mass balance is calculated. Right click "Derived Values" under Results, choose "Line Integration." Then, in turn, integrate the velocity over boundaries 2, 3 (input) and 9 (output). Do this by selecting a boundary, + to put it in the Selection box, then = (at the top). Taking into account the direction of flow (one of the integrals is negative), the sum of flow rates is 0.9914 in and 0.9726 out. This is accurate to within 1.8%.

Step 9 (Improve the Solution) There are a few ways to improve the accuracy. The first one is to change the trial functions for velocity; the default choice is linear functions, but we can change them to quadratic functions, which roughly doubles the number of unknowns. Do that by first selecting the Show icon and clicking Discretization (if you have not already set it in Preferences). Then choose the "Laminar Flow" node, and then the Discretization option (near the bottom). If it is not showing, use the arrow at the top of the Model Builder to say Show More Options. Change from "P1 + P1" to "P2 + P1". The flow rate in is 0.9972 versus 0.9971 out, for an error of 0.01%. These solutions were obtained using 1213 elements; for P1 + P1 there were 2260 degrees of freedom whereas for P1 + P2 there were 8150 degrees of freedom. This is a good lesson: for laminar flow at low Reynolds numbers, using quadratic functions for velocity is much better than the default functions (linear).

Step 10 (Solve for Fully Developed Flow In) If we wanted fully developed flow into the device, we would use Eq. (10.30) for the inlet profile rather than 1.0 (with the appropriate sign at the top and bottom):

$$v = -24(x + 1)\left(x + \tfrac{1}{2}\right) \tag{10.30}$$

To do that, change the two inlets to Velocity field and insert the formula for the y-velocity. Right click Study =; now the streamlines coming in are straight (see Figure 10.9). The flow rates in sum to 1.000 and that is the flow rate out, too, so this is very accurate.

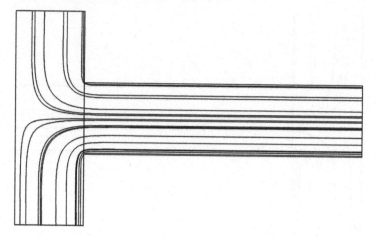

FIGURE 10.9 Streamlines in a T-sensor when the inlet velocity is parabolic.

EXAMPLE: TURBULENT FLOW IN A PIPE

Next, consider flow in a pipe at a flow rate high enough for turbulence to occur. Generally, this occurs whenever the Reynolds number is greater than 2200, where the Reynolds number is defined in terms of the average velocity and the pipe diameter:

$$Re = \frac{\rho <v> D}{\mu} = \frac{<v> D}{\nu} \tag{10.31}$$

You can solve the problem with an inlet velocity that is flat; thus, you find the entry length it takes to achieve fully developed turbulent flow, and the velocity profile downstream is the fully developed one. When you solve for a kinematic viscosity of 10^{-6} m²/s (water), diameter of 0.05 m (about 2 inches), and a velocity of 2 m/s (a common optimal velocity), you will get a Reynolds number of 10^5.

The turbulence model in Comsol Multiphysics is in dimensional (SI) units. It is contained in the CFD module, which may not be available in your system. If it is, turbulence is modeled using the k-ε model. In this model, the turbulent kinetic energy is represented by k, and the rate of dissipation of turbulent kinetic energy is represented by ε. Furthermore, the viscosity is augmented by a turbulent eddy viscosity, which is a function of k and ε. Special equations have been developed for both variables, and these must be solved along with the momentum equation which has the turbulent eddy viscosity in it as well. All these equations are included in Comsol Multiphysics.

$$(\mathbf{u}\cdot\nabla)\mathbf{u} = -\nabla p/p + \nabla\cdot[(v + v_T)\nabla\mathbf{u}]$$

$$(\mathbf{u}\cdot\nabla)k = \tau_{ij}\partial x_j - \varepsilon + \nabla\cdot[(v + v_T/\sigma_k)\nabla k] \tag{10.32}$$

$$(\mathbf{u}\cdot\nabla)\varepsilon = C_{\varepsilon 1}\frac{\varepsilon}{k}\tau_{ij}\partial u_i/\partial x_j - C_{\varepsilon 2}\frac{\varepsilon^2}{k} + \nabla\cdot[(v + v_T/\sigma_\varepsilon)\nabla\varepsilon]$$

Although $C_{\varepsilon 1}$ and $C_{\varepsilon 2}$ are dimensionless, the turbulent viscosity, v_T, is calculated in the SI system. Thus, all dimensions must be in meters, meter per second, and so on.

The main caution is that these equations are only an approximation of reality, and the formulas and equations have been chosen to model experimental data. This means that they are only as good as the data they were derived from, and the data are usually found in relatively simple flows. The experiments are definitely not easy to carry out, but they are for idealized situations, such as fully developed flow in a pipe or past a flat plate or in a jet. Thus, the equations should be used with caution. Other methods in turbulence are much more time consuming and may require banks of computers running for hours. That works for research, but is not suitable for day-to-day engineering work.

Step 1 (Open and Set Preferences)   Open Comsol Multiphysics and choose 2D, axisymmetric. Then choose Fluid Flow/Turbulent Flow/Turbulent Flow, *k-ε*, Finish.

Step 2 (Define the Geometry)   In "Model 1," right click Geometry and select Rectangle; make it a width of 0.025 and a length of 0.1 and click "Build All." We are going to solve the problem for fully developed flow, so the length is not important. If it is too long, though, lots of elements would be required, leading to longer solution times.

Step 3 (Define the Materials)   To define the materials, right click Materials and choose Open Material Browser. Then open Build-In and choose water and + to add the material. Rather than using that here, we will use a density of 1000 kg/m³ and a viscosity of 0.001 Pa s, set in "Turbulent Flow, User defined density and viscosity."

Step 4 (Define the Problem)   Click "Turbulent Flow" and change the Compressibility to "Incompressible Flow."

Step 5 (Set the Boundary Conditions)   Right click "Turbulent Flow" and choose Wall; click the wall boundary, + to add it to the Selection window. Right click "Turbulent Flow and choose "Periodic Flow Condition." Set the "Pressure Difference" to 60 Pa. Right click "Turbulent Flow" and choose "Pressure Point Constraint." Click one node at the exit, +, to add it to the Selection box, and set it to zero. Since periodic boundary conditions are being used, no velocity boundary conditions are needed except for zero velocity along the wall.

Step 6 (Create a Mesh)   Click the Mesh node and "Build All." A mapped mesh will require the fewest number of elements.

Step 7 (Solve the Problem)   Then right click "Study 1," =. The solution may take awhile (depending on the mesh you have chosen). The streamlines are straight indicating the flow is fully developed throughout.

Step 8 (Examine the Solution)   The exit velocity profile and centerline velocity are shown in Figure 10.10. The peak velocity is about 2.06 m/s. The centerline velocity fluctuates a bit about the value 2.06 (note the scale). The profiles of *k* and *ε* are shown in Figure 10.11. The pressure along the centerline is shown in Figure 10.12.

Step 9 (Examine the Solution Again)   The most important check, though, is to compare with experimental data. To get the average velocity, right click "Derived Values" under

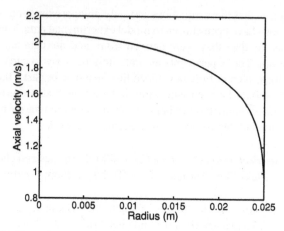

FIGURE 10.10   Velocity profile for fully developed turbulent entry flow in a pipe.

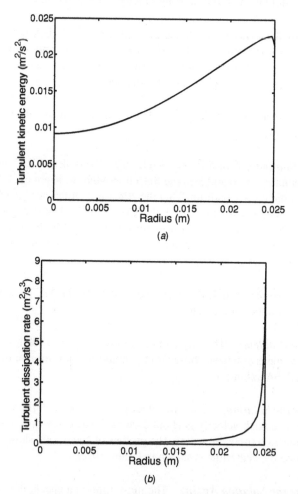

FIGURE 10.11   Turbulent kinetic energy (*a*) and rate of dissipation (*b*) for fully developed turbulent flow in a pipe.

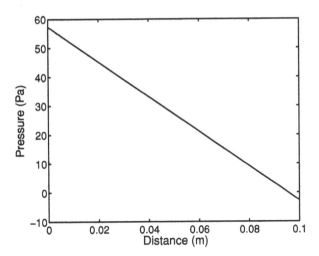

FIGURE 10.12 Pressure drop in turbulent flow.

the Results node. Select "Line Integration 1." In the "Line Integration" panel, open "Integration Settings" and choose "Compute surface integral (axial symmetry)." Just to test the integration, change the Expression to 1.0 and choose =. The value is 0.0019635, which is appropriate if one is integrating $2\pi r dr$. Thus, the cross-sectional area is 0.0019635 m^2. With the Expression replaced by spf.U, one gets 0.0034715. The average velocity is then 0.0034715/0.0019635 = 1.768.

Step 10 (Compare with Experimental Data)   The pressure drop per length is 60 Pa/ 0.1 m with a radius of 0.025 m. The wall stress is $\tau_w = 7.5$ Pa. The friction factor is then

$$\tau_w = \frac{\Delta p}{2}\frac{R}{L}, \quad f = \frac{2\tau_w}{\rho <v>^2} = \frac{\Delta p}{L\rho}\frac{R}{<v>^2} = 4.8 \times 10^{-3} \qquad (10.33)$$

This is a bit higher than the value for a smooth pipe at a Reynolds number of 8.84×10^4, where $f = 4.6 \times 10^{-3}$ (Perry and Green, 2008, pp. 6–10).

EXAMPLE: START-UP FLOW IN A PIPE

A classic problem in fluid mechanics is the transient development of flow in a pipe. In this idealized problem, the fluid is initially at rest with no applied pressure gradient. At time zero, a pressure gradient is applied, and flow begins. The velocity is slow enough that the flow is laminar. After a certain amount of time, the flow becomes steady.

Step 1 (Open and Set Preferences)   Open Comsol Multiphysics and choose "2D, axisymmetric." Then choose "Laminar Flow;" next, "Time Dependent," finish.

Step 2 (Define the Geometry)   In "Model 1," right click Geometry and select Rectangle; make it a width of 0.0025 m and a length of 0.025 m and click "Build All." We are going to solve the problem for fully developed flow, so the length is not important. If it is too long, though, lots of elements would be required, leading to longer solution times.

Step 3 (Define the Problem)   Click "Laminar Flow," then choose Discretization and change to "P2 + P1." (Use the Show icon if Discretization is not showing.) Click on "Fluid Properties" node and set the density and viscosity to User defined, 1000 kg/m^3 and 0.001 Pa s.

Step 4 (Set the Boundary Conditions)   Right click "Laminar Flow" and choose "Periodic Flow Condition." Set the pressure difference to 2.5 Pa. Right click on "Laminar Flow" and choose "Pressure Point Constraint." Choose a point at the outlet and set the pressure p_0 to 0.

Step 5 (Create a Mesh)   Choose Mesh and "Build All."

Step 6 (Solve the Problem)   Under "Study 1," click on "Step 1: Time Dependent" to set the times at which the solution will be plotted. We want a solution from 0 to 7, but want a finer discretization initially. Thus, use for Times: 0 0.1 0.2 0.3 0.4 0.5 0.6 0.7 0.8 0.9 1.0 2 3 4 5 6 7. Right click Study = to get the solution.

Step 7 (Examine the Solution)   Right click Results and choose "1D Plot Group." Right click it and choose "Line Graph." The velocity is chosen automatically. In the Settings window, open Legends and mark the "Show legends" box. Click on the plot icon to get Figure 10.13. The figure shown was replotted using MATLAB.

Step 8 (Plot the Solution Versus Time)   Right click on Results and choose "Point Plot." Select the point at $r = 0$, $z = 0$, +, to add it to the Selection box. Click on the plot icon to obtain Figure 10.14. This shows that steady state is achieved by $t = 7$.

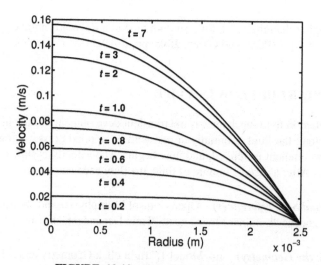

FIGURE 10.13   Transient velocity profiles.

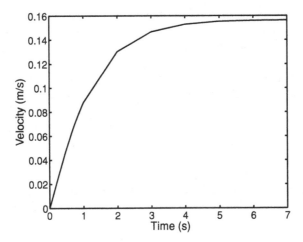

FIGURE 10.14   Transient laminar flow in a pipe, velocity at $r = 0, z = 0$.

Step 9 (Check Reynolds Number)   Next, check the Reynolds number, which is 390, ensuring that the flow is laminar. The peak average velocity is 0.078 m/s, and in laminar flow the peak velocity should be twice that, or 0.156 m/s, which it is.

EXAMPLE: FLOW THROUGH AN ORIFICE

The next problem is flow through an orifice. If the pipe or channel is very small, the flow may be laminar. This is usually the case in microfluidic devices. If the channel is small, the orifice would not be infinitesimally small, and the width of the orifice may have an effect. You will solve for the flow here, whereas in Chatpter 11 you will extend the problem to solve for the temperature rise due to viscous dissipation. In this chapter, you assume the viscosity is constant. This problem was solved by a senior chemical engineering student, Febe Kusmanto, while studying the effect of orifices with a finite thicknesses (Kusmanto et al., 2004).

Step 1 (Open and Set Preferences)   Open Comsol Multiphysics and choose "2D, axisymmetric"; next, "Laminar Flow"; next, Stationary, finish. Set the units to "none" in the root node, and change the discretization from "P1 + P1" to "P2 + P1."

Step 2 (Define the Geometry)   Right click on Geometry and choose Rectangle; make one of width 3, height 16, base corner at 0,0; click "Build All." Right click geometry again and choose a rectangle; set the width to 2.5, height to 0.15, base corner at $r = 0$, $z = 3.925$. This will make the diameter of the orifice hole, D, equal to 1, and thickness of the orifice plate/diameter of the hole, L, is 0.15. Right click Geometry again, choose "Boolean Operations" and click Difference. The domain we want is the first one minus the second one.

Step 3 (Define the Problem)   Click on the "Laminar Flow" node, then the "Fluid Properties" node. Under "Fluid Properties," change the density to "User defined" and set the value to 1.0. Do the same for the viscosity and set the viscosity to $1/Re$; this corresponds to a nondimensional equation. We will make the Reynolds number depend on x, so right click Definitions, choose Variables, and define Re as $10^{\wedge}x$.

Step 4 (Set the Boundary Conditions)   Right click on the "Laminar Flow" node and choose Inlet. Select the bottom boundary, +, to add it to the "Boundary Selection" box. Enter the z-velocity as $(2/36)*(1-(r/3)^{\wedge}2)$. This makes the average velocity through the orifice 1.0. Right click "Laminar Flow" and choose Outlet. Select the top boundary, + to add it to the Selection box, and set the boundary condition to $p = 0$ (not the one with zero shear stress).

Step 5 (Create a Mesh)   Click on the Mesh node and "Build All" to get the mesh shown in Figure 10.15. This has 4556 elements and 25,023 degrees of freedom.

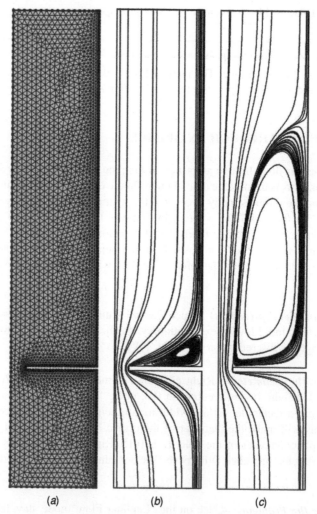

(a)   (b)   (c)

FIGURE 10.15   Flow through an orifice: (a) mesh; (b) $Re = 1$; (c) $Re = 10^{1.5} = 31.6$.

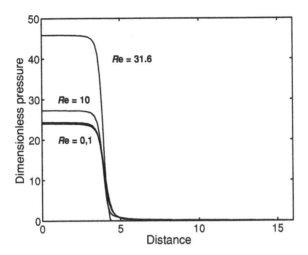

FIGURE 10.16   Pressure profiles for Re from 0.01 to 31.6.

Step 6 (Solve the Problem)   Right click "Study 1" and choose "Parametric Sweep." We want the Reynolds numbers to go from 1 to about 30; make the Reynolds number $= 10^{\wedge}x$, with x going from 0 to 1.5. Thus, set the parameter as x in "Global Definitions/Parameters" and define the parameter values as range (0,0.1,1.5) (see Chapter 9, Section "Parametric Solution"). Then right click "Study 1" $=$ to get the solutions.

Step 7 (Examine the Solution)   To see the streamlines, right click Results and choose "2D Plot Group." Then right click it and choose Streamlines. The solution shown is for $Re = 10^{1.5}$ (the last parameter. To see the solution for some other Re, change the parameter value in the "2D Plot Group panel"; click the plot icon to create the plot. Two solutions are shown in Figure 10.15. Figure 10.15b and c shows the streamlines at two Reynolds numbers. This figure illustrates the fact that recirculation occurs, and the region of recirculation increases as the Reynolds number increases, even though the flow is still laminar. Figure 10.16 shows the corresponding pressure profiles. The pressure changes occur very near the orifice, and the pressure drop is almost entirely caused by the orifice with only a small contribution from fully developed flow in the tube. As the Reynolds number increases, the pressure reduces after the orifice. Then, it gradually returns to a higher value in the region known as the vena contracta.

Figure 10.17 shows the pressure drop as a function of Reynolds number. These results are plotted in two ways, corresponding to the two different ways to nondimensionalize the equations. The one in Figure 10.17a is most suitable for slow speed laminar flow, because the pressure term plotted is constant with Reynolds number over quite a range. The one in Figure 10.17b is most suitable for high Reynolds numbers because that form of the pressure term is constant once the Reynolds number gets high enough. Figure 10.17b also looks like a typical friction factor plot. Both curves, however, give the same value of pressure drop in pascals, and the plotted coefficients are related by Eq. (10.34):

$$\frac{\Delta p}{\frac{1}{2}\rho <v>^2}\frac{Re}{2} = \frac{\Delta p}{\rho <v>^2}\frac{\rho <v> D}{\mu} = \frac{\Delta p D}{\mu <v>} \tag{10.34}$$

FLUID FLOW IN TWO AND THREE DIMENSIONS

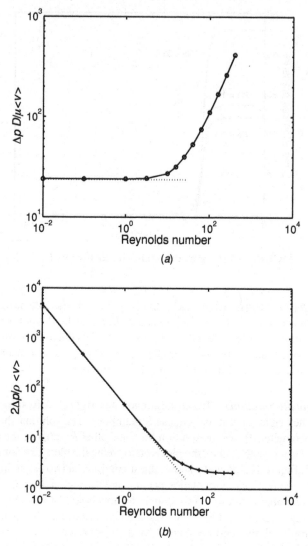

FIGURE 10.17 Dimensionless pressure drop in orifice: (*a*) preferred plot for low Reynolds number; (*b*) preferred plot for high Reynolds number.

The dotted line is the analytical solution to Stokes flow by Dagan et al. (1982):

$$\frac{2\Delta p}{\rho <v>^2} = \frac{12\pi}{Re} + \frac{64}{Re}\frac{L}{D}, \text{ or } \frac{\Delta p D}{\mu <v>} = 6\pi + 32\frac{L}{D} \qquad (10.35)$$

As the Reynolds number increases, you must decide if the length of the domain is sufficiently long that the velocity profiles and pressure profiles have returned to their fully developed values. Another way to determine this is to solve the problem on a longer domain and compare the answers from the problems with different lengths. Because the pressure drop should be the same, as long as the pressure drop from the fully developed flow is negligible or taken into account, this is a valid test.

FIGURE 10.18   Mesh used to model the serpentine mixer.

EXAMPLE: FLOW IN A SERPENTINE MIXER

Slow flow in a serpentine mixer is used to mix two chemicals that come in at different sides of the port to the left. The geometry and mesh are shown in Figure 10.18; there were 70,008 elements and 170,598 degrees of freedom to solve for the three velocity components and pressure using a "P1 + P1" discretization (the default). It was solved using "P2 + P1" discretization in an earlier version of Comsol Multiphysics by Zachery Tyree when he was a senior chemical engineering student at the University of Washington (Neils et al., 2004). A few flow lines are plotted in Figure 10.19, showing how the mixing of the fluid streams occurs as the fluid moves along the channels, turns, and goes up or down a chimney. To solve this problem in Comsol Multiphysics, download the geometry from the book website. Then set the density and viscosity, inlet ($\mathbf{u} \cdot \mathbf{n} = -1$) and outlet ($p = 0$) boundary conditions,

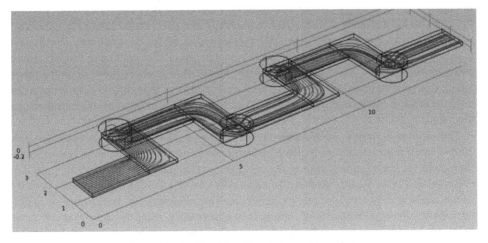

FIGURE 10.19   Flow lines in serpentine mixer.

and solve. Plots are made as in 2D problems with the minor change of choosing 3D Plots rather than 2D Plots. Of course you can create 2D Plots from the 3D simulation, too.

MICROFLUIDICS

The field of microfluidics involves flow and diffusion in small devices, usually defined as less than 1 mm in width/height. There are two important technologies that use microfluidics: microreactors and a lab-on-a-chip. The lab-on-a-chip applications include both analytical devices for the chemistry laboratory and medical devices that are used in point-of-care medical treatment. A sophisticated measurement can be made wherever you are, the results can be electronically transmitted to your doctor, who then has the finished lab results and can act on them immediately. The technologies provide interesting problems for flow in the laminar region. They also provide important challenges for diffusion, since sometimes the diffusion layer is very thin and requires a very fine mesh to resolve. Provided here are such problems, along with simpler versions that allow the engineer to determine how sophisticated the model needs to be in order to design the device. Sometimes your basic understanding of fluid mechanics, heat transfer, and diffusion will allow you to make reasonable estimates for the design. References involving microreactors are Koch et al. (2007)and Gokhale et al. (2005); a references dealing with micro- and nanoscale transport are Kirby (2010)and Finlayson et al. (2007); a good reference about biological applications is Gomez (2008).

Why would you want to have a chemical reactor that you can hold in your hand or put on your desk? Well, you would not want to hold it or put it on your desk, but microreactors have several advantageous features. First, you can control the reaction time, temperature, and pressure very closely in a way you cannot in a packed bed or large tube. This may enhance safety since you can control the temperature and reactive material better. Atom efficiency is also improved, allowing you to make more of the material you want and less of the by-products that have to be disposed of. The results are often more reproducible, too. How, then, do you make more than a laboratory beaker of material? You scale up by what is called "numbering up": you simply make many replicas and combine the outputs. One advantage of this is scale up: you do not have to study a larger version of the reactor, in which new phenomena may be relevant, since the duplicative reactors are the same as the one you have already studied. This also means that you do not have to build a huge plant; build a modest plant, and then as the market grows, add to it with similar reactors. Of course you lose the benefit of the sixth-tenth rule: the capital cost goes up as the sixth-tenth power of the capacity. This leads to very large plants. But, in return, you may have lower labor costs (one estimate 50–75% less). As with any reactor, though, you have to solve for the flow in a specific geometry, find the pressure drop, and then solve for concentration and temperature.

For medical purposes, microfluidic devices can be used in a variety of ways (Gomez, 2008). Small fluid streams with different chemicals can be mixed in precise ways. Sometimes cells are sorted to provide just the ones you want to analyze, thus improving the discrimination. Point-of-care diagnostics is especially useful in developing countries, because sometimes the device can be made cheaply, and the chemicals do not require refrigeration. Rather than collecting blood, taking it to a laboratory, using multiple reagents to do biochemical analyses, a microfluidic device can take a drop of blood, add small amounts of the appropriate chemicals and do the analysis "on the spot." For example, Professor

Paul Yager's laboratory at the University of Washington designed a device to differentiate between six different diseases whose main symptom is a fever. Without the blood test, the doctor or nurse may not know which disease is causing the fever, and the treatment may not be focused as it should be. Thus, Chapters 10 and 11 have problems involving the flow and diffusion of chemicals in microfluidic geometries.

When the flow is slow, as in microfluidic devices, it is convenient to present results in a slightly different way than for turbulent flow. Consider the flow of an incompressible fluid in a straight pipe. The pressure drop is a function of the density and viscosity of the fluid, the average velocity of the fluid, and the diameter and length of the pipe. The relationship is usually expressed by means of a friction factor defined as follows:

$$f \equiv \frac{1}{4} \frac{D}{L} \frac{\Delta p}{\frac{1}{2}\rho <v>^2} \tag{10.36}$$

The Reynolds number is

$$Re \equiv \frac{\rho <v> D}{\eta} \tag{10.37}$$

The usual curve of friction factor is given in Figure 10.20a in a log–log plot. For any given Reynolds number, one can read the value of friction factor, and then use Eq. (10.36) to determine the pressure drop. In the turbulent region, for $Re > 2200$, one correlation is the Blasius formula:

$$f = 0.0791/Re^{0.25} \tag{10.38}$$

In the laminar region, for $Re < 2200$, the correlation is

$$f = 16/Re \tag{10.39}$$

When the definitions are inserted into Eq. (10.39) for laminar flow, one obtains for the pressure drop

$$\Delta p = 32 \frac{L}{D} \frac{\mu <v>}{D} \tag{10.40}$$

Thus, in laminar flow in a straight channel, the density of the fluid does not affect the pressure drop. To use the correlation for friction factor, or Figure 10.20a, one must know the density. This problem is avoided if one plots the product of friction factor and Reynolds number versus Reynolds number, since from their definition

$$f Re = \frac{\Delta p D^2}{2L\mu <v>} \tag{10.41}$$

This is shown in Figure 10.20b. Notice that for $Re < 1$ the only thing needed is the numerical constant, here 16. For a fully developed pipe flow, the quantity fRe does not change as the Reynolds number increases to 2200, but it does in more complicated cases. This covers the

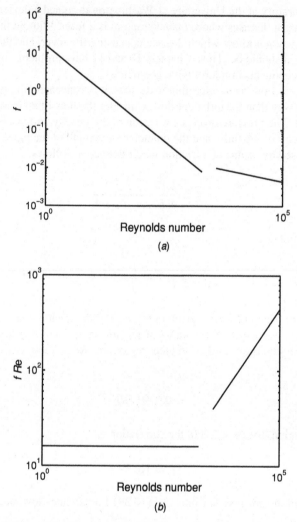

FIGURE 10.20   Friction losses (pressure decreases) for fully developed pipe flow. (*a*) Standard presentation Reynolds number. (*b*) Low Reynolds number presentation.

range in which most microfluidic devices operate. To predict the pressure drop, all that is needed is the single number for the geometry and flow situation in question. The way these numbers are determined is best illustrated by looking at the mechanical energy balance for a flowing system.

Consider the control volume shown in Figure 10.21, with inlet at point 1 and outlet at point 2. In the following, Δ variable means the variable at point 2 (outlet) minus the variable at point 1 (inlet). The mechanical energy balance is (Bird et al., 2002)

$$\Delta \left(\frac{1}{2} \frac{<v^3>}{<v>} \right) + g\Delta h + \int_1^2 \frac{dp}{\rho} = \hat{W}_m - \hat{E}_v \qquad (10.42)$$

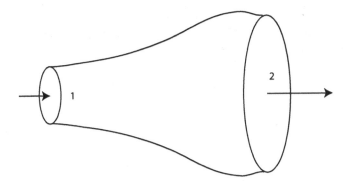

FIGURE 10.21   Geometry and notation for mechanical energy balance.

For no height change $g\Delta h = 0$; for no pump in the control volume $\hat{W}_m = 0$; for an incompressible fluid with constant density:

$$\int_1^2 \frac{dp}{\rho} = \frac{1}{\rho}(p_2 - p_1) = \frac{1}{\rho}\Delta p \qquad (10.43)$$

Thus, under all these assumptions we have

$$\Delta \left(\frac{1}{2}\frac{<v^3>}{<v>} \right) + \frac{1}{\rho}\Delta p = -\hat{E}_v \qquad (10.44)$$

By continuity the mass flow rate is the same, $w_1 = w_2$. The definition of the viscous losses is

$$E_v = -\int_V \boldsymbol{\tau} : \nabla \mathbf{v} dV = \int_V \mu[\nabla \mathbf{v} + (\nabla \mathbf{v})^{\mathsf{T}}] : \nabla \mathbf{v} dV \geq 0, \ \hat{E}_v = \frac{E_v}{w}, \ w = \rho \int \mathbf{v} \cdot \mathbf{n} dS$$

$$(10.45)$$

Rewrite Eq. (10.44) in the form

$$\Delta \left(\frac{1}{2}\frac{<v^3>}{<v>} \right) + \hat{E}_v = -\frac{1}{\rho}\Delta p = \frac{p_1 - p_2}{\rho} \qquad (10.46)$$

Mechanical Energy Balance for Laminar Flow

When the flow is slow and laminar, we use the nondimensionalization for the pressure identified as Option One earlier. Then the nondimensional mechanical energy balance is

$$Re\Delta \left(\frac{1}{2}\frac{<v'^3>}{<v'>} \right) + \hat{E}'_v = (p'_1 - p'_2) \text{ where } \hat{E}'_v = \frac{\rho x_s}{\mu v_s}\hat{E}_v = \frac{\rho x_s}{\mu v_s}\frac{E_v}{w} \qquad (10.47)$$

For turbulent flow, a reasonable approximation is that the velocity is constant with respect to radial position in the inlet and outlet ducts, and then $< v'^3 > / < v' >=< v'^2 >$. This is not true for laminar flow, though.

For fully developed flow through a tube with a constant diameter, D, the kinetic energy does not change from one end to the other. We already have a formula for the pressure drop in terms of the friction factor, so that

$$p_1' - p_2' = \frac{\Delta p D}{\mu <v>} = 32\frac{L}{D} \tag{10.48}$$

Thus, the viscous dissipation term in Eq. (10.47) for laminar flow in a straight channel is

$$\hat{E}_v' = 32\frac{L}{D} \tag{10.49}$$

Thus, the dimensionless viscous dissipation is proportional to L and does not depend on the Reynolds number as long as the flow is laminar in a straight pipe. The reason for using the nondimensionalization of Option One is because of this result: it covers a wide range of flow rates using only a single number.

Note that in Eq. (10.47) the Reynolds number multiplies the term representing the change of kinetic energy, and the complete term may be small because the Reynolds number is small. Consider the case when the Reynolds number is extremely small (i.e., zero). Then any flow situation is governed by Stokes equation, that is, Eq. (10.17) with $Re = 0$:

$$\nabla'^2 \mathbf{v}' = -\nabla' p' \tag{10.50}$$

Thus, the solution to Eq. (10.50) does not depend on the Reynolds number. In this case, with $Re = 0$, the dimensionless viscous dissipation term in Eq. (10.47) is a constant and the kinetic energy term is negligible. To correlate the pressure drop in laminar flow, the main task is to calculate the viscous dissipation term, $\hat{E}_v'$. This is usually the case in microfluidic devices.

A convenient way to correlate data for excess pressure drop when there are contractions, expansions, bends and turns, and so on, when the flow is turbulent is to write them as

$$\Delta p_{excess} = K\frac{1}{2}\rho <v>^2 \tag{10.51}$$

Then, to calculate the excess pressure drop for any given fluid and flow rate, one only needs to know the values of K, which are tabulated in Perry and Green (2008).

As shown earlier, for slow flow the pressure drop is linear in the velocity rather than quadratic, and a different correlation is preferred. Write

$$\Delta p_{excess} = K_L\frac{\mu <v>}{D} \tag{10.52}$$

Values of K_L for different geometries are calculated in several problems. The two coefficients can be related by equating the Δp_{excess} formulas:

$$K_L = \frac{K}{2}Re \tag{10.53}$$

Thus, to estimate the Δp in a device one merely needs to solve one flow problem to get K_L. We see below that this works well for $Re < 1$.

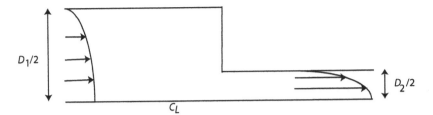

FIGURE 10.22   Contraction flow geometry.

Pressure Drop for Contractions and Expansions

Consider a large circular tube emptying into a smaller circular tube, as shown in Figure 10.22. The pressure drop in this device is due to viscous dissipation in the fully developed regions of the large and small channels, plus the extra viscous dissipation due to the contraction, plus the kinetic energy change. For the flow illustrated in Figure 10.22, we write the excess pressure drop for the contraction as the total pressure drop minus the pressure drop for fully developed flow in the large and small channels. The pressure drop for fully developed laminar flow is known analytically, of course [Eq. (10.40)]:

$$\Delta p_{\text{excess}} = \Delta p_{\text{total}} - \Delta p_{\text{large channel}} - \Delta p_{\text{small channel}} \tag{10.54}$$

The excess pressure drop is correlated by using $\Delta p_{\text{excess}} = K_L \eta <v>/D$ and K_L can be determined from finite element calculations. Note that one must specify which average velocity and distance are used in this correlation, and the usual choices are the average velocity and the diameter or total thickness between parallel plates, all at the narrow end. The value of K_L depends on the contraction ratio, too, and some of them are calculated in the problems. Typical values range from 8 to 15. Figure 10.23 shows the total pressure drop

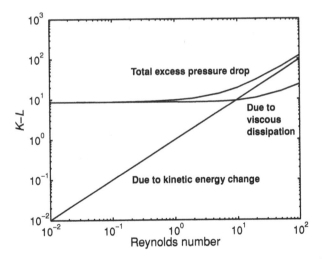

FIGURE 10.23   Laminar flow excess pressure drop for a 3:1 contraction in a circular channel. K_L is the total pressure minus the fully developed pressure drop in the two regions, expressed as $K_L = \Delta p_{\text{excess}} D_2/\eta <v_2>$.

and the contributions due to viscous dissipation and kinetic energy changes. The viscous term predominates at low Reynolds numbers ($Re < 1$, i.e., in microfluidics) whereas the kinetic energy term predominates at high Reynolds number (>100).

When the Reynolds number is not vanishingly small, K_L depends on Reynolds number, too, and the dependence is easily determined using CFD. For example, for a 3:1 contraction in a pipe the results are shown in Figure 10.23. The departure from a constant K_L is due partially to extra viscous dissipation at higher Reynolds number, but mostly due to kinetic energy change for $Re > 10$. Formulas for the kinetic energy change for a variety of axisymmetric or channel contractions or expansions are given by Finlayson et al. (2007).

GENERATION OF TWO-DIMENSIONAL INLET VELOCITY PROFILES FOR THREE-DIMENSIONAL SIMULATIONS

Consider flow in the channel shown in Figure 10.24. In the square corner, the channel is 1 dimensionless unit square, and each leg is 3 units long. The inlet velocity is fully developed flow with an average flow rate of 1.0. The excess pressure drop is calculated. To obtain the fully developed flow at the inlet, it is necessary to learn how to use Linear Extrusion in Comsol Multiphysics. To figure the pressure drop in a straight channel with a square cross section, use $fRe = 14.2$ (Finlayson, 1972). The D in Eq. (10.36) is the hydraulic diameter, $D = 4A/C$. The area is W^2 and the circumference is $4W$, so the hydraulic diameter is W.

Comsol Multiphysics is opened for a 3D simulation using Laminar, steady flow. In the root icon, set the units to "none." Be sure that discretization is chosen to be displayed. In the "Model 1"/"Laminar Flow," open Discretization and change the discretization from

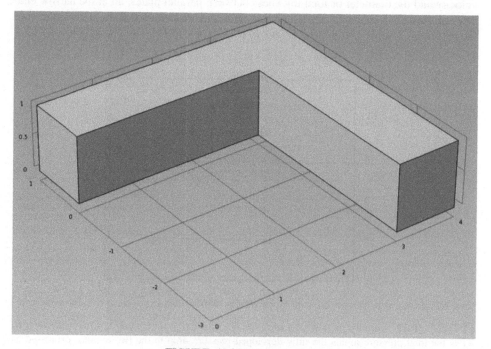

FIGURE 10.24 Square corner.

P1 + P1 to P2 + P1. In Geometry, choose "Work Plane" and draw the projection on the surface $z = 0$, choose "Boolean operations/Union" (and press Build) to connect them without internal boundaries. Then choose extrude, for a distance of 1. In the "Model 1/Laminar Flow/Fluid Properties" set density to 0.001 and viscosity to 1 (for the low Reynolds number case). For the first simulation, set the inlet velocity to 1. Set the pressure at the outlet to zero (not the one with no shear stress). Under "Model 1/Mesh" choose "normal" mesh and build; it uses 48,863 elements and 381,536 degrees of freedom. The average velocity at the inlet is 0.9934. The pressure is zero at the outlet, and the average pressure at the inlet is 215.882. Using the formula involving friction factor and Reynolds number gives

$$\frac{W^2 \Delta p}{L4W \frac{1}{2} \rho < v >^2} \frac{\rho < v > W}{\mu} = 14.2, \text{ or } \Delta p = \frac{28.4 \mu < v >}{W^2} L \qquad (10.55)$$

Here, $W = 1$, $\mu = 1$, $<v> = 1$, $L = 2 \times 3 = 6$ so that $\Delta p = 28.4 \, L = 170.4$. Thus, the excess pressure drop is 45.4. Some of this pressure drop is due to the rearrangement of the velocity profile at the inlet.

Next, solve the problem when the inlet velocity is the proper one, proceed as follows.

Step 1 (Define a 2D Problem)   satisfying the Navier–Stokes equations by adding a second "Model for 2D, Poisson's equation"; this will be "Model 2." Make the domain be the square from 0 to 1 in each dimension. Set the value of $u2 = 0$ on all four boundaries, and solve for some value of f. Then adjust the f so that the average is 1.0; the f is 28.4543.

Step 2 (Identify Vertices)   Now go to "Model 2/Definitions" and right click to get "Model Couplings/Linear Extrusion." For Source, use the solution for $u2$ on the square. For Destination, choose the solution on Geometry 1. Set the vertices as follows: for the source, make them appear as in Table 10.1; do this for the destination, too. The fourth one is, of course, defined by the other three. (Note: It is confusing to only specify three vertices for the square, but that is the way it is; the program will not work if you specify the fourth one.) In the same settings windows, for the geometric entity, for the 2D problem select the entire Domain and for the 3D problem select Boundary and pick the inlet boundary.

Step 3 (Create a Function)   Then go to "Model 1/Definitions/Variables" and define a variable u_fully_dev as mod2.linext1(mod2.u2).

Step 4 (Assign Boundary Conditions)   In the boundary conditions, set the inlet velocity to u_fully_dev.

Step 5 (Connect the problems)   You also must choose "Step 1:Stationary" and open "Values of Variables not solved for." Be sure to click the box and identify the solution to be used (Solution 2).

TABLE 10.1   Vertex Choice for Linear Extrusion; 2D Problem; 3D Inlet

3	4		1	2
2	–		4	–

FIGURE 10.25   Solution of Poisson's equation on a square.

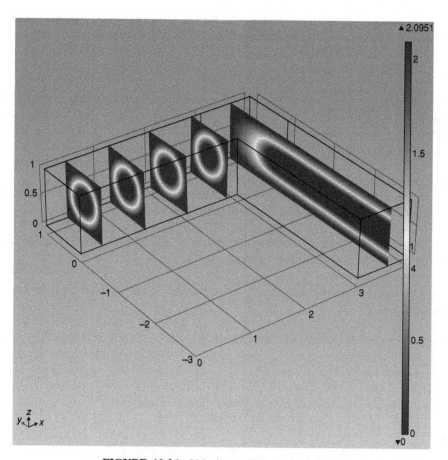

FIGURE 10.26   Velocity profile in a square turn.

Step 6 (Solve the 3D Problem)   For "Study 1" (the 3D case), you can turn off "Stationary, Poisson's Eq." For "Study 2" (the 2D case), you can turn off "Stationary, Laminar Flow." Then solve "Study 2," getting the solution shown in Figure 10.25, with a peak value of 2.0956; then solve "Study 1." The x-velocity is shown in Figure 10.26. Now the average pressure on the inlet is 190.65. Thus, the excess pressure is 190.65 – 170.4 = 20.25. When the density is changed to 1, the average pressure is almost the same, 190.68, so the excess pressure is 20.48. As expected, there is very little difference between those solutions at low Reynolds number, provided the nondimensional form is used.

CHAPTER SUMMARY

This chapter illustrates how to use Comsol Multiphysics to solve the Navier–Stokes equations in a variety of situations. Some of these problems are classic, such as entry flow into a pipe and transient start-up of pipe flow. Most examples are for laminar flow in two dimensions, but one model is turbulent flow into a pipe, and other models are for complicated 3D geometries.

PROBLEMS

10.1₁ Solve for the entry flow of a Newtonian fluid into a pipe using the parameters in the example. Reproduce Figure 10.2 through Figure 10.4.

10.2₁ Solve for the entry flow of a power–law fluid into a region between two flat plates. The viscosity is given by

$$\eta = 0.2\dot{\gamma}^{-0.6} \tag{10.56}$$

The plates are a distance of 0.005 m apart, and the applied pressure gradient is 60 Pa in a length of 0.3 m. The density of the fluid is that of water. (Hint: follow the steps in the example of entry flow in a pipe, except use a power–law formula for viscosity and use two flat plates rather than a pipe.)

10.3₁ Solve Problem 10.2 except use the Carreau function, Eq. (10.29) with $\eta_\infty = 0.05$, $\eta_0 = 0.492$, $\lambda = 0.1$, and $n = 0.4$. (Hint: follow the steps in the example of entry flow in a pipe, except use the Carreau function for viscosity and use two flat plates rather than a pipe.)

10.4₁ Compare the Carreau viscosity as a function of shear rate (Figure 10.5 and Problem 10.3) with that of the power-law fluid in Problem 10.2.

10.5₁ Solve for the start-up flow of the power-law fluid and problem described in Problem 10.2. How long does it take to reach steady state?

10.6₁ Solve for the start-up flow of the Carreau fluid and problem described in Problem 10.3. How long does it take to reach steady state?

10.7₁ Solve for the flow through an orifice. Reproduce Figure 10.15 for Reynolds numbers of 31.6 using the parametric sweep from 1 to 31.6. Plot the pressure coefficient versus Reynolds number, as in Figure 10.16.

10.8₁ Solve the example of a microfluidic T-sensor, except have the average velocity coming in the bottom twice as high as in the example. Compare the flow rates in both streams and out to ensure that the total flow in equals the total flow out.

10.9₂ Consider laminar flow past a flat plate. Create a geometry as a long rectangle, 1 m long and 0.1 m high. Use density of 1000 kg/m³ and viscosity of 0.001 Pa s. Set the velocity to zero on the plate, and set it to 0.1 m/s on the left. Use an open boundary with zero normal stress at the top and an outlet with $p = 0$ at the right. You may have to refine the mesh considerably to see the boundary layer. Compare your solution with the Blasius solution in your textbook.

10.10₂ Solve for the drag on a sphere in a flowing stream with a uniform velocity profile upstream. Solve for zero Reynolds number (Stokes flow). Compare the solution with an analytical solution in your textbook. (Hint: set the density to a small number to simulate Stokes flow. The drag is obtained by integrating the stress over the boundary of the sphere. Use the Expression button.)

10.11₁ Solve for the drag on a sphere in a flowing stream with a uniform velocity profile upstream. Solve for Reynolds number from 1 to 100. Put the sphere in a cylindrical tube, but use an open boundary with zero normal stress on the tube (this will mimic an infinite domain). How does the qualitative behavior of the solution change with Reynolds number?

10.12₂ Solve for the drag on a sphere in a flowing stream with a uniform velocity profile upstream. Solve for Reynolds number from 1 to 100. Put the sphere in a cylindrical tube with a diameter of 5. For this problem, use zero velocity on the cylindrical tube, which mimics the effect of the wall.

10.13 **Project.** Use the problem set up in Problem 10.12 and deduce the wall effect when measuring the terminal velocity by dropping a sphere into a fluid. Compare with Perry and Green (2008).

10.14₂ Solve for the drag on a cylinder in a flowing stream with a uniform velocity profile upstream. Solve for Reynolds number = 1, 5, 10, 20, 40. Use the nondimensional form with radius = 0.5, upstream velocity = 1, $\rho = Re$, and $\eta = 1$. [The other method of nondimensionalization ($\rho = 1$ and $\eta = 1/Re$) works best as the Reynolds number gets high. But use the one with $\rho = Re$ and $\eta = 1$ when calculating the drag force.] Far from the cylinder use an open boundary with zero normal stress (this will mimic an infinite domain.)

(a) How does the qualitative behavior of the solution change with Reynolds number? Prepare graphs that display interesting features of the solution for $Re = 10$.

(b) Plot the dimensionless drag (what you get by integrating the total force on the cylinder) versus Reynolds number. The drag coefficient can be determined from your forces using the formula $C_D = 2F'/Re$. The force F' can be calculated by integrating the total force on the surface of the cylinder. Then multiply it by 2 (to account for the other side, assuming you solve the problem using symmetry).

(c) Compute the drag coefficient at $Re = 1$ and 10 and compare with Perry and Green (2008).

10.15₂ Solve for turbulent flow in a pipe using the parameters in the example (if you have access to the CFD module). Use velocities of (a) 0.2; (b) 0.6; (c) 1.4 m/s. Compare results with Figures 10.10 and 10.11.

10.16₂ Do the microfluidic problem in Figure 10.7 but in three dimensions for the following parameters. The size of the channels are 100 μm in cross section, the inlet ports and the exit ports are 100 × 100 μm. The section after the inlet is 500 μm long. The inlet velocities are fully developed with an average velocity of 0.1 mm/s. The density and viscosity are those of water (density of 1000 kg/m³, viscosity = 0.001 Pa s). What is the Reynolds number in each inlet; in the section between the inlet and outlet? Show the streamlines. What is the pressure drop in the device?

10.17₁ (All parts as a Project or one part as a problem.) Solve for the 2D flow in channels of various shapes as shown in Figure 10.27.

   (a) In the sharp bend, the channel is 1 dimensionless unit wide, and each leg is 3 units long.

   (b) In the smooth bend, the radius along the centerline is one gap width.

   (c) In the smooth bend, long radius, the radius is 1.5 times the gap size.

   (d) In the T, each leg is 3 units long. For all cases use a fully developed flow profile for a channel at the inlet with a dimensionless average flow rate of 1.0.

   Use Comsol Multiphysics in a nondimensional form with a density of 0.001 (to get basically low Reynolds number) to determine the pressure drop; then calculate it for flow in a straight channel that is as long as the centerline distance of the device. Subtract the straight channel result from the total to obtain the excess pressure drop. Comment. What is K_L? For one of the cases recalculate the pressure drop with a dimensional density of 1 ($Re = 1$).

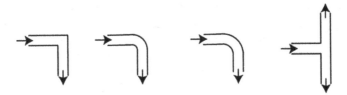

FIGURE 10.27 Geometries for pressure drop calculations.

10.18₂ Solve for the 3D flow in a 90 degree turn of a round pipe with $r/d = 1$ (Figure 10.28). Use an average inlet velocity of 1.0 and solve for Reynolds numbers from 0.01 to 100. Plot K and K_L versus Reynolds number from 0.01 to 100 for laminar flow and from 10^4 to 2×10^5 for turbulent flow. For turbulent flow, use correlations in Perry and Green (2008, Eq. 6-97, pp. 6-18–6-19) to get $K = 0.18$ at a Reynolds number of 106, and $Re > 2 \times 105$. For Reynolds numbers between 10^4 and 2×10^5, multiply this number by the correction factor given there:

$$C_{Re} = 2.2 + \frac{1 - 2.2}{\log_{10}(2 \times 10^5) - \log_{10}(10^4)}(\log_{10} Re - \log_{10}(10^4))$$

$$C_{Re} = 2.2 + \frac{1 - 2.2}{5.3 - 4}(\log_{10} Re - 4)$$

(10.57)

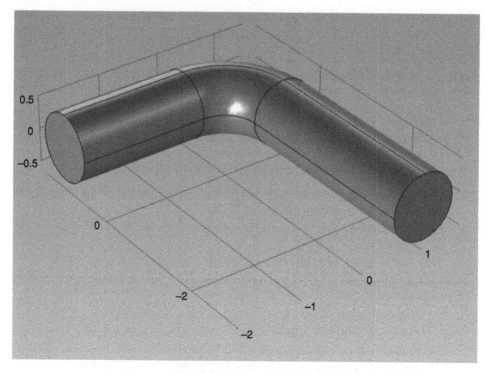

FIGURE 10.28   Pipe corner.

10.19_1 Project. The pressure drop for a contraction is represented as the difference between the total pressure drop minus the theoretical pressure drop for fully developed flows in the large and small section. Determine the K_L for slow flow and a 2:1; 3:1; and 4:1 contraction for both channel flow and pipe flow. For the 3:1 pipe flow case (Figure 10.22), solve for Reynolds numbers from 0.01 to 100. Base the definition of K_L on the velocity and diameter/thickness of the smaller region, since the pressure drop there is the largest.

10.20_2 Compute the excess pressure drop in a 3:1 contraction of a square channel (Figure 10.29). The excess pressure drop is the total pressure drop minus theoretical pressure drop for fully developed flow along the centerline. Express the result in terms of K_L.

10.21_1 Compute the excess pressure drop in the 2D T-sensor and H-sensor at $Re = 1$. The excess pressure drop is the total pressure drop minus the theoretical pressure drop for fully developed flow along the centerline. Express the result in terms of K_L.

10.22_1 Solve for flow in a regular porous media, as shown in Figure 10.30. The domain is a square from -1 to $+1$ and each circle has a radius of 0.5. The top and bottom boundaries are symmetry boundaries, the left boundaries are inlet boundaries, and the right boundaries are outlet boundaries. There is no slip on the surfaces of the cylinders. Solve for slow flow in dimensionless terms when the normal stress on the right side is 0 and on the left side is 1.

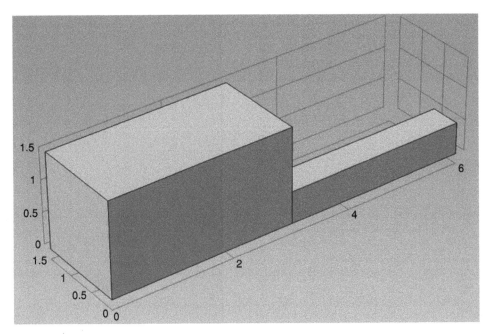

FIGURE 10.29   3:1 contraction using symmetry on the $y - x$ plane ($z = 0$) and the $z - x$ plane ($y = 0$).

10.23$_2$ Consider flow in a cross as shown in Figure 10.31, first for channel flow and then for 3D flow when the cross section is a square. The length of the legs is 7 and the channel width is 1, in dimensionless terms. Oftentimes a microfluidic device will have a flow segment like this, but with much longer channels. In those channels the flow is fully developed and we know what it is. However, at the cross, the flow is disturbed. What distance does it take (in channel widths) to return to fully developed flow? Do this for small Reynolds number.

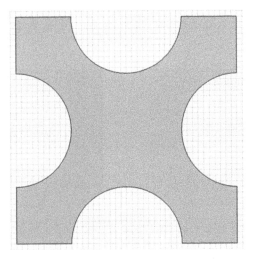

FIGURE 10.30   Periodic flow between cylinders.

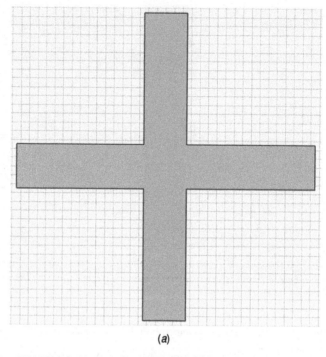

(a)

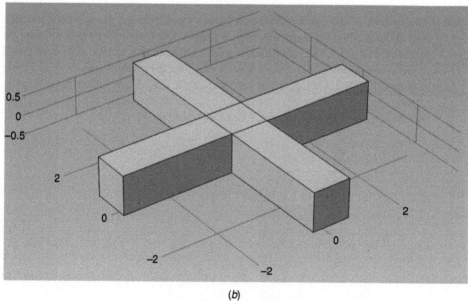

(b)

FIGURE 10.31   Flow in a cross: (a) channel flow; (b) 3D flow.

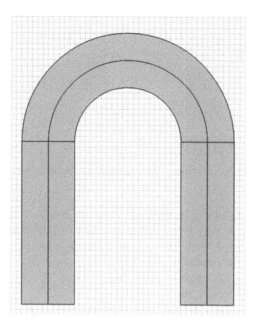

FIGURE 10.32 Flow in a bend.

10.24₂ Solve for the flow in a bend, first in two dimensions and then in three dimensions as shown in Figure 10.32. The length of the straight channels is 3, with width 1. The circular regions have radii 1, 1.5, and 2. For the 3D problem, extrude the geometry to a height of 1.0.

10.25₂ Solve for the flow in an H-sensor in three dimensions. Extrude the H-sensor in Problem 10.21 to one unit height. You will have to use two linear extrusion couplings, one for each inlet branch, but you only need to solve Poisson's equation once. Determine the excess pressure drop.

10.26 Project. Compute the pressure drop for the serpentine mixer at $Re = 1$. By using the cut-plane option, determine the pressure drop for 1, 2, 3, and 4 segments: straight channels, chimneys. Then correlate your results in terms of the number of straight channels and chimneys in order to provide an analytical formula for a serpentine of any number of straight channels and chimneys.

10.27 Project. A microfluidic device was designed by Micronics (Redmond, WA) and adapted for metabolic fingerprinting applications by Drs. Colin Mansfield and R. Anthony Shaw at the NRC Institute for Biodiagnostics in Canada. The purpose was to take a sample of serum (comprising both metabolites and proteins) and use diffusion across a serum/water interface to partly extract metabolites for diagnostic or analytical characterization. The efficacy of this process was assessed and optimized by specifically targeting the extraction of serum creatinine (a representative metabolite) at the expense of serum albumin (a representative protein). With that accomplished, infrared spectroscopy could be used to characterize the metabolite-rich aqueous stream with minimal interference from serum proteins. The convection and diffusion problem is specified in Chapter 11 (Problem 11.28);

FIGURE 10.33 Geometry for a laminar fluid diffusion interface device.

here, consider just the flow problem. First consider flow into the device shown in Figure 10.33. The height is 330 mm and the two inlet portions each are 110 mm high. The depth is 4.5 mm, but solve the problem in two dimensions. The flow rates are 2 μL/s for the receiver (water; top stream) and 1 μL/s for the sample. For the flow problem in this chapter, take the viscosity as 1.6 mPa s for both fluids; this is changed in Problem 11.28.

(a) Solve for the streamlines at the inlet in the geometry shown below.

(b) Compare that with a geometry that has a knife edge at a height of 165 mm. Further details about the device can be obtained from the Comsol website (Finlayson and Shaw, 2010), the book website, and Schattka et al. (2011).

10.28_1 When fluid is flowing between two flat plates and there is a hole or slit in the bottom plate, there is also a difference between the pressure at the bottom of the hole/slit and the top plate. This is called the hole pressure and the effect can be used to measure the first normal stress coefficient of polymers (a measure of elasticity). Here consider a Newtonian fluid and solve the hole pressure problem in two dimensions (i.e., for a slit) when $W = H = L$ for $Re = 1$ (see Figure 10.34). Plot the pressure on the bottom of the hole and the top just across it.[1]

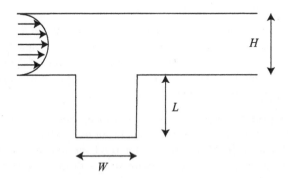

FIGURE 10.34 Two-dimensional hole pressure geometry.

---

[1] This problem was the first problem solved by the Galerkin finite element method by the author and his students (Jackson and Finlayson, 1982a,b). At the time, we wrote our own finite element code, computers were much slower and very much smaller, so only 12, 29, and 110 elements were used for a 2D problem. Despite that limitation, comparison with experiment was quite reasonable for Newtonian fluids.

10.29 **Project.** Expand Problem 10.28 to three dimensions. For the 3D case, make the entrance be 1×8, make the length be 4, and put the cylinder of diameter 1 centered at a distance of 1.5 from the inlet. Then the geometry from the side is the same as the 2D case. The cylinder is made to extend through to the top surface so that integrals can be computed on the circle on the top surface to compare with the same integrals on the bottom of the cylinder. The problem was solved in nondimensional form, with $\rho = 1$, $\mu = 1$. Be sure to change from compressible flow to incompressible flow (in the model), use no dimensions, and change to quadratic velocity (Discretization). As an approximation, take the inlet velocity as $u = 6^*z(1 - z)^* y^{1/7} {}^* (8 - y)^{1/7}/1.328$. Compare the results for three dimensions with those obtained in two dimensions. Rationalize any differences.

11

HEAT AND MASS TRANSFER IN TWO AND THREE DIMENSIONS

Chemical reaction and mass transfer are two unique phenomena that help define chemical engineering. Chapter 8 describes problems involving chemical reaction and mass transfer in a porous catalyst, and how to model chemical reactors when the flow is well defined, as in a plug-flow reactor. Those models, however, did not account for the complicated flow situations sometimes seen in practice, where flow equations must be solved along with the transport equation. Microfluidics is the chemical analog to microelectromechanical (MEMS) systems, which are small devices with tiny gears, valves, and pumps. The generally accepted definition of microfluidics is flow in channels of size one millimeter or less, and it is essential to include both distributed flow and mass transfer in such devices.

Microfluidic devices often are smaller than a penny, so a new name has been invented for them: lab-on-a-chip. The pharmaceutical industry uses these devices for drug design, delivery, and detection, and the biomedical industry uses them for sensors and drug delivery. Industrial applications include the study of new reaction mechanisms and pathways that lead to improved selectivity with fewer pollutants. Because of the small size of microfluidic devices, a precise control of temperature is possible, which improves kinetic information, especially with catalysts. Some elements of a fuel cell are small enough to be microfluidic devices, too. If large-scale production is required, engineers use "number up" rather than "scale up" to achieve high throughput. The techniques of the electronic industry can be used to produce many small devices cheaply, and engineers simply add "clones" to increase the production rate of chemicals.

This chapter uses Comsol Multiphysics to solve problems involving two-dimensional (2D) or three-dimensional (3D) flow and diffusion of mass and/or heat. Convection and diffusion of mass and heat is often governed by the convective diffusion equation. The flow problems are treated in Chapter 10, but sometimes the parameters, such as viscosity and density, depend upon the concentration and temperature. In that case, you have to

Introduction to Chemical Engineering Computing, Updated Second Edition. Bruce A. Finlayson.
© 2014 John Wiley & Sons, Inc. Published 2014 by John Wiley & Sons, Inc.

solve all the partial differential equations together. This chapter illustrates how to use Comsol Multiphysics to solve problems in the geometries discussed in Chapter 10: the T-sensor, an orifice, and the serpentine mixer. You will learn how to solve the equations either sequentially, one after the other, or simultaneously; make the transport properties such as viscosity functions of temperature or concentration; and learn to assess the relative importance of convection and diffusion, which is essential for interpreting numerical results.

Instructional Objectives: After working through this chapter, you will be able to

1. Model diffusion and heat conduction in 2D and 3D situations, with and without flow.
2. Solve the flow and convective diffusion problems sequentially to save time.
3. Apply various boundary conditions, including periodic boundary conditions.
4. Handle variable viscosity depending on concentration and wall reactions.
5. Understand the importance of the Peclet number and use that information adroitly.
6. Calculate properties of the solution (mixing cup average concentrations and variances).
7. Check your solution for continuity of mass.

In addition, you will be introduced to mass transfer problems in the field of microfluidics.

CONVECTIVE DIFFUSION EQUATION

The equation governing mass transfer with flow is

$$\frac{\partial c}{\partial t} + \mathbf{u} \cdot \nabla c = \nabla \cdot (D \nabla c) \tag{11.1}$$

When the diffusivity is constant it is

$$\frac{\partial c}{\partial t} + \mathbf{u} \cdot \nabla c = D \nabla^2 c. \tag{11.2}$$

In 2D rectangular coordinates, Eqs. (11.1) and (11.2) become

$$\frac{\partial c}{\partial t} + u \frac{\partial c}{\partial x} + v \frac{\partial c}{\partial y} = \frac{\partial}{\partial x} \left(D \frac{\partial c}{\partial x} \right) + \frac{\partial}{\partial y} \left(D \frac{\partial c}{\partial y} \right)$$
$$= D \left(\frac{\partial^2 c}{\partial x^2} + \frac{\partial^2 c}{\partial y^2} \right) \quad \text{when } D \text{ is constant.} \tag{11.3}$$

Heat transfer is similar, but with different parameters. The vector representation is

$$\rho C_p \left(\frac{\partial T}{\partial t} + \mathbf{u} \cdot \nabla T \right) = \nabla \cdot (k \nabla T)$$
$$= k \nabla^2 T \quad \text{when } k \text{ is constant.} \tag{11.4}$$

In 2D rectangular coordinates, the equation is

$$\rho C_p \left(\frac{\partial T}{\partial t} + u\frac{\partial T}{\partial x} + v\frac{\partial T}{\partial y} \right) = \frac{\partial}{\partial x}\left(k\frac{\partial T}{\partial x} \right) + \frac{\partial}{\partial y}\left(k\frac{\partial T}{\partial y} \right)$$

$$= k\left(\frac{\partial^2 T}{\partial x^2} + \frac{\partial^2 T}{\partial y^2} \right) \quad \text{when } k \text{ is constant.}$$

(11.5)

The velocity in the equation comes from the solution to the Navier–Stokes equation described in Chapter 10 (or another flow equation).

NONDIMENSIONAL EQUATIONS

Diffusion in liquids can be slow, and this makes the problems hard to solve, as shown below. The best way to learn how important diffusion is for any problem is to calculate the Peclet number. Equation (11.3) is made nondimensional using the techniques illustrated in Eq. (10.13). Let $c'(x', y', t')$ be the nondimensional concentration, and $c = c_s c'(x', y', t')$. Eq. (11.3) for a constant diffusivity becomes

$$\frac{c_s}{t_s}\frac{\partial c'}{\partial t'} + \frac{u_s c_s}{x_s}\left(u'\frac{\partial c'}{\partial x'} + v'\frac{\partial c'}{\partial y'} \right) = \frac{Dc_s}{x_s^2}\left(\frac{\partial^2 c'}{\partial x'^2} + \frac{\partial^2 c'}{\partial y'^2} \right)$$

(11.6)

Multiplying by x_s^2/Dc_s gives

$$\frac{x_s^2}{t_s D}\frac{\partial c'}{\partial t'} + \frac{u_s x_s}{D}\left(u'\frac{\partial c'}{\partial x'} + v'\frac{\partial c'}{\partial y'} \right) = \left(\frac{\partial^2 c'}{\partial x'^2} + \frac{\partial^2 c'}{\partial y'^2} \right)$$

(11.7)

You may already have picked the time standard in the flow problem. If you choose $t_s = x_s/u_s$, the coefficients of the first and second terms are the same, and the coefficient is called the Peclet number:

$$Pe = \frac{u_s x_s}{D}$$

(11.8)

The equation is then

$$Pe\frac{\partial c'}{\partial t'} + Pe\left(u'\frac{\partial c'}{\partial x'} + v'\frac{\partial c'}{\partial y'} \right) = \left(\frac{\partial^2 c'}{\partial x'^2} + \frac{\partial^2 c'}{\partial y'^2} \right)$$

or $$\frac{\partial c'}{\partial t'} + u'\frac{\partial c'}{\partial x'} + v'\frac{\partial c'}{\partial y'} = \frac{1}{Pe}\left(\frac{\partial^2 c'}{\partial x'^2} + \frac{\partial^2 c'}{\partial y'^2} \right)$$

(11.9)

Typical boundary conditions include setting the concentration (temperature) to a specific value, having no flux through a boundary, and sometimes having the flux through the boundary, to be proportional to the difference between an interface concentration (temperature) and an external value. At an outflow boundary, choose convective flux.

EXAMPLE: HEAT TRANSFER IN TWO DIMENSIONS

The first problem considered is heat conduction in a square, which is a classical problem:

$$\frac{\partial^2 T}{\partial x^2} + \frac{\partial^2 T}{\partial y^2} = 0 \quad \text{in} \quad 0 \le x \le 1, 0 \le y \le 1 \tag{11.10}$$

$$\begin{aligned} T &= 1 \text{ on } x = 0 \\ T &= 1 \text{ on } y = 0 \\ T &= 0 \text{ on } y = 1 \\ \mathbf{n} \cdot \nabla T &\equiv \frac{\partial T}{\partial n} = 0 \text{ on } x = 1 \end{aligned} \tag{11.11}$$

Step 1 (Open and Set Preferences) Open Comsol Multiphysics and choose two dimensions and click the next arrow. Then select Heat Transfer/Heat Transfer in Solids and next; Finally, click Stationary and the finish flag. (See the icons displayed in Figures 9.2–9.4.) Click the root node and change units to "none" at the bottom of the settings window.

Step 2 (Define the Geometry) Right click Geometry and choose a square. Make it 1 by 1 with base point at (0,0); click "Build All."

Step 3 (Define the Problem) Right click "Heat Transfer" and select "Heat Transfer in Solids" if it is not already there. Open "Heat Transfer in Solids" and change the thermal conductivity to "User defined," value 1.

Step 4 (Set the Boundary Conditions) Right click "Heat Transfer" and choose Temperature. Select boundaries 1 and 2 (the lowest one and the left most one), + to add it to the Selection box, and set the value to 1. Do this again and select boundary 3 (top one) and set the temperature to 0. The right-hand boundary is insulated, as can be seen by clicking on "Thermal Insulation"; the other boundaries are shown as overridden.

Step 5 (Create a Mesh) Click on Mesh and "Build All"; the mesh is shown in Figure 11.1.

Step 6 (Solve the Problem) Right click on "Study 1" and = to get the solution.

Step 7 (Examine the Solution) The solution will be plotted in a color plot. Contours of temperature are also made automatically. If they are not, they can be made by right clicking Results, choosing "2D Plot Group," and then right clicking on "2D Plot Group" and choosing Contour. Click on the plot icon to see the plot, Figure 11.2. Save it by clicking the camera.

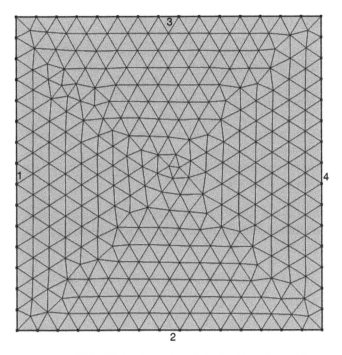

FIGURE 11.1   Finite element mesh for heat transfer problem.

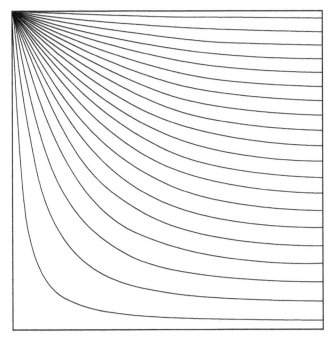

FIGURE 11.2   Solution to heat transfer problem.

EXAMPLE: HEAT CONDUCTION WITH A HOLE

Next, introduce a hole in the middle of the same diagram. The boundary conditions on the sides of the hole are thermal insulation.

Step 1 (Open and Set Preferences) Open the previous solution.

Step 2 (Define the Geometry) Right click on Geometry and choose Square. Make the sides of the square 0.4, and put the base position at (0.3,0.3); click "Build All." Right click on Geometry; choose "Boolean Operations," and choose Difference. Add the first square and subtract the second one. Then choose "Build All."

Step 3 (Set the Boundary Conditions) To check that the inside boundaries are insulated, click on Heat Transfer/Thermal Insulation. The new boundaries are listed, so there will be no heat flux across them.

Step 4 (Create a Mesh) Click on Mesh, "Build All" to get the mesh, which has 572 elements.

Step 5 (Solve the Problem) Right click "Study 1" and choose = to get the solution.

Step 6 (Examine the Solution) The contour plot is shown in Figure 11.3; the settings are the same as in the first case. Notice that the contours of constant temperature are perpendicular to the thermally insulated boundaries, which is appropriate because there is

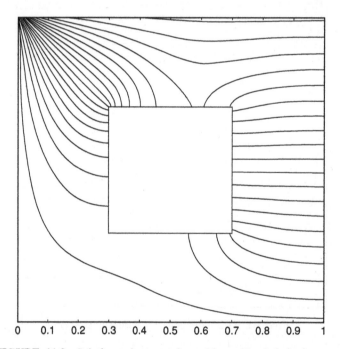

FIGURE 11.3 Solution to heat transfer problem with a hole in the center.

no gradient normal to the boundary. To find the heat flux on the upper surface of the hole, right click the Results/Derived Values' node and choose "Line Integration." Then change the expression to $-Ty$, which is the vertical heat flux. The value is 0.046. It should be zero. To see the effect of mesh refinement, choose two finer discretizations, fine and finer. The values of this integral are then 0.0414 for fine, 798 elements; 0.0325 for finer. Choosing two finer meshes gives 0.0213 for extra fine, 1598 elements; 5434 elements; and for extremely fine, 0.0133, 43644 elements. Clearly the integrated heat flux is converging to zero as the mesh is refined.

If we wanted to solve this problem with physical properties instead of in a nondimensional form, we would need to open "Model 1" and "Open Materials Browser." Choose "Built-in" and select Aluminum and right click "Add Material" to put it in the simulation. Be sure that the domain number, 1, appears in the box. Then go back to "Heat Transfer in Solids" and change the thermal conductivity from "User defined" to "From material." Change the temperatures to 296 and 273 and solve again.

EXAMPLE: CONVECTIVE DIFFUSION IN MICROFLUIDIC DEVICES

The T-sensor is described in Chapter 10, but its key use is to transfer a chemical from one flowing stream to the other. Thus, the convective diffusion equation must be solved, too. A fluid (such as water) comes in the top and bottom, but the top stream contains a dissolved chemical that needs is to be transferred. The bottom stream may contain a different chemical that will react and fluoresce, thus permitting a visual detection. Your goal is to predict how fast the transfer takes place. The T-sensor was developed by Wiegl and Yager (1999) and Hatch et al. (2001).

Step 1 (Open and Set Preferences) Open the Comsol Multiphysics file for the fluid flow problem in Chapter 10. First, we add physics by right clicking "Model, 1" choosing "Transport of Diluted Species," next; Stationary, finish.

Step 2 (Set the Variables) Right click Definitions and choose Variables, and set *Pe* to 1.

Step 3 (Define the Problem) Click on the "Transport of Diluted Species" node and choose Discretization. Change from Linear to Quadratic. If Discretization does not appear, go to the Show button (see Figure 9.2) and choose Discretization. Right click on "Transport of Diluted Species" and choose "Convection and Diffusion." Select the total domain, +, to add it to the Selection box. Change the velocity to u and v. The diffusivity is a stand-in for $1/Pe$, so change it to $1/Pe$.

Step 4 (Set the Boundary Conditions) To set the boundary conditions, right click on"Transport of Diluted Species" and choose Inlet. Choose the top boundary, +, to add it to the Selection box. Keep the value as 0. Do the same thing for the bottom boundary, but set the Concentration to 1.0. Right click on "Transport of Diluted Species," choose Outlet; select the outlet boundary, +, to add it to the Selection box.

Step 5 (Solve the Problem) To solve the problem, right click on "Study 2" and choose = .

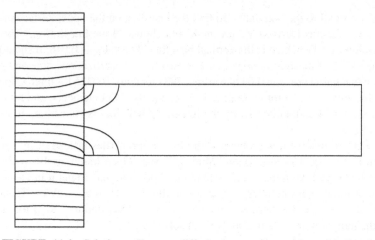

FIGURE 11.4   Solution to T-sensor diffusion/convection problem with $D = 1$.

Step 6 (Examine the Solution) The velocity shows first in the graphics window. To see the concentration, right click on Results and choose "2D Plot Group." Right click on "2D Plot Group" and choose Surface. The default expression is spf.U. Change it to c. Also change the Data set to "Solution 2." Click the plot icon to see the plot of concentration. In Figure 11.4 we do the same thing but choose Contour rather than Surface. To check the mass balance, right click on the Derived Values under Results and choose "Line Integration." Integrate v^*c over the top and bottom boundary (0.0 and 0.5000000). Integrate u^*c over the outlet (0.4999998). Thus, the mass balance is quite good.

Step 8 (Change the Peclet Number) Change the Peclet number to 50 and repeat. Click the Variables node under Definitions and change 1 to 50. Right click "Study 2" and choose =. Change to the plot of concentration, Figure 11.5. Now there is good mixing, but not as good as when the Peclet number is smaller. The flow rates in and out are 0.499959 and 0.5000000, which gives a quite accurate mass balance. Change the Peclet number to 500

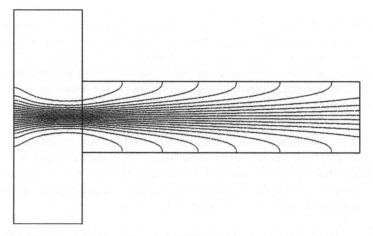

FIGURE 11.5   Solution to T-sensor diffusion/convection problem with $D = 1/50$.

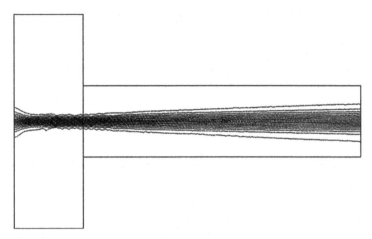

FIGURE 11.6   Solution to T-sensor diffusion/convection problem with $D = 1/500$.

and repeat, giving Figure 11.6. Now the mixing is not nearly as good. The mass balance numbers are the same as before, giving a good overall mass balance. Note, however, that the concentration (in the scale) goes from -0.019 to 1.0212. This indicates that there are oscillations in the solution, but they are on the order of 2%. This error can be decreased if desired by refining the mesh.

Step 9 (Plot Two Solutions)   To plot outlet concentration profiles for $Pe = 50$ and 500, right click Global Definitions, choose Parameters, and use Pe (you can set a value). Now right click on "Study 2" and choose "Parametric Sweep." When you click on the $+$, the Pe will appear in the window; select it and Add, then OK. Click on the parameter icon and set the range from 50 to 500 with a step of 450. Right click "Study 2" and choose $=$. Right click on Results, choose "1D Plot Group"; right click it and choose "Line Graph." The two concentration profiles are plotted on one graph, Figure 11.7.

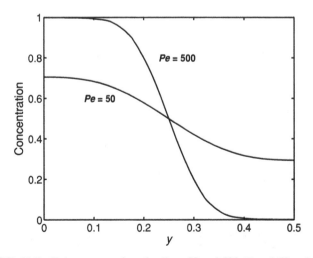

FIGURE 11.7   Exit concentrations for $Pe = 50$ and 500 ($D = 1/50$ and $1/500$).

EXAMPLE: CONCENTRATION-DEPENDENT VISCOSITY

The viscosity of a mixture depends upon its concentration, although often that dependence is small. To resolve the previous problem and allow the viscosity to depend upon concentration use the following (arbitrary) formula

$$\mu = 1 + c \tag{11.12}$$

Then the viscosity of the material coming into the top of the T-sensor is 2, whereas the viscosity coming into the bottom is 1. To modify the model, perform the following steps.

Step 1 (Set the Variables) We want to change the viscosity from 1 to 1 + alpha*c, starting with the previous solution. The alpha is used so that it can be a parameter, either 0 or 1. Click on the Parameters node under Global Definitions. Define a parameter alpha and set it to 0. Be sure the *Pe* is 500. (The alpha can also be set in Variables in "Model 1".)

Step 2 (Define the Problem) Click on the node "Fluid Properties" under "Laminar Flow" and change the viscosity from 1 to 1 + alpha*c.

Step 3 (Solve the Problem) Right click "Study 1" and choose another Parametric Sweep; use alpha as the parameter: click on the parameter icon, select alpha, Add, then OK, then set the value to range (0,1). Also right click on the "Parametric Sweep" and delete the one using Peclet number. Click on "Study 1" and notice that there are two equations listed there, the flow equation and the concentration equation. Make sure that both are selected to be solved (green check mark), since the viscosity depends on the concentration, which depends on the flow, which depends on the viscosity. Right click "Study 1," choose =.

Step 4 (Examine the Solution) Click on the "Line Graph" node under "1D Plot Group" to see the concentration profile at the outlet for the two solutions: both with *Pe* = 500 but one with viscosity = 1 the other 1 + alpha*c. Clearly there is not much difference; see Figure 11.8.

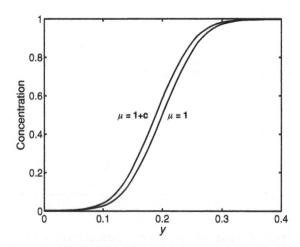

FIGURE 11.8 T-sensor diffusion/convection problem with $D = 1/500$ and $\mu = 1, \mu = 1 + c$.

EXAMPLE: VISCOUS DISSIPATION

So far the examples have assumed that viscous dissipation does not cause an appreciable temperature rise. Viscous dissipation, of course, always increases the temperature, but often the increase is negligible. However, it is important to know the magnitude or provide an upper bound. To see the effect of viscous dissipation, a heat generation term is added to the energy equation. In 2D problems this heat generation term is

$$\Phi_v = \mu \left\{ 2 \left[\left(\frac{\partial u}{\partial x} \right)^2 + \left(\frac{\partial v}{\partial y} \right)^2 \right] + \left[\frac{\partial v}{\partial x} + \frac{\partial u}{\partial y} \right]^2 \right\}, \quad \text{Cartesian geometry} \qquad (11.13)$$

and

$$\Phi_v = 2\mu \left[\left(\frac{\partial u}{\partial z} \right)^2 + \left(\frac{\partial v}{\partial r} \right)^2 + \left(\frac{v}{r} \right)^2 \right] + \mu \left[\frac{\partial u}{\partial r} + \frac{\partial v}{\partial z} \right]^2, \quad \text{cylindrical geometry}$$

$$(11.14)$$

The effect of viscous dissipation is illustrated here for the orifice problem solved in Chapter 10. The notation uses u as the radial velocity and w as the axial velocity:

$$Q = 2 * Br * (wz^\wedge 2 + ur^\wedge 2 + (u/r)^\wedge 2) + Br * (wr + uz)^\wedge 2 \qquad (11.15)$$

The Brinkman number is

$$Br = \Pr \frac{u_s^2}{C_p T_s}, \quad \Pr = \frac{C_p \mu}{k} \qquad (11.16)$$

The parameters used in this example are

$$\rho = 1, \eta = 1/Re, \Pr = 7, Re = 31.62, C_p = \Pr * Re = 221.3, Br = 0.141 \qquad (11.17)$$

There are two options for the temperature boundary conditions. In the first option, the system can be adiabatic where all walls use the boundary condition of no flux. In the second option, the walls can be made of metal, with plenty of coolant outside, and then the boundary condition would use a constant temperature. Here, adiabatic boundary conditions are used. Thus, the temperature out of the device (minus the inlet temperature) gives the temperature rise due to viscous dissipation.

Step 1 (Open and Set Preferences) Open the Comsol Multiphysics file for the orifice problem solved in Chapter 10. Right click "Model 1" and choose "Add Physics." Select "Heat Transfer/Heat Transfer in Fluids"; next, Stationary, finish.

Step 2 (Set the Variables) Click on Variables and add the following: $Re = 31.62$, $Pr = 7$, $Cp = Pr*Re$, $Br = 0.141$.

Step 3 (Define the Problem) Click on the node "Heat Transfer in Fluids" under "Heat Transfer." Change the Velocity to "Velocity field (spf/fp1)." Set the physical properties to $k = 1$, density $= 1$, $Cp = Cp$, gamma $= 1$. The flow problem needs to be changed to use a viscosity of 1/31.62. Under Study 1, delete the "Parametric Sweep." Under "Study 1," click on "Step 1: Stationary" and make sure the Physics interface selected is "Laminar flow." Right click on "Study 1" and choose = . This solves the flow problem. To create the source, right click "Heat Transfer" and choose "Heat Source." Select the domain, +, to add it to the Selection box. The Q is defined as $2*Br*(wz^\wedge2 + ur^\wedge2 + (u/r)^\wedge2) + Br*(wr + uz)^\wedge2$ to agree with Eq. (11.15). For Heat Transfer, choose Discretization and choose Quadratic if it is not already Quadratic.

Step 4 (Set the Boundary Conditions) The thermal problem is solved with adiabatic boundaries. The default conditions are axial symmetry on the centerline and adiabatic boundaries everywhere else. Right click on -Heat Transfer-, choose Inlet, select the inlet boundary, +, and set the Temperature to 1. Then Right click on -Heat Transfer-, choose Outlet, select the outlet boundary, +.

Step 5 (Solve the Problem) Click on "Step 1: Stationary" under "Study 2" and make sure the heat transfer option is selected. Right click on "Study 2" and choose = .

Step 6 (Examine the Solution) Then right click on Results, choose "2D Plot Group." Right click on it and choose Contour. Change the expression to T2 and click the plot icon, giving Figure 11.9. The effect of viscous dissipation is shown in Figure 11.9 when the flow field is that shown in Figure 10.15c. The peak temperature is only 1.045°C, which is only 0.045°C higher than the input temperature. It is also possible to make the viscosity depend upon temperature. Before doing that, you can solve the problem with a constant viscosity to determine the maximum temperature rise. If the temperature rise is small, you have verified that the assumption that the viscosity is constant. If the temperature rise is large, then the problem can be solved allowing the viscosity to depend upon temperature. Let the viscosity depend upon temperature, as in Eq. (11.18):

$$\mu = \mu_0 \exp(E/RT) \tag{11.18}$$

Equation (11.18) is inserted for the viscosity for the Physics/Subdomain Settings for the Navier–Stokes equation:

$$\mu = mu0 * \exp(E/(R * T)) \tag{11.19}$$

Naturally, you must use the correct units. You can also use the Materials Browser to get the viscosity variation with temperature if desired.

EXAMPLE: CHEMICAL REACTION

It is easy to model the microfluidic device as a chemical reactor. You prepare the flow problem as illustrated in Chapter 10, add the convective diffusion equation, and enter the reaction rate. Suppose the rate of reaction is

$$\text{Rate} = kc^2, \text{ where rate is in mole per volume per time} \tag{11.20}$$

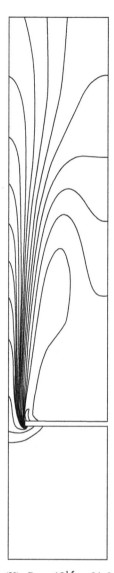

FIGURE 11.9   Temperature rise (K), $Re = 10^{1.5} = 31.6$, 20 contour levels from 1 to 1.048.

Step 1 (Open and Set Preferences)   Open the microfluidic simulation with both flow and diffusion.

Step 2 (Define the Problem)   To add a reaction rate, right click "Transport of Diluted Species" and choose Reactions. Then select the domain, +, to get it into the Selection box. Set R_C to rate. Right click the Definitions node under "Model 1" and choose Variables. Define rate $= -k^*c^*c$. Right click "Global Definitions" and choose Parameters and set $k = 0.1$.

Step 3 (Solve the Problem)   Right click on "Study 1" and choose = to get the solution. The material reacts very quickly.

EXAMPLE: WALL REACTIONS

A wall reaction can be included as well. Suppose rate of reaction on the wall is

$$\text{Rate} = -k_b c^2, \text{ where rate is in mole per area per time} \qquad (11.21)$$

The flux boundary condition is

$$-\mathbf{n} \cdot \mathbf{N} = N_0, \quad \mathbf{N} = -D\nabla c + \mathbf{u}c \qquad (11.22)$$

Next, you need to augment the flux equation on the wall. The normal velocity on the wall is set to zero for the flow problem. Thus, the diffusive flux equals the rate of reaction. Enter into the window for N_0 the following expression:

$$N_0 = -kb * c * c \qquad (11.23)$$

Step 1 (Define the Problem) For a surface reaction, right click on "Transport of Diluted Species" and choose Flux. Choose a boundary, +, to add it to the Selection box. Check the Species c box and set $N_{0,c} = -kb*c*c$. Choose the Parameters subnode under "Global Definitions" and set $kb = 0.1$.

Step 2 (Solve the Problem) Right click "Study 1," and choose = .

Step 3 (Examine the Solution) Choose a contour plot of c; now the reaction interface is near the dividing streamline and near the wall, too.

EXAMPLE: MIXING IN A SERPENTINE MIXER

To illustrate what is possible in three dimensions, consider the serpentine mixer that is modeled in Chapter 10. The idea is that at the inlet there are two streams entering, and one of them contains a different chemical. The objective is to mix the chemicals in as short a distance as possible. Because mixing by diffusion is slow, chimneys and turns are inserted to enhance the mixing and make it occur in a shorter distance (Neils et al., 2004).

Step 1 (Open and Set Preferences) Open the Comsol Multiphysics file for the serpentine fluid flow problem in Chapter 10. First, we add physics by right clicking "Model 1," choosing "Transport of Diluted Species," next; Stationary, finish.

Step 2 (Define the Problem) Click on the "Transport of Diluted Species" node and choose Discretization. Change from Linear to Quadratic. Right click on Transport and choose "Convection and Diffusion." Select the total domain, +, to add it to the Selection box. Change the velocity to "Velocity field (spf/fp1)." The diffusivity is a stand-in for $1/Pe$, so change it to $1/Pe$. Go back to Definitions, right click Variables, and set Pe to 1000.

Step 3 (Set the Boundary Conditions) To set the boundary conditions, right click on"Transport of Diluted Species" and choose Inlet. Choose the left inlet boundary, +, to add it to the Selection box. Set the value to 1. To get this boundary you may have to change the orientation of the model. Do the same thing for the right inlet boundary, but leave the Concentration as 0. Right click on Transport, choose Outlet; select the outlet boundary, +, to add it to the Selection box.

Step 4 (Solve the Problem) To solve the problem, right click on "Study 2" and choose = .

Step 5 (Examine the Solution) The velocity shows first in the graphics window. To see the concentration, right click on Results and choose "2D Plot Group." Right click on "2D Plot Group" and choose Surface. The default expression is spf.U. Change it to c. Also change the "Data set" to "Solution 2." Click the plot icon to see the plot of concentration, Figure 11.10.

Step 6 (Make More Plots) A plot of the concentration profile inside the domain (Figure 11.11) is obtained by opening "Data sets" under Results, and right clicking to get "Cut Plane." In the settings window you can determine where the plane will be, and choose the plot icon to see the plot. Notice that there are some small errors, since the maximum concentration is 1.081 and the minimum value is −0.0219, but these are small enough to be ignored. The region where they occur can be found by plotting only values of concentration greater than 1.0; next plot only values of concentration less than 0.0.

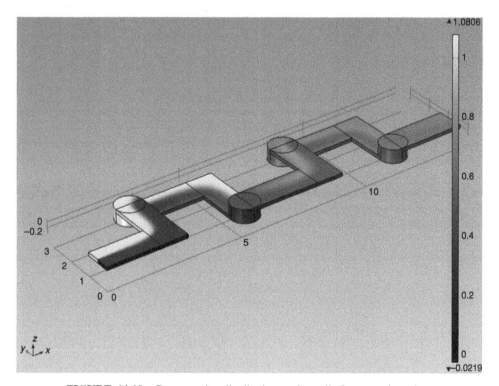

FIGURE 11.10 Concentration distribution on the wall of a serpentine mixer.

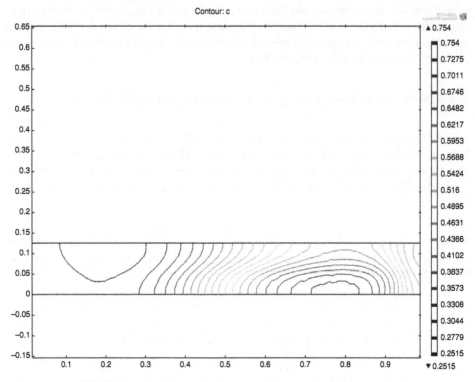

FIGURE 11.11   Concentration distribution in the yz-plane at $x = 8$.

MICROFLUIDICS

Many microfluidic devices involve diffusion of mass as well as fluid flow. The following articles give details of these applications: Jensen (1999), Stone and Kim (2001), Chow (2002), Freemantle (2005), Henry (2005), and a useful website is www.Lab-on-a-Chip.com. The velocity crossing a plane normal to the flow is seldom uniform, so the diffusion must be examined in the midst of nonuniform velocity profiles. The simplest illustration of this effect is with Taylor dispersion.

For simplicity, consider flow between two flat plates as shown in Figure 11.12. The flow is laminar and the velocity profile is quadratic. Now suppose the entering fluid has a second chemical with the concentration shown, 1.0 in the top half and 0 in the bottom half. As this fluid moves downstream (to the right), the velocity is highest in the center, so the second chemical moves fast there. However, it can also diffuse sideways, and the fluid element that does first diffuse sideways then moves slower down the flow channel. At a point not on the

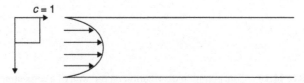

FIGURE 11.12   Diffusion in a fully developed planar flow.

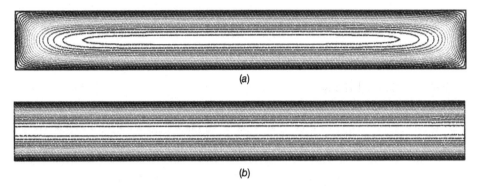

(a)

(b)

FIGURE 11.13 Fully developed velocity profile in (a) 1×8 rectangular channel; (b) 1×8 rectangular channel with slip on the side walls.

centerline, the concentration is determined by this slower velocity in the flow direction and diffusion sideways because the concentration on the centerline is always higher than the concentration in adjacent fluid paths. This is called Taylor dispersion (Deen, 1998). Taylor (1953) and Aris (1956) showed how to model the average concentration as a function of length using a simpler equation (dispersion in the flow direction only) and an effective dispersion coefficient given by (where D_h is the hydraulic diameter)

$$K = D + \frac{<v>^2 D_h^2}{192D} \quad \text{or} \quad \frac{K}{D} = 1 + \frac{<v>^2 D_h^2}{192D^2} = 1 + \frac{Pe^2}{192} \tag{11.24}$$

Next, consider fully developed flow in a narrow channel, which is typical of a microfluidic device. The normal view is shown in Figure 11.13 with flow going into the paper. The velocity varies in x and y according to the equation

$$\eta \left(\frac{\partial^2 v}{\partial x^2} + \frac{\partial^2 v}{\partial y^2} \right) = -\frac{\Delta p}{L} \tag{11.25}$$

A typical solution is shown in Figure 11.13a. If one uses a model ignoring the edge effects, one is essentially using the velocity profile shown in Figure 11.13b.

When diffusion is added into the problem, these two flow situations (a and b) are different. Take the 3D flow situation shown in Figure 11.14, with the velocity profile being the same for all x. Let the material on one side have a dilute concentration of some species

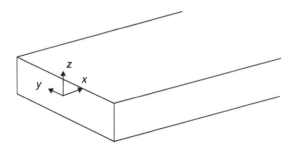

FIGURE 11.14 Three-dimensional rectangular channel.

(taken $= 1$), and on the other side there is no concentration of that species ($c = 0$). As the material flows down the channel, diffusion perpendicular to flow occurs, so we must characterize the mixing.

Characterization of Mixing

To evaluate mixing, it is necessary to have a quantitative measure. The variance from the mean is appropriate, but in a flowing system some fluid elements have a higher velocity than others. The concept in chemical engineering that is used to account for this is the mixing cup concentration. This is the concentration of fluid if the flow emptied into a cup that was well stirred. In mathematical terms, the mixing cup concentration is

$$c_{\text{mixing cup}} = \frac{\int\limits_A c(x, y, z)v(x, y)\mathrm{d}x\mathrm{d}y}{\int\limits_A v(x, y)\mathrm{d}x\mathrm{d}y} \tag{11.26}$$

Note that the average velocity times the mixing cup concentration gives the total mole/mass flux through an area at position z. Then it is natural to define the variance of the concentration from the mixing cup concentration:

$$c_{\text{variance}} = \frac{\int\limits_A [c(x, y, z) - c_{\text{mixing cup}}]^2 v(x, y)\mathrm{d}x\mathrm{d}y}{\int\limits_A v(x, y)\mathrm{d}x\mathrm{d}y} \tag{11.27}$$

Average Concentration along an Optical Path

Optical measurements are frequently made through the thin layer of a microfluidic device. The average concentration is then determined by the integral along the path, such as

$$c_{\text{optical}} = \int\limits_0^L c(x.y, z)\mathrm{d}y / \int\limits_0^L \mathrm{d}y \tag{11.28}$$

When the velocity is variable in the y-direction some regions of the fluid are moving faster than others. Thus, the optical measurement, or integration along the optical path, may not be comparable to the mixing cup concentration along the same path. Indeed, it is entirely possible that several different flow and concentration distributions can give the same average along an optical path. In that case, computational fluid dynamics can be used to interpret the optical measurements.

Peclet Number

The dimensionless convective diffusion equation is Eq. (11.9), where the Peclet number is a ratio of the time for diffusion, x_s^2/D, to the time for convection, x_s/v_s. Typically, *Pe* is very large, and this presents numerical problems. The second form of Eq. (11.9) shows that the coefficient of the term in the differential equation with highest derivative goes to zero; this makes the problem singular. For large but finite Pe the problem is still difficult to solve. For a simple one-dimensional (1D) problem it can be shown (Finlayson, 1992) that the solution will oscillate from node to node unrealistically unless

$$\frac{Pe\Delta x'}{2} \leq 1, \quad \text{or} \quad \frac{v_s x_s}{2D}\frac{\Delta x}{x_s} \leq 1 \tag{11.29}$$

This means that as Pe increases, the mesh size must decrease. Since the mesh size decreases, it takes more elements or grid points to solve the problem, and the problem may become too big. One way to avoid this is to introduce some numerical diffusion, which essentially lowers the Peclet number. If this extra diffusion is introduced in the flow direction only, the solution may still be acceptable. Various techniques include upstream weighting (finite difference; Peyret and Taylor, 1983) and Petrov–Galerkin finite element; Hughes and Brooks, 1979). Comsol Multiphysics includes a variety of such techniques, called stabilization, and you can tweak the parameters. Basically, if a numerical solution shows unphysical oscillations, either the mesh must be refined, or some extra diffusion must be added. Since it is the *relative* convection and diffusion that matter, the Peclet number should always be calculated even if the problem is solved in dimensional units. The value of Pe will alert the chemist, chemical engineer, or bioengineer whether this difficulty would arise or not. Typically, v_s is an average velocity, x_s is a diameter or height, and the exact choice must be identified for each case.

Sometimes an approximation is used to neglect axial diffusion since it is so small compared with axial convection. If the flow is fully developed in a channel then one solves (for steady problems)

$$w(x, y)\frac{\partial c}{\partial z} = D\left(\frac{\partial^2 c}{\partial x^2} + \frac{\partial^2 c}{\partial y^2}\right) \tag{11.30}$$

This is a much simpler problem since the D can be absorbed into the length, z,

$$w(x, y)\frac{\partial c}{\partial(zD)} = \left(\frac{\partial^2 c}{\partial x^2} + \frac{\partial^2 c}{\partial y^2}\right) \tag{11.31}$$

In nondimensional terms the equation is

$$w'(x', y')\frac{\partial c'}{\partial(z'/Pe)} = \left(\frac{\partial^2 c'}{\partial x'^2} + \frac{\partial^2 c'}{\partial y'^2}\right) \tag{11.32}$$

Then all results should depend upon z'/Pe not z' and Pe individually. The velocity is not a constant; it depends upon x and y. Thus, the problem is different from the following formulation:

$$<w'>\frac{\partial c'}{\partial(z'/Pe)} = \left(\frac{\partial^2 c'}{\partial x'^2} + \frac{\partial^2 c'}{\partial y'^2}\right) \text{ or } \frac{\partial c'}{\partial t} = \left(\frac{\partial^2 c'}{\partial x'^2} + \frac{\partial^2 c'}{\partial y'^2}\right), \text{ where } t' = \frac{z'}{<w'> Pe} \tag{11.33}$$

where $<w'> = \int_A w'(x', y')dx'dy' / \int_A dx'dy'$.

These phenomena are explored in the problems. Figures 11.4–11.6 illustrate the concentration profiles that can occur.

EXAMPLE: CONVECTION AND DIFFUSION IN A THREE-DIMENSIONAL T-SENSOR

Next, we investigate a 3D T-sensor.

Step 1 (Open and Set Preferences) Prepare a problem for flow in a 3D T-sensor that is extruded from the 2D version by 0.5 in the vertical direction.

Step 2 (Define the Problem) Right click on "Model 1" and choose"Add Physics." Select "Transport in Diluted Species" and then Stationary, and finish. In that model, set the diffusion coefficient to $1/Pe$ and insert the velocity components u, v, and w. Change the Discretization to Quadratic.

Step 3 (Set the Boundary Conditions) For the two inlet boundaries, make one have concentration zero and the other one. Select the outflow boundary.

Step 4 (Solve the Problem) A "Study 3" will be generated, and deselect the flow and Poisson's equation so that only the convective diffusion equation is solved. 165,692 degrees of freedom are used.

Step 5 (Examine the Solution) For $Pe = 50$, the integral of c is 0.5006, the integral u is 1.937, the integral of c^*u is 0.9686, the mixing cup concentration, integral of $c^*u/1.937$, is 0.50008, and the variance is 0.02008. The limits on the plot of concentration are 0–1, so the solution has no visible oscillations; isoconcentration surfaces are shown in Figure 11.15.

Step 6 (Examine Another Solution) For $Pe = 500$, the integral of c is 0.5012, the integral u is 1.937, the integral of c^*u is 0.9684, the mixing cup concentration, integral of $c^*u/1.937$, is 0.500004, and the variance is 0.1594, indicating that mixing is minimal. The limits on the plot of concentration are -0.013 to 1.0025, so the solution has small oscillations; isoconcentration surfaces are shown in Figure 11.16.

Step 7 (Examine the Solution on a Cut-Plane) Under Results/Data Sets, right click and choose 2D Cut Plane. Define the plane as the yz-plane at $x = 0.51$. The value 0.51 was used so that it was just inside the turn; this avoids involving all the side boundaries of the inlet ducts. The similar numbers for $Pe = 500$ were: the integral of c is 0.500004, the integral u is 1.943, the integral of c^*u is 0.9716, the mixing cup concentration, integral of $c^*u/1.943$, is 0.5002, and the variance is 0.2207. In a 2D problem, the maximum variance is 0.25, so this indicates that little mixing has occurred up to that point.

Step 8 (Examine the Solution on a Line Probe) Sometimes it is convenient to measure the concentration by shining a light of a specific wavelength through the device and determining the absorption. The analog of this is using a line through the device. Choose Results, right click on "Cut Line 3D" and identify the coordinates of the line. For example, for the solutions shown in Figures 11.15 and 11.16, at the point $x = 2.5$ (the end) and $y = -0.075$

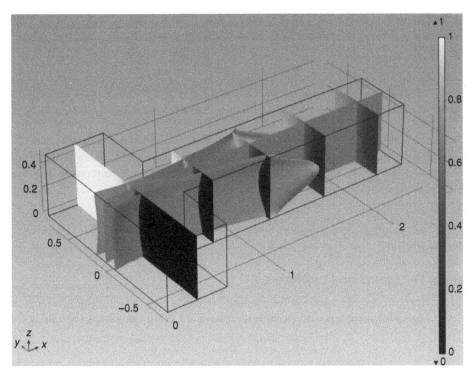

FIGURE 11.15   Concentration for $Pe = 50$.

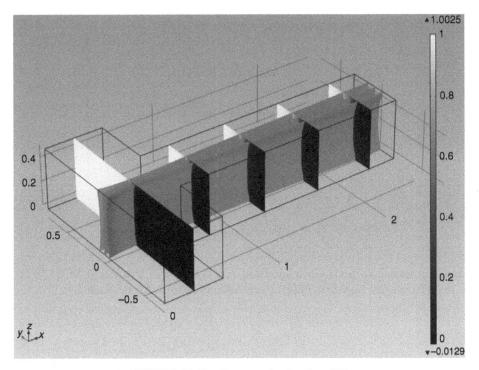

FIGURE 11.16   Concentration for $Pe = 500$.

the integration of c from $z = 0$ to 0.5 was 0.1948 for $Pe = 50$ but only 0.0571 for 500. This indicates that little of the material from the top half got through to the lower half when Pe was 500.

CHAPTER SUMMARY

This chapter illustrated the use of Comsol Multiphysics for problems of heat conduction, heat conduction and convection, and mass diffusion and convection. These problems are complicated, but they do follow conditions in real-life that engineers must tackle. Knowing how to do this will allow you to work in a variety of modern industries.

PROBLEMS

Steady, Two-Dimensional Problems

11.1₁ Heat transfer takes place in the geometry shown in Figure 11.17. Boundaries A are insulated; along boundary B the temperature is 1.0; the boundary condition at C is

$$-k\frac{\partial T}{\partial n} = hT \tag{11.34}$$

Use $k = 1$, $h = 3$ and solve for the temperature profile. (b) Use $k = 0.01$, $h = 3$ and solve for the temperature profile; compare the results.

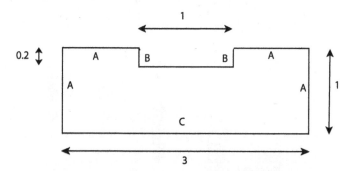

FIGURE 11.17   Heat transfer in a microchemical reactor.

11.2₁ A slotted-electrode electrochemical cell is shown in Figures 11.18–11.19 (see Orazem and Newman, 1984). The governing equation for potential is

$$\nabla^2\phi = 0, \quad \text{or} \quad \frac{\partial^2\phi}{\partial x^2} + \frac{\partial^2\phi}{\partial y^2} = 0 \tag{11.35}$$

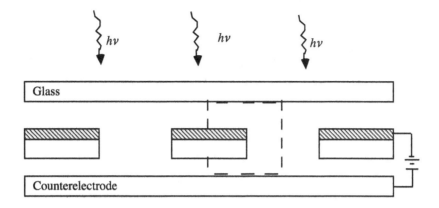

Semiconductor

FIGURE 11.18   Slotted-electrode electrochemical cell.

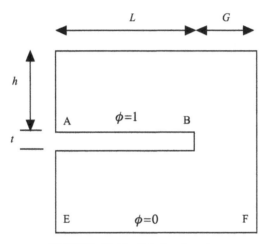

FIGURE 11.19   Solution domain.

Determine the current distribution $(\partial\phi/\partial n)$ along the faces AB and EF. The variables are $t = 0.25$, $h = 1$, $L = 2.5$, and $G = 1$.

11.3₁ Project. Solve Laplace's equation [Eq. (11.33)] in the region shown in Figure 11.20 with the following boundary conditions:

$$u = 0 \text{ along } \theta = 0, \quad \frac{\partial u}{\partial n} = \frac{\partial u}{\partial \theta} = 0 \text{ along } \theta = \alpha, \quad u = \sin\left(\frac{\pi\theta}{2\alpha}\right) \text{ along } r = 1$$

$$(11.36)$$

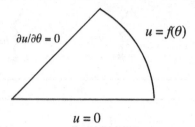

FIGURE 11.20   Solution domain.

(a) $\alpha = \pi/4$; (b) $\alpha = 3\pi/4$; (c) $\alpha = \pi$; (d) $\alpha = 3\pi/2$. Comment on the singularity that occurs at the corner, despite the smooth boundary conditions for u. The exact solution is

$$u = r^{\pi/2\alpha - 1} \sin\left(\frac{\pi\theta}{2\alpha}\right) \tag{11.37}$$

In Comsol, you need to convert the (x, y) values of a point on the boundary into θ values, using the arccos function (acos).

11.4₁ Solve Problem 9.2 in a cylindrical pellet whose length equals its diameter. Compare the average reaction rate in the cylindrical pellet with that in a spherical pellet.

11.5₁ Solve Problem 9.6 in a cylindrical pellet whose length equals its diameter. Compare the average reaction rate in the cylindrical pellet with that in a spherical pellet.

11.6₂ It is desired to measure the thermal conductivity of a ferrofluid, a suspension of magnetic particles in an oil-based solvent (Rosensweig, 1985). The fluid is contained in a thin layer between two thick copper plates, with insulation at the edges, as shown in Figure 11.21. The dimensions are radius of inner circle is 0.02 m;

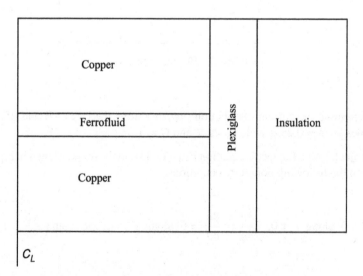

FIGURE 11.21   Thermal conductivity measurement device.

thickness of plexiglass is 0.005 m; thickness of insulation layer is 0.01 m. The height of the copper cylinders is 0.01 m, and the thickness of the internal layer containing the ferrofluid is 0.002 m. Solve for the heat transfer problem in Comsol using the material properties in the Material Browser. Use nylon for the plexiglass (it has the same thermal conductivity). Create a new material named "Insulation" and use $k = 0.09$ W/mK, which is a value for asbestos. The estimate of the thermal conductivity is $k = 0.25$ W/mK. In the experiment, an electric heater will supply heat to the top copper layer and the temperature would be measured. Here, take the top of the top copper at 30°C and the bottom of the bottom copper cylinder 25°C and determine the heat flux. Use an insulating boundary on the top and bottom of the plexiglass and insulation, and a temperature on the outside of the insulation of 25°C. Determine the accuracy of the measurement: use the heat flux and temperature drop across the ferrofluid from Comsol and the thickness to determine the measured thermal conductivity. What is the error? How can the measurement be improved?

Heat Transfer with Flow

11.1₁ Solve for the heat transfer from a sphere maintained at temperature 1 when a Newtonian fluid is flowing past it with a temperature of zero. Use constant viscosity and integrate the flux over the surface to find the rate of heat transfer. (a) $Re = 0$ (Stokes solution) and $Pe = 1$, 10, and 100. (b) repeat part (a) with $Re = 10$.

11.2₁ Solve the following Graetz problem using the convection and conduction module in Comsol Multiphysics. When there is no axial conduction, you can absorb the Peclet number into the distance, z, and solve the problem once for all Pe. Solve with a radius of 1.0 and a length of $z/Pe = 1$; use "Outlet" as the outlet boundary condition:

$$2(1 - r^2)Pe\frac{\partial T}{\partial z} = 2(1 - r^2)\frac{\partial T}{\partial(z/Pe)} = \frac{1}{r}\frac{\partial}{\partial r}\left(r\frac{\partial T}{\partial r}\right), \quad \left.\frac{\partial T}{\partial r}\right|_{r=0} = 0,$$

$$T(1, z) = 1, \quad T(r, 0) = 0 \tag{11.38}$$

Compute the Nusselt number, which is given by

$$Nu = \frac{-\left.\frac{\partial T}{\partial r}\right|_{r=1}}{<T> -1}, \quad <T> = \text{average temperature}, \quad Pe = \frac{u_{avg}R}{D} \tag{11.39}$$

11.3₁ Solve the same Graetz problem as in Problem 11.8, but add in axial conduction. Solve the equations in the form of Eq. (11.40) with a radius of 1.0, a length of 6, and $Pe = 20$:

$$2(1 - r^2)\frac{\partial T}{\partial z} = \frac{1}{Pe}\left[\frac{1}{r}\frac{\partial}{\partial r}\left(r\frac{\partial T}{\partial r}\right) + \frac{\partial^2 T}{\partial z^2}\right], \quad \left.\frac{\partial T}{\partial r}\right|_{r=0} = 0, \quad \left.\frac{\partial T}{\partial z}\right|_{z=L} = 0,$$

$$T(1, z) = 1, \quad T(r, z = 0) = 0 \tag{11.40}$$

11.4₁ The boundary conditions in Problem 11.9 are not really appropriate since there can be conduction upstream into the unheated region. Solve the problem from $z = -2$ to 6, including the upstream region. Use $Pe = 20$. The equations are

$$2(1 - r^2)\frac{\partial T}{\partial z} = \frac{1}{Pe}\left[\frac{1}{r}\frac{\partial}{\partial r}\left(r\frac{\partial T}{\partial r}\right) + \frac{\partial^2 T}{\partial z^2}\right], \quad \frac{\partial T}{\partial r}\bigg|_{r=0} = 0, \quad \frac{\partial T}{\partial z}\bigg|_{z=L} = 0,$$

$$T(r, z = -2) = 0 \tag{11.41}$$

$$T(1, z) = 0, \quad z \geq 0, \quad T(1, z) = 1, \quad z < 0 \tag{11.42}$$

Compare the answer with the answer in Problem 11.9.

11.5₁ Solve the Graetz Problem 11.9 but use a boundary condition of $T = 1$ at the outlet. Solve with $Pe = 30$ on a short domain, $L = 6$; solve on a long domain, $L = 30$. This problem illustrates the importance of having the correct boundary conditions (see Chang and Finlayson, 1980).

Reaction with Known Flow

11.12₂ Convection, diffusion, and reaction in a packed bed reactor are governed by the following equations in a cylindrical domain with length 2 and radius 0.1:

$$\frac{\partial c}{\partial t} + u\frac{\partial c}{\partial z} = \frac{D}{r}\frac{\partial}{\partial t}\left(r\frac{\partial c}{\partial r}\right) - kc$$

$$c(r, z, 0) = c_0, \quad c(r, 0, t) = c_1, \quad \frac{\partial c}{\partial r}\bigg|_{r=0} = 0, \quad -D\frac{\partial c}{\partial r}\bigg|_{r=R} = k_m\left(c|_{r=R} - c_2\right)$$

$$\tag{11.43}$$

Solve with the following parameters. You should integrate to about 0.2:

$$c_0 = 0, \ c_1 = 1, \ c_2 = 0, \ u = 0, \ w = 10, \ D_{rr} = 0.1, \ D_{zz} = 0, \ k = 10, \ k_m = 1$$

$$\tag{11.44}$$

11.13₂ Equation (11.45) shows the dimensionless equations for a nonisothermal packed bed reactor in a cylindrical domain:

$$\frac{\partial c}{\partial z} = \alpha\frac{1}{r}\frac{\partial}{\partial r}\left(r\frac{\partial c}{\partial r}\right) + \beta R(c, T), \quad \frac{\partial T}{\partial z} = \alpha'\frac{1}{r}\frac{\partial}{\partial r}\left(r\frac{\partial T}{\partial r}\right) + \beta' R(c, T)$$

$$\frac{\partial c}{\partial r}\bigg|_{r=0} = \frac{\partial T}{\partial r}\bigg|_{r=0} = 0, \quad \frac{\partial c}{\partial r}\bigg|_{r=1} = 0, \quad -\frac{\partial T}{\partial r}\bigg|_{r=1} = Bi_w[T(1, z) - T_w]$$

$$c(r, 0) = 0, \quad T(r, 0) = 1$$

$$\tag{11.45}$$

Here, c represents the conversion, not the concentration. Solve this problem using a 1D Axisymmetric option with Time Dependent. Think of z as time, t, and integrate to $t = 3$:

$$R(c, T) = (1 - c)e^{y-y/T}, \quad \alpha = \alpha' = 1, \quad \beta = 0.3, \quad \beta' = 0.2, \quad \gamma = 20, \quad T_w = 0.92$$

$$(11.46)$$

and (a) Biw $= 1$; (b) Biw $= 20$. The dimensionless variables are defined as

$$\alpha = \frac{L d_p}{R^2 Pe_m}, \quad \alpha' = \frac{L d_p}{R^2 Pe_h}, \quad Pe_m = \frac{G d_p}{\rho D_e}, \quad Pe_h = \frac{C_p}{G d_p k_e}$$

$$\beta = \frac{k_0 L \rho}{G}, \quad \beta' = \frac{(-\Delta H_{rxn}) k_0 c_0 L}{C_p G T_0}, \quad Bi_w = \frac{h_w R}{k_e}$$

$$(11.47)$$

where d_p is the particle diameter, G is the mass flux, D_e is the effective diffusivity, k_e is the effective thermal conductivity, and h_w is the heat transfer coefficient at the cylindrical wall. Comment on the radial distribution of temperature and concentration in the two cases. Based on this limited experience, give your recommendations about when the temperature and concentration will vary significantly in the radial direction.

Reaction with No Flow

11.14$_1$ Equation (11.48) governs the transient reaction and diffusion with a Michaelis–Menten reaction in a porous media. Integrate the equations to steady state using the parameters $\phi^2 = 10$, $v = 2$, $Bi_m = 20$:

$$\frac{\partial c}{\partial t} = \frac{1}{r^2} \frac{\partial}{\partial r} \left(r^2 \frac{\partial c}{\partial r} \right) - \phi^2 R(c), \quad R(c) = \frac{c}{v + c}$$

$$\left. \frac{\partial c}{\partial r} \right|_{r=0} = 0, \quad -\left. \frac{\partial c}{\partial r} \right|_{r=1} = Bi_m[c(1, t) - 1], \quad c(r, 0) = 0$$

$$(11.48)$$

11.15 Project. Equation (11.49) governs the reaction of carbon monoxide in a thin porous material with platinum as the catalyst:

$$\frac{\partial c}{\partial t} = \frac{1}{r^2} \frac{\partial}{\partial r} \left(r^2 \frac{\partial c}{\partial r} \right) - \phi^2 R(c, T), \quad \frac{\partial T}{\partial t} = \frac{1}{r^2} \frac{\partial}{\partial r} \left(r^2 \frac{\partial T}{\partial r} \right) + \beta \phi^2 R(c, T)$$

$$\left. \frac{\partial c}{\partial r} \right|_{r=0} = \left. \frac{\partial T}{\partial r} \right|_{r=0} = 0, \quad -\left. \frac{\partial c}{\partial r} \right|_{r=1} = Bi_m[c(1, t) - c_w],$$

$$-\left. \frac{\partial T}{\partial r} \right|_{r=1} = Bi_h[T(1, t) - T_w], c(r, 0) = 0, \quad T(r, 0) = 1,$$

$$R(c, T) = \frac{c}{(1 + Kc)^2} \exp\left[\gamma \left(1 - \frac{1}{T} \right) \right]$$

$$(11.49)$$

The parameters are

$$\phi^2 = 625, \ \beta = 0.02, \ K = 20, \ \gamma = 30, \ Bi_m = 200, \ Bi_h = 5, \ c_w = 1, \ T_w = 1$$
$$(11.50)$$

What is the maximum temperature rise in the catalyst? (Note: If the standard temperature is $600°$K, used in the definition of the dimensionless activation energy, then the actual temperature is 600 times the dimensionless temperature.) Where does the maximum reaction rate occur?

Solve for Concentration and Flow

11.16₁ Solve the flow and diffusion problem for the T-sensor as illustrated in the example. Then consider a similar flow problem with water coming in both inlets. However, one inlet contains one chemical, identified as A, and the other inlet contains another chemical, identified as B. The two react with the reaction rate

$$A + B \rightarrow C, \ rate = -kc_Ac_B \qquad (11.51)$$

To solve this problem, you need to use the Navier–Stokes equation and two convection and diffusion equations. Use the same parameters as in the example, plus $k = 3$, c_A in $= 1$, c_B in $= 1$. Compare solutions with $Pe = 1$, 10, and 100.

11.17₂ Solve for the diffusion in a 2D H-sensor. Take the T-sensor geometry and add on a section that mimics the input section. Determine the fraction of material going out the top outlet when it all comes in the top inlet; also compute the variance for $Pe = 50, 100, 200, 500, 1000$.

11.18₁ Reproduce Figures 11.4–11.6. Also compute the variance of the concentration coming out for $Pe = 50$ and 100.

11.19₁ Solve the problem in Example: Viscous dissipation.

11.20₂ Consider flow into a 3D channel that is one unit high, eight units across, and ten units long for a Reynolds number of 1.0. Solve for the flow coming in the 1×8 cross section. Use "P1 + P1" discretization for flow. First, use a uniform velocity in, to check for any errors in your representation. Next, apply linear coupling (see Chapter 10) to have a fully developed flow coming in. Then solve the convective diffusion equation with the concentration $= 1$ on the right-hand side of the inlet and zero on the left-hand side of the inlet. Use linear discretization for concentration. Compute the variance at the outlet for $Pe = 100, 200, 500, 1000$. Discuss the accuracy of your calculations and indicate steps that could be taken to improve the accuracy. Also solve the problem when the velocity profile is the fully developed velocity in a channel, which is equivalent to having slip on the side walls. To achieve this, just take the velocity in the x direction as $6^*z^*(1-z)$. Although the problem with slip could be solved in two dimensions, solve it in the 3D geometry with the same mesh used with no-slip.

11.21₁ Solve the same problem as in 11.20 except have the concentration $= 1$ in the lower half of the inlet and 0 in the top half.

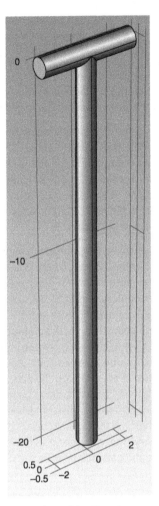

FIGURE 11.22   Pipe T-sensor.

11.22   Project. Solve for flow and diffusion in a pipe-T consisting of two pipes coming together and branching off into one (Figure 11.22). The diameter of the pipes is 1.0, the inlet lengths are 2.0 long and the length of the single pipe is 20.0. Compute the variance for a Peclet number of 500 and use the cut-plane option to calculate variances at several values of z'.

11.23   Project. Solve for the flow and diffusion in a 3D H-sensor where the channel has a square cross section. Take the T-sensor geometry and add on a section that mimics the input section. Determine the fraction of material going out the top (or left) outlet when it all comes in the top (or left) inlet. Solve for $Re = 1$ (based on the inlet to one port), $Pe = 100, 200, 500, 1000$.

11.24$_2$ Solve for diffusion in the serpentine mixer for $Pe = 100, 300$, and 1000. Determine the variance at different lengths using cut-planes. Plot the results on one graph as a function of z'/Pe.

11.25₂ For the geometry of the T-sensor and conditions of Problem 10.16, solve the concentration problem in a 3D T-sensor when the diffusivity is 10^{-10} m²/s and 10^{-11} m²/s.

11.26₂ Solve for convection and diffusion for the problem 10.24 in two dimensions. Use a Peclet number of 1000. Construct the mesh in strips so that you can use boundary elements along the centerline to improve the accuracy.

11.27₂ Suppose one has done experiments with water flowing in straight channels measuring 300 × 120 mm. The average flow rate is 0.01 m/s. The inlet concentration takes the value 1.0 in the bottom half and 0.0 in the top half and the diffusivity is 10^{-8} m²/s. It is desired to scale up to 2.5 × 1 mm, that is, a scale-up by a factor of 8.5, keeping the average velocity at 0.01 m/s. What are the Reynolds number and Peclet number in both devices? Do you expect the solutions to be the same? Calculate the variance out of each device.

11.28 **Project.** The laminar fluid diffusion interface device introduced in Problem 10.27 is to be used when both a metabolite (taken here as creatinine) and protein (taken here as albumin) enter in the sample. Using the same specifications from Problem 10.27, but extending the dimensionless length by 10, solve for the concentration profiles of both creatinine and albumin when both enter with a dimensional concentration of 1.0. Now the viscosity depends upon the albumin concentration: $\eta = 1 + 0.6cA$ mPa s (cA = 1 is the maximum albumin concentration). Since the viscosity depends upon the albumin concentration, even in fully developed flow the velocity profile will not be quadratic (Bird et al., 2002, pp. 56–68), and the analytical solution there (when the interface is halfway between the two sides) is extended to more general cases in "Poiseuille Flow of Two Immiscible Fluids Between Flat Plates with Applications to Microfluidics" on the book website. The diffusivities of creatinine and albumin in both serum and water are taken as $9.19 \ 10^{-10}$ m²/s and $6.7 \ 10^{-11}$ m²/s, respectively. Determine the enhancement factor if the stream in the upper half at the outlet is collected.

11.29₂ One use of serpentine mixers is to create a gradient in concentration across the device by careful arrangement of successive mixers. Gomez (2008) describes devices developed by Chung et al. (2005) to generate gradients for studying proliferation and differentiation of human neural stem cells. They can also be used to differentiate stem cells (Gomez, 2008). Consider the device shown in Figure 11.23; the geometry is available on the book website. Model it for flow and concentration in two dimensions when $Re = 1$. The bottom inlet contains water and BSA (bovine serum

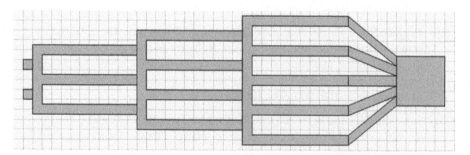

FIGURE 11.23   Device to generate concentration gradients.

albumin) with a dimensionless concentration of 1; the top one contains water and Laminin, with dimensionless concentration of 1.0. Although both of these have similar diffusivities (6×10^{-11} m^2/s), use a Peclet number of 1000 for the Laminin in the top inlet and 100 for the BSA in the bottom inlet to demonstrate the effect of the Peclet number. As the two streams enter the cross-flow channel, the center outflow has a concentration of both BSA and Laminin. This is repeated further downstream. Determine the concentration gradient after the streams have divided into three, then four, then five streams and recombined. The long segments are 10 units long and the width of the channel is always 1 unit long. The mesh can be obtained from the book website.

Numerical Problems

11.30$_1$ Solve the problem of heat transfer shown in Figure 11.2 using the finite difference method in Excel. Use $\Delta x = \Delta y = 0.1$ with $Q = 0$ in Eq. (E.53). The points go from 1 to $n + 1$ in x and from 1 to $m + 1$ in y. For the boundary conditions on temperature, make the cell value be the boundary condition. For the boundary condition of zero flux, in the y direction, set

$$T_{n+2,j} = T_{n,j} \tag{11.52}$$

Then the equation for a point on that boundary is

$$T_{n+1,j} = \tfrac{1}{4}\left(T_{n+1,j+1} + T_{n+1,j-1} + 2T_{n,j}\right) \tag{11.53}$$

Plot the temperature, too.

11.31$_2$ Solvethe problem of heat transfer shown in Figure 11.3 using the finite difference method in Excel. Use $\Delta x = \Delta y = 0.1$ with $Q = 0$ in Eq. (E.53). In other words, add the hole. For the points along the internal boundary, $x = 0.3$, the equation will be

D5 = 0.25*(D4+D6+2*C5) for points 5–7 in the y-direction
H5 = 0.25*(H4+H6+2*I5) for points 5–7 in the y-direction
E4 = 0.25*(D4+F24+2*E3) for points 5–7 in the x-direction
E8 = 0.25*(D8+F28+2*E9) for points 5–7 in the x-direction
D4 = 0.25*(D3+C4+D5+E4), similarly for each of the corners

These equations are derived by noting that for zero flux, the points on either side of the boundary have the same value (with a false boundary). Plot the temperature, too.

11.32 Project. Solve for flow between flat plates with an inlet concentration of 1.0 in one half and 0.0 in the other half. Use the finite difference method for both a flat velocity profile and a quadratic velocity profile. Plot the variances versus z'/Pe; compare with the approximation solution (see book website)

$$\sigma^2 = \begin{cases} 0.25\left(1 - 1.476\sqrt{2z'/Pe}\right), & z'/Pe \le 0.05 \\ 0.220\exp(-10z'/Pe), & z'/Pe > 0.05 \end{cases} \tag{11.54}$$

APPENDIX A

HINTS WHEN USING EXCEL®

This appendix provides hints when using Microsoft Excel.[1] Appendix A includes general features that are useful in all the applications solved with Excel in this book. Other features are illustrated in the context of specific examples, and a list of these is provided at the end of this appendix for handy reference. This appendix assumes you have some experience with Excel; special features useful for calculation are described. For a more detailed description of Excel, consult other books devoted to Excel or the book website.

INTRODUCTION

When Excel opens, there are several tabs across the top, as listed in Table A.1. It is easy to find out what each one does: simply put your cursor above one of the commands and an explanation appears, along with a link to more information. You can also use the question mark in the upper-right corner. Notice particularly the File/Options near the bottom. Click on it and then click on Add-Ins. You want to be sure that the following applications are Active Applications: Analysis Toolpak, Aspen Plus V7.3 Excel Calculator, Aspen Simulation Workbook V7.3, and Solver Add-In. If they are not active, select them and choose Go. If they do not appear, use your original DVD

Each tab has a number of groups. The home tab has the following groups: clipboard, font, alignment, number, and so on. To copy cells down, click on a cell and choose Fill down. The contents of the cell above are copied down. If the cell above has a formula, the formula is also copied down, incrementing the row number by one. You can also select the lower right corner of a cell and copy down, too. Suppose the original formula used B2. If

[1]Excel is a registered trademark of Microsoft Corporation, Inc.

Introduction to Chemical Engineering Computing, Updated Second Edition. Bruce A. Finlayson.
© 2014 John Wiley & Sons, Inc. Published 2014 by John Wiley & Sons, Inc.

TABLE A.1   Tabs in Excel

File	Home	Insert	Page Layout	Formulas	Data Review	View	Aspen

the formula is dragged to the right, B2 is changed to C2. If the formula is dragged down, the B2 is changed to B3. If you do not want that to happen, that is, you want the formula in all columns to use B2, then use absolute addresses. In the formula, change B2 to $B$2. Options such as $B2 and B$2 are also possible. If you select a cell tag in the equation and press F4, it turns into an absolute address. On the Macintosh, selecting the cell tag and pressing command-T (apple-T) does the same thing; if you continue to press it, the tag cycles between B2, $B$2, B$2, $B2, B2. To insert a new row, select a row on the left-hand side and choose Insert/Row. A new row is placed above the one you have selected. The cells are renumbered, and all the formulas are changed, too, to reflect the new numbering system. Use this same method to insert a column; the new column will appear to the left of the one you have selected.

CALCULATION

When doing calculations, it is convenient to name the cells so that they can be referred to in a way that means something to you, for example, T rather than B2. To name a cell, select the cell and enter the name of the variable in the upper left-hand corner of the spreadsheet and then press Return or Enter. Then you can use T in your formulas rather than B2.

In the Formulas Tab, in *Function Library Group*, you can find a number of functions in Math&Trig. These include cos, cosh, acosh, ln, log10, and so on. "More Functions" has engineering functions such as Bessel functions. "Show Functions" (to the right of "Trace Precedents") display all formulas rather than the calculated results. *Calculation Options* allow you to choose Automatic (the spreadsheet recalculates every time you change one cell) or Manual (the spreadsheet will wait till you choose Calculate Now (for the whole workbook) or Calculate Sheet (for only the sheet showing). This is sometimes useful when solving a problem in which the formulas are intertwined and the formula in one cell depends on the answer in another cell, which depends on the value in the first cell. In these cases, it is convenient to turn off iteration while preparing the formulas in the spreadsheet, and then turn it on once the entire spreadsheet is complete.

You can use Goal Seek to find the value of a variable to make an equation zero (or some other number. For example, suppose we wish to solve $f(x) = x*x - 2*x = 0$. The formula will be put in one cell, and another cell will be identified as x and so named. Open What-If Analysis in the *Data Tools Group,* Goal Seek and set the formula cell to zero by changing the x cell. The answer may not be as good as you like, but Preferences can be used to reduce the error criterion.

If Solver has been supplied as an Add-In (see above), the word Solver is seen in the *Analysis Group*. This program is a sophisticated program to solve nonlinear equations. This operates in the same way as Goal Seek but it is much more robust and gives smaller errors. You have a choice of methods; usually GRG Nonlinear will work, but sometimes it is useful to use Evolutionary first. If the answer is not accurate enough, you can change the Constraint Precision to a value you think is reasonable for your problem. (The default parameters are set for business applications, where tenths of a cent do not matter.) When

TABLE A.2   Matrix Multiplication in Excel

	A	B	C
1	10	5	2
2	1	11	4
3	3	7	12
4			
5			
6	0.105263	-0.046559	-0.002024
7	0.000000	0.115385	-0.038462
8	-0.026316	-0.055668	0.106275
9			
10			
11	1	0	0
12	-1.38E-17	1	0
13	0	0	1

the solver window is open, choose Options, and click All Methods. You can also limit the number of iterations. Examples are given in Chapters 2–5.

If the spreadsheet iterates to meaningless numbers, you can reset it by using CTRL-Z or using the undo typing key.

You can solve multiple equations by using Solver, too. For example, if the variables x and y are placed in cells A1 and A2, respectively, you can put equations into cells B1 and B2 that use x and y. Then choose Solver and set B1 to zero by varying A1 through A2, subject to the constraint that B2 = 0. Convergence may be difficult and this might not work.

It is possible to use Excel with Aspen. The Aspen Tab is an add-in (see File/Options/Add-Ins); using it allows you to use Excel to calculate certain things in your Aspen model. This procedure is beyond the scope of this book, but more information is available in Chapter 5 of the Aspen Plus User Model.

Two of the functions under Math&Trig are MINVERSE and MMULT. These functions have the obvious use. The inverse function, of course, must act on a square matrix, and creates a square matrix. Consider a 3 × 3 matrix. Create the matrix, say in cells A1:C3 (see Table A.2). Then in cell A6 type =MINVERSE(A1:C3). You can also insert the A1:C3 cell by selecting the cells. Click on cell A6, press the shift key, and select the other corner of the matrix, C8. Press Function Key F2, then CTRL-Shift-Enter (CTRL-Shift-Return on the Macintosh). The inverse appears in A6:C8 cell. You can check by multiplying these two matrices together; the result should be the identity matrix. Click on cell A11, type =MMULT(A1:C3,A6:C8). Click on cell A11, press the shift key, and select the other corner of the matrix, C13. Press F2, the CTRL-Shift-Enter. The matrix multiplication appears in A11:C13. Indeed, it is the identity matrix.

PLOTTING

Plotting in Excel is easy. Choose the cells you wish to plot, such as the first two columns. Then choose the *Insert Tab*, choose *XY* Scatter, and click on the graph style you want. To label the axes, select the figure, choose Layout, Axis Titles. Click on the title you wish to change and type the new values. To put multiple lines on the graph, select more columns.

IMPORT AND EXPORT

There are several ways to do this. The first method works when you have a single column to transfer. To export columns from Excel, select the column of cells, choose the Home Tab and then Copy, open an application that accepts text files, and choose Paste. To get a single column of numbers into Excel, copy the column in another application (one with Tabs), in Excel select one cell, and choose Paste. To get several columns (with tabs or spaces) into Excel, use the Import function.

You can export the contents of the spreadsheet to a text file that can be used as the input to another program. This is a nice way to have a user interface for a program that requires significant computing capability (beyond what Excel offers). You could put the physical parameters for the case you are studying into the Excel spreadsheet, allowing them to be changed easily. You then export the information and have the other computer program read the text file to get the data. To export the information (the contents of the cells, not the figures from the spreadsheet), choose Save As, change the format to a text formats, such as Text (Tab delimited), and save it. If you open this file in a word processor or text editor, you will see the information displayed in exactly the same format as in the spreadsheet. It is not trivial to get these numbers into other programs, which have to read the spaces appropriately, and some knowledge of the C computer language may be necessary, but it does make a nice interface for a program that you will use many times. For example MATLAB® can read such files.

PRESENTATION

The insert tab allows you to put figures and charts into your spreadsheet. One important option is Text Box in the *Text Group*. You can draw a box on your spreadsheet (and resize it) and type into it whatever you want. It can be anywhere on the spreadsheet, covering multiple cells. In this tab, in the *Sheet Options Group*, you can choose to show the gridlines (or not) and the headings (1, 2, 3, ..., A, B, C, ...) (or not).

In the Review Tab you have a chance to check your spelling, translate to a new language, and so on. You can protect your sheet or workbook so that others cannot open it without a password, and you can share the workbook with others on a network so that several people can use it. The New Comment command allows you to fill a text box with some information that will be hidden until you click on it. A small red arrow is created in the upper right location of the cell to indicate there is a comment there. Put your cursor above the arrow and the comment is displayed.

In the View Tab, the *Workbook Views Group* provides a variety of ways that the sheet can be displayed. In the *Show Group*, Gridlines works the same as in the Page Layout Tab: if it is checked, lines are shown dividing the cells, and they are not displayed if it is unchecked. In the *Zoom Group*, Zoom allows you to zoom in or out; Zoom to Selection allows you to choose a section of the sheet and expand it to the entire screen. Choose 100% to return to the original view.

In the *Window Group*, there are provisions for breaking the sheet into separate parts. You may wish to display one column while you scroll through other columns. Put that column (e.g., column A) at the leftmost column shown, and choose Freeze Panes. Then as you scroll to see columns B → Z → AA → AX, and so on, column A continues to be displayed. This is helpful for a case in which the column represents the chemical names and properties and each of the other columns refers to one stream. There might be hundreds of streams and you would like the first column identifying rows to remain visible at all times.

To make a professional-looking spreadsheet, it is important to provide documentation for the reader. You can establish different sections, surrounded by boxes (see Figures 2.1).

Title, your name and date

Data Input—these are values that the user may change

Parameter Values—these are values that the user will not change, but are needed

Formulas used—this defines the problem

Results

To make the boxes, select the cells to be enclosed, in the Home Tab, Number Group, choose the arrow at the bottom right, choose Border, and then Outline (or some other format.

The Split command allows you to divide the Sheet into two vertical panels or two horizontal panels, or four, 2×2. Simply put your cursor in the upper left-hand cell of the lower right-hand panel and choose Split. To undo this, choose Split again.

You can insert pictures, equations, and hyperlinks (to websites), too. Be sure to try the hyperlink while connected to the internet to ensure that it works.

Sometimes you want to reproduce part of your spreadsheet in another spreadsheet or in another area of your existing one. You can select the area of the spreadsheet, choose Copy, move to the upper, left-most cell of the new area and choose Paste. Be careful because this will copy all formulas, too. If you want only the values from a spreadsheet, copy the part you want and choose Paste Special; then choose the desired option (such as Value). You can also paste into Word. If you simply use Paste, you can manipulate the spreadsheet inside Word (resize it, change the font, etc.). If you use Paste Special, one option is to paste the spreadsheet as a picture. In that case the whole image is frozen before it is put into Word.

APPENDIX B

HINTS WHEN USING MATLAB®

This appendix provides hints and tips when using MATLAB®.[1] It assumes that you are a beginner in using MATLAB, but not an absolute beginner in computer programming. Most likely, you remember concepts from a computer programming class taken earlier. Included in Appendix B are general features that are useful in all the applications solved with MATLAB. Other features are illustrated in the context of specific examples; a list of examples is provided at the end of the appendix for handy reference. You will probably want to skim this appendix first, then start working some of the problems that use MATLAB to gather experience, and finally come back and review this appendix in more detail. That way you will not be burdened with details that do not make sense to you before you see where and how you need them. A book with more detail is by Gilat (2010), and MATLAB itself has excellent help menus that are referred to in this Appendix.

Outline of Appendix B:

1. General features
2. Programming options: input/output, loops, conditional statements, timing, matrices
3. Finding and fixing errors
4. Eigenvalues of a matrix
5. Evaluation of an integral
6. Interpolation: splines and polynomials
7. Solve algebraic equations
8. Integrate ordinary differential equations that are initial value problems
9. Plotting

[1] MATLAB is a registered trademark of The Math Works, Inc. The version discussed here is R2012a.

Introduction to Chemical Engineering Computing, Updated Second Edition. Bruce A. Finlayson.
© 2014 John Wiley & Sons, Inc. Published 2014 by John Wiley & Sons, Inc.

10. Other applications
11. Import/export data
12. Programming graphical user interfaces (GUIs)
13. MATLAB help
14. Applications of MATLAB

GENERAL FEATURES

Screen Format

To start, open MATLAB. The screen will appear as in Figure B.1. On the upper line, note the question mark that leads to excellent help information. To the right of that is the local folder that MATLAB is pointing to. If the picture were larger, it would have the words "Current Folder" in front of it. When you ask MATLAB to find a file, it looks in that folder first. To move to one of your folders, click on the . . . and browse to the desired folder. The large areas on the screen are as follows. On the left is a list of the files that are in your current folder, and below it would be information about a file you selected from the list. In the middle is the Command Window, which is where you will issue commands interactively. To the right is the Workspace that lists variables you have defined and their values, and below it is the Command History, which records your commands in sequence. Many of these items you can drag from the Workspace or Command History into the Command Window to enter their values into the command workspace or to execute a command again. For a quick illustration of using MATLAB, look at the "Getting Started with MATLAB" video. Click on the "New to MATLAB?, Watch this Video" link at the top of the command window. Alternatively, choose the question mark, then open MATLAB, and double click on

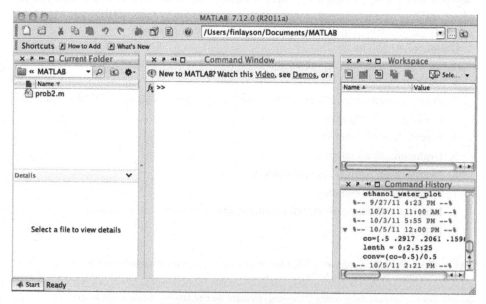

FIGURE B.1 MATLAB screen.

Demo. The video is the first one, and it is either on your computer (if the entire MATLAB was loaded originally) or on the internet.

Look under the Desktop menu; you can choose which windows you want open. If you made a mistake, choose Desktop Layout/default to get back to Figure B.1.

Sometimes you have files in more than one folder, and you need to access them even if that is not your "current folder." Use the path command to point to an area where you store your programs and data. First enter path in the command window to see the list of MATLAB folders that will be searched. Then enter the command (modified to identify the new folder, perhaps with a C:\:

```
path(path,'/Users/yourname/foldername').
```

When MATLAB looks for an m-file (a program you write), it will look in all these directories, from top to bottom until it finds one with the same name. Then, it will use that m-file.

Stop/Closing the Program

When you stop work, save your m-files in a folder, so you can use them next time. You can also save the workspace, which will save all the numerical values that have been calculated. This is convenient if you have to interrupt your work for a short time and do not want to start over. Of course, most of your work is contained in your m-files, which you always want to save. If a simulation continues for a long time and you want to halt it, press Control+Break or Control+C on the PC or Control+C on the Macintosh. This will stop the program as a forced quit.

m-files and Scripts

In MATLAB you write computer programs that are called MATLAB function files (or m-files) as well as scripts, which run in the command window. The m-files are saved on your computer and can be used by simply typing their name in the command line. In addition one m-file can call another m-file. Scripts are a list of commands that can also be saved; if they are pasted into the command window they will be executed. You will want to include comments in your m-files and scripts that explain what the file does. This is done by inserting a %; everything after % on a line is considered a comment. These comments remind you of details when you come back to the program at a later time. You can execute an m-file that is listed in your current folder by double clicking it; then edit it if desired, save it, and finally run it by pressing the green arrow. MATLAB has many functions of its own; you can see a list of them by pressing the function button (*fx*). Then you can search either by category or by key words. To see how to use one of them, issue the command: help functionname. When you write your own m-files, the line may extend too far. By putting . . . at the end, the command continues in the next line. Simple MATLAB programs are illustrated in Chapters 2–9. As you become more experienced using MATLAB you should look at the Videos "Working in the Development Environment" and "Writing a MATLAB Program." Access these via the Help question mark, MATLAB, Demos

Workspaces and Transfer of Information

When you use MATLAB, your results will only be as good as the programs you write. Thus, it is important to understand the structure of workspaces and m-files, as well as how to display intermediate results, transfer parameters from one m-file to another, and debug your code. Most of the problems encountered by beginning users involve errors in these areas, and only careful checking will help you find them.

The command workspace is an area of computer memory that is reserved for parameters you define from the command line in MATLAB. Each m-file has its own unique workspace, and this workspace does not communicate with the command workspace unless you make it happen. It is sort of like Facebook: you and your friends have to agree to be friends to see what the others are doing (transfer information). There are two ways to set up the communication: using the global command or calling arguments of the m-files.

"Global" Command

Figure B.2 shows a command window and two m-files. The global commands in each one determine which variables are available in the m-files. Assume you have typed the command.

```
global a b c
```

in the command line. Then give a value to a, b, and c.

$$a = 2$$
$$b = 3$$
$$c = 4$$

The question is then: how can you get these values of a, b, and c into an m-file? You do this by putting a global command in the m-file; the command lists the variables that you want to be available. In Figure B.2, the parameters $a = 2$ and $b = 3$ are available in "m-file one" and $a = 2$ and $c = 4$ are available in "m-file two." If a is changed in "m-file one," it

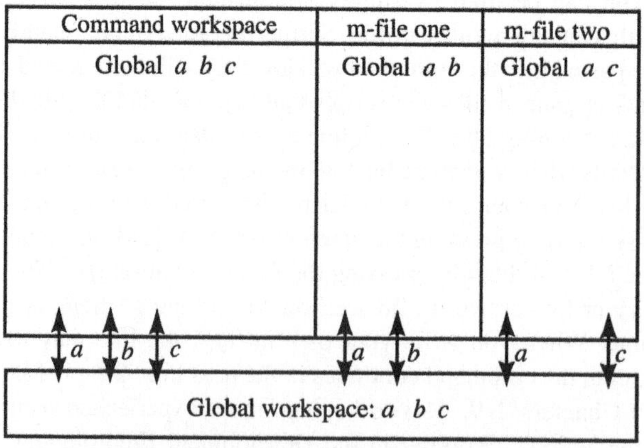

FIGURE B.2   Global and local variables.

is changed for all the programs. In other words, the change is made in the central, global location and thus, the exchange goes two ways. If the variable c is used in "m-file one," it must be defined there since it is a local variable in "m-file one." Whatever its value, it does not change the global value of c because c is not listed in the global command of "m-file one."

The second way to pass parameters is to use "list the parameters" in the function statement of the m-file. For example, the m-file

```
function y=name(x,parameters)
```

expects values of x and *parameters* to be supplied when the m-file is called. The variable *parameters* can be a vector, that is, containing several elements.

Display Tools

When you are checking your programming, you will want to examine the value of various parameters. The ";" represses the display of the result of the calculation. Remove the ";", and you will see the result of each line of your code in the Command Window as it is executed. It is essential that you look at this to verify your code.

You can also cause a variable to be displayed using one of the following three commands:

disp(x) gives:	$x = 2.34$
x gives:	$x = 2.34$
disp("now is the time") gives:	now is the time

You can issue these commands from the command window to see the value of any variable in the command workspace. If you want to know the value of a local variable in an m-file you must put one of these commands in the m-file and execute the m-file. Also look at the Workspace. The variables are listed there along with their values. This is another way to check your results.

If you need more significant digits, try

```
format long: 1.234567890e + 02
format long e: 1.234567890e + 02
format short: 123.4568
format short e: 1.2346e+02
```

To eliminate all variables

```
clear all
```

This is helpful when debugging because it eliminates variables that may have been set incorrectly in previous calculations. MATLAB then starts from scratch. MATLAB may be saving a number that you do not realize, which makes it hard to find your mistake.

Classes of Data

Most of your work will involve numbers, but sometimes you will want to use text or words. A number is usually a double precision number, using 64 bits. Its class is double, which can be seen by issuing the command: whos. A character string is of class char. A number can be converted to a character using the num2str command:

»a = 1.234		
»b = num2str(a)		
»whos		
Name	Bytes	Class
a	8	double
b	10	char

The reverse command is str2num(b). There are other classes as well; see the help command and search on class.

PROGRAMMING OPTIONS: INPUT/OUTPUT, LOOPS, CONDITIONAL STATEMENTS, TIMING, AND MATRICES

Input/Output

You can ask the user for input with the following command.

```
viscosity = input('What is the viscosity (Pa s)?')
```

When you see "What is the viscosity (Pa s)?," type the value and press return. You can display in a specified format.

```
fprintf('%5.3f %10.5f\n',x,y)
2.345 234.56789
```

These are C-commands, and the notation means

```
f - floating point
\n is a carriage return
e - exponential format
5.3 means five characters, three after the decimal point.
```

To write a text file, you must open it, write it, and close it. The following command writes a text file that has two columns, x and y, with as many entries as defined by the vectors x and y (which have the same length).

```
fid = fopen('ChEData','w') % w for write
fprintf(fid,'%5.3f %10.5f\n',[x,y])
fclose(fid)
```

To read a text file, you must open it, read it, and close it.

```
fid = fopen('ChEData','r') % r for read
[x y] = fscanf (fid,'%5.3f %10.5f')
fclose(fid)
```

Loops

It is sometimes useful to execute a command over and over, putting each result in one element of a vector. The format is

```
for i=1:10
   y(i) = ...
end
```

If you want to stop before the end

```
for i=1:10
 y(i) = ...
 if y(i) > 25 break,end
end
```

Conditional Statements

Sometimes the program needs to execute different instructions depending on a calculated result. Here are some examples.

```
if (condition1)  % examples of condition1: i==3, i<3
                              % (note double equal sign)
  ...
else
  ...
end
```

Another option:

```
if (condition1)
  ...
elseif (condition2)
  ...
else
  ...
end
```

Still another option:

```
switch lower(num)
case {1} % execute when num = 1
  y = ...
case {2, 3} % execute when num = 2 or 3
  y = ...
otherwise % execute when num is neither 1, 2 or 3
  y = ...
  display ('there is an error')  % this is optional if num
                                      should be 1, 2, or 3

end
```

Timing Information

Turn on clock at one place in the program—tic. Turn off clock at another place in the program—toc—and the elapsed time is displayed—elapsed_time = 2.34.

Matrices

Vectors are row (1×3)

$$x = [2 \quad 3 \quad 4]$$

or column (3×1)

```
x = [ 2
      3
      4]
```

The transpose of x is x'.
Set matrix values in different ways:

```
A = [ 1 2 3 4 carriage return
      5 6 7 8]
A = [ 1 2 3 4; 5 6 7 8]
```

size(A) gives dimensions of A

```
size(A) = 2 4
```

Operate on all elements at once. If $T(:)$ is a vector of T (K) you might want it in °C:

$$t = T - 273$$

MATLAB will do this for all elements in $T(:)$ and make a new vector $t(:)$. Operate on one element of a matrix

$t(i) = T(i) - 273$ % Note : MATLAB is case sensitive so that $t(i)$ and $T(i)$ are different.

Obtain rows from matrix

$$a = A(1, :) \quad \text{or} \quad a = [1 \, 2 \, 3 \, 4]$$

You can use a or A(1,:) in calculations or plotting.

Matrix Multiplication

Take the matrix $A = \begin{vmatrix} 2 & 4 & 6 \\ 8 & 10 & 12 \\ 14 & 16 & 18 \end{vmatrix}$ and the vector $x' \begin{vmatrix} 2 \\ 3 \\ 4 \end{vmatrix}$.

The 3 × 3 matrix A is formed by

$$A = [2\,4\,6;\ 8\,10\,12;\ 14\,16\,18]$$

Matrix multiplication is achieved with the asterisk.

$$\begin{vmatrix} 2 & 4 & 6 \\ 8 & 10 & 12 \\ 14 & 16 & 18 \end{vmatrix} \begin{vmatrix} 2 \\ 3 \\ 4 \end{vmatrix} = \begin{vmatrix} 40 \\ 94 \\ 148 \end{vmatrix} \tag{B.1}$$

```
A * x'
ans =  [40
        94
        148]
```

is [3 × 3][3 × 1] = [3 × 1] matrix. Note the apostrophe on x; is a column vector, which is needed for matrix multiplication. If you had used A*x, MATLAB would have told you that the inner matrix elements have to agree. Multiply terms element by element:

$$u(i) = [x(i)]^2 \tag{B.2}$$
$$u = x.*x = [4\,9\,16]$$

Note the period before the asterisk. The multiplication $\sum_{i=1}^{3} [x(i)]^2$ is made by

$$x^*x' = 29$$

This is a single number because the matrix multiplication is [1 × 3][3 × 1] = [1 × 1]

```
x' * x = [ 4 6 8
           6 9 12
           8 12 16]
```

is [3 × 1][1 × 3] = [3 × 3].
    $x * x$ is meaningless.

$$[1 × 3][1 × 3] =?$$

Element by Element Calculations

If you want to calculate something for every element in the vector, you need to precede the +, -, *, /, ^, and so on, command with a period. For example, if you have a series of x and y values:

$$x = 0:0.1:1;\ y = 1:0.1:2$$

the command

$$z = x./y$$

computes $z(i) = x(i)/y(i)$ for each i.

More Information

To learn more about matrices, view the Video "Working with Arrays" (under help, MAT-LAB, Demos). Also look at Mathematics (under help, MATLAB, Users Guide).

FINDING AND FIXING ERRORS

An important lesson you can learn about using MATLAB is how to find errors. One common error message from MATLAB is

```
???Undefined function or variable 'q'.
```

This means that q is either an undefined function or an undefined variable. If q is a function or m-file, then MATLAB cannot find it. Check the working directory or use set path and make sure that the path is set to the folder where you stored the m-file and that you have added that path to the list. Do not forget to save your pathlist so that it is there the next time you use MATLAB. If q is a variable, it has not been set or was not transferred correctly.

One source of confusion is when you think you have defined a variable, but have not really. Use disp(x) to see the current value of x in the command workspace. If the variable is supposed to be in an m-file, run the m-file with a command added to display the variable, and that will give you the value of x in the m-file workspace. That should give you a hint about possible sources of the error. The most common cause of this mistake (and one that is hard to find unless you check the code systematically) is forgetting to pass a parameter to the m-file either using "global" or as a parameter. The reason it is difficult to find these mistakes is that in your mind you have set the value of the parameter in the program somewhere but failed to transfer it to the appropriate m-file.

Another typical error message is

```
???  Index exceeds matrix dimensions
```

This message means that a matrix like A(i,j) was called with either i or j too big (i.e., beyond the definition of A), or a function uses a matrix that has not been defined yet.

The programs illustrated in this book are fairly simple, and they are easy to debug. All programmers make mistakes; good programmers learn to find them! Finding them is easy if you take the time. First consider a line in a code in the command window. Set all the parameters used in the line and execute the line. Compare the answer with a calculation done with a calculator or one you did by hand. Be sure that all the parameters you use are different, and do not use zero or one, because some mistakes will not be picked up if the variable is zero or one. The parameters you use for testing can be single-digit numbers that are easy to calculate in your head, and they can be completely unrelated to the problem you are solving. You can also copy a line from an m-file and paste it into the command window to check it, too.

If the code you wish to check is in an m-file, the first check is to see that all the parameters are passed correctly and have been set in the workspace available to the m-file. The easiest way to do this is to request that the m-file print all the variables you expect to be transferred to the m-file, at the start of the m-file, and check that they are correct. If a variable has not been set, MATLAB will give you an error message without running the code. It tells you the line and column where the undefined variable is. The problem is probably improper passing of parameters or incorrect use of the global command or, sometimes, a typing error.

MATLAB provides several commands to help you debug the program. Check the value of a variable by looking in the workspace. The command "whos" will list the variables and their type, as well as whether they are local or global variables. This is also a command that is useful inside an m-file (when checking), since it lists all the variables that the m-file can access, and if the one needed is not there you can then search for the reason.

You can make the program stop at a certain point and return to the keyboard mode. Then you can access values of the parameters. To stop the program, issue the command "dbstop at linenumber in filename." "Filename" is the name of the m-file, and "linenumber" is the line number in the m-file where you wish to stop. Alternatively, put the command "keyboard" into the m-file where you wish to stop. The command cursor changes to K» and you can issue commands. For example, typing the name of a variable will cause its value to be printed. This is a convenient way to access the value of variables in an m-file. If you then want to access variables in the program that called the m-file, say dbup. To continue computing, type "return" and press the return key. To stop the calculation altogether, type "dbquit."

EIGENVALUES OF A MATRIX

Eigenvalues are the polynomial roots to Eq. (B.3).

$$|A_{ij} - \lambda \delta_{ij}| = 0 \qquad (B.3)$$

They are easy to calculate in MATLAB using the "eig" command.

```
» lambda=eig(A)
lambda = 32.2337
        -2.2337
        -0.0000
```

EVALUATE AN INTEGRAL

To evaluate the integral, Eq. (B.4), define a set of points in the interval to be integrated, as the vector x (need not be uniformly spaced). Then evaluate the function at each of these points. Calculate the integral using the trapezoid rule.

$$\text{area} = \int_0^2 x^2 \, dx \qquad (B.4)$$

$$x = [0 : 0.1 : 2]$$
$$y = x. * x$$
$$\text{area} = \text{trapz}(x, y)$$

SPLINE INTERPOLATION

Make a cubic B-spline pass through all the data points: $[x(i), y(i), i = 1,\ldots,m]$. A cubic spline is a cubic function of position, defined on small regions between data points. It is

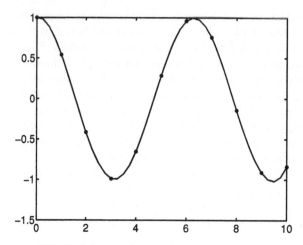

FIGURE B.3   Spline fit of a set of data points.

constructed so the function and its first and second derivatives are continuous from one region to another. It usually makes a nice smooth curve through the points. The following commands create Figure B.3.

```
x  =  0:1:10;  y  =  cos(x); % evaluate cos at a set of x values
xx = 0:.25:10;              % set up a finer mesh for plotting
yy = spline (x,y,xx);       % evaluate the spline function at the
                             points xx
plot(x,y,'*',xx,yy)         % plot the original points with*
                             and the spline curve
                           % x(i),z(i),i = 1, ... ,m
```

Interpolate Data, Evaluate the Polynomial, and Plot the Result

For data points $x(i)$, $z(i)$, $i = 1, \ldots, m$ fit a polynomial of degree n in the least squares sense.

$$y = p_1 x^n + p_2 x^{n-1} + \cdots + p_n x + p_{n+1} \tag{B.5}$$

```
p = polyfit(x,z,n)
```

The command "v = polyval(p,w)" evaluates the polynomial p at the w points (w can be a vector). Plot the data points (y) and the polynomial (v)

```
w = [0:.1:10]
v = polyval(p,w)
plot(x,y,'ro',w,v,'b-')
```

SOLVE ALGEBRAIC EQUATIONS

Using "fsolve"

Solve $f_i(\{y_j\}) = 0, i = 1, \ldots, n$ for the vector $\{y_j\}$. There are three steps.

Step 1 Set the initial guess of the solution.

$$y0 = [0 \quad 0]$$

Step 2 Name and construct the m-file "prob2.m," which evaluates the function.

Step 3 Call "fsolve," which is in the optimization toolbox in MATLAB, which all versions do not have.

```
z = fsolve(@prob2,y0)
z =   - 1.4456  - 2.4122
```

You need to make several checks. The number of unknowns is set by the number of elements in the initial guess, y0. The m-file "prob2" calculates f_i, $i = 1, \ldots, n$, and there have to be as many elements in f as there are in y. The m-file needs to be checked, of course. The only way you can make MATLAB find the solution to your problem is to make sure the m-file gives the correct set of fs when given a set of ys. If the "fsolve" does not work well, try making an initial value problem and solving it using implicit methods, integrating to a longer time.

$$\frac{dy_i}{dt} = f_i(\{y_j\}), \quad y_i(0) = \text{initial guess} \tag{B.6}$$

Solve Algebraic Equations Using "fzero" or "fminsearch" (Both in Standard MATLAB)

If you have a single function of one variable, you can use the "fzero" function in place of "fsolve," and it is called in a similar way. If you have several fs that you want to make zero, you can make the norm of the vector f zero. The norm can be the square root of the sum of the squares of the elements, or the maximum of the absolute value of all elements. Then "fminsearch" uses optimization methods to find the vector y that makes the norm a minimum. This may work for simple problems, especially ones with constraints, but it is less successful on large problems.

INTEGRATE ORDINARY DIFFERENTIAL EQUATIONS THAT ARE INITIAL VALUE PROBLEMS

To solve a single ordinary differential equation

$$\frac{dy}{dt} = f(t, y), \quad y(0) = y_0 \tag{B.7}$$

we use the "ode" set of functions in MATLAB. There are four steps.

Step 1 Set the initial condition, y0.

$$y0 = 3$$

Step 2 Set the time span for the integration.

$$\text{tspan} = [0 \quad 10]$$

Step 3 Name and construct the m-file rhs.m, which evaluates the function $f(t,y)$.

Step 4 Call ode45.

```
[t, y] = ode45(@rhs, tspan, y0) or [t, y] = ode45(@rhs, [0 10], 3)
```

The m-file is

```
function ydot=rhs(t,y)
ydot = ...
```

Since MATLAB will integrate whatever equation you give it, and the m-file "rhs" is going to be used many times, you must insure that it is correct: given t and y, it computes the correct $f(t,y)$ and puts the value into the variable *ydot*, which is then used by MATLAB to construct the solution. Once you have checked the program and run it, it is easy to plot the solution.

```
plot(t,y)
```

If you want data output at specific times, use

```
tspan = [0:1:10] for data at t = 0, 1, 2, ..., 10.
```

If the integration is taking an interminable amount of time, it is possible that you have defined a stiff system of equations (see Appendix E). In that case, change to an implicit method (for stiff equations, hence the s-designation).

```
[t , y] = ode15s(@rhs, tspan, y0)   or ode23s(@rhs, tspan, y0)
```

If you want the solution to be plotted as it is computed, leave off the $[t, y]$.

```
ode45(@rhs, tspan, y0)
```

To solve a set of ordinary differential equations

$$\frac{dy_i}{dt} = f_i(\{y_j\}), \quad y_i(0) = y_{i0} \tag{B.8}$$

we do the same steps except that the initial condition is now a vector with N elements, the m-file "rhs" must compute N functions using the vector y with N elements, and we have to return the result as a column vector.

```
y0 = [0  1  3]
tspan = [0  10]
[t , y] = ode45(@rhs, tspan, y0)
```

The m-file is

```
function ydot=rhs(t,y)
% now there are y(1), y(2) and y(3) and t, which can be used
% to evaluate fi.
ydot(1) = ...        term1 = ...
ydot(2) = ...   or   term2 = ...
ydot(3) = ...        term3 = ...
ydot =ydot'          ydot = [term1;term2;term3]
```

Now the output from "ode45" is a vector, t, and a matrix, y. The vector y has as many columns as there are unknowns (here called N), and it has as many rows as there are elements in the vector t. The output has to be a column vector, hence the $'$. Once the solution is found, all components of it can be plotted.

```
plot(t,y)
```

If you wish to solve the problem more accurately than is done with the default parameters, you can change them. The following commands show how to change the relative tolerance and absolute tolerance from the default values of 10^{-3} to 10^{-9}. You can place different accuracy limits on different variables, too. To see all the options, say help odeset.

```
OPTIONS=odeset('RelTol',1e-9,'AbsTol',1e-9)
[t , y] = ode45(@rhs, tspan, y0,OPTIONS)
```

Differential–Algebraic Equations

Some problems may have ordinary differential equations and algebraic equations, too. These are not always easy to solve, but MATLAB has methods for them. Open the help MATLAB/User's Guide/Mathematics/Calculus/Ordinary Differential Equations and move down to Differential–Algebraic Equations. Examples are given there, but they require careful adjustment of the error tolerances, which depend on the type of problem being solved.

Checklist for Using "ode45" and Other Integration Packages

When using "ode45," your m-file for the right-hand side must meet these conditions:

- The name in the calling command must be the same as the filename; and name in the function is immaterial. If "ode45" identifies @rhs, then the filename of the m-file has to be "rhs.m." The first line of the m-file can be function ydot=nothing(*t,y*).
- The variable tspan (or whatever it is called) must have at least two values.
- The number of entries in the vector for the initial conditions must be the same as the number of right-hand sides calculated in the m-file.
- Whether you call the output from the m-file "ydot" or something else is your choice, but if the output is called "ydot," the m-file must computer all elements of the vector *ydot*.
- The output of the m-file must be a column vector.

- Variables can be used in the m-file provided they are (a) global variables, (b) passed as parameters in the calling argument, or (c) set in the m-file. Check their value inside the m-file by running the m-file.
- The m-file will be called many times by the "ode45" function. However, you only have to check the calculation once. Check the m-file by giving it t and all the $y(i)$; compute what you expect the right-hand sides to be and see that the computer gives those values. This is the only way to ensure that MATLAB is solving the problem you want solved.

PLOTTING

Simple Plots

For a vector x with n entries, and a vector y with n entries, plot them in one of four ways:

1. plot(x,y)	2. loglog(x,y)	3. semilogx(x,y)	4. semilogy(x,y)

To plot more than one variable:

```
plot(x, y1, x, y2)    % if the y1 and y2 are known at the same
                        values of x
plot(x1,y1,x2,y2)     % if y1 is known at x1 and y2 is known at x2
```

Add titles:

```
title('This is the title')
xlabel('x')
ylabel('y')
```

Add a legend: `legend(''first curve,'' ''second curve'')`.

Plot only one column of a matrix, here the second column: `plot(t,y(:,2))`.

Add Data to an Existing Plot

Issue the command: hold on and continue plotting. Further lines, symbols, and so on will be added to the existing figure. When done, type: hold off.

Dress Up Your Plot

Plot in different colors using: `plot(x, y, 'r', x, y2, 'b')`.

Use different markers: `plot(x, y, 'or', x, y2, '*b')`.

Use different line options: `plot(x, y, '-r', x, y2, ': b')`.

To get both the symbols and the lines, use both line styles and markers: `plot(x, y, '-or', x, y2, ':*b')`.

Limit the plot to $x_{min} \leq x \leq x_{max}$, $y_{min} \leq y \leq y_{max}$, regardless of the data: `axis([xmin xmax ymin ymax])`.

FIGURE B.4 Multiple plots.

Limit the plot to $x_{min} \leq x \leq x_{max}$, regardless of the data: `xlim( [xmin xmax] )`.
Limit the plot to $y_{min} \leq y \leq y_{max}$, regardless of the data: `ylim( [ymin ymax] )`.
Get more information: `help axis`.

To change the width of the lines add "LineWidth",2 to the plot calling argument.

Multiple Plots

To get multiple plots on the same screen use the command subplot. Figure B.4 shows six plots on one screen.

The command: `subplot(2,3,1)` makes a 2 × 3 array of plots in one figure and the plotting commands following this command apply to the first plot (upper left-hand corner). For the second figure, say `subplot(2,3,2)` and the next plot will be in the middle of the top row of three figures. The final number is the plot sequence, numbered from left to right and then top to bottom.

Sometimes you want to prepare several plots, examine them in turn, and decide whether to save them or print them. One way to do this is to insert the following command after the plot command: pause. This causes the program to stop at that point in the code, and you can export the figure or print it. Then press any key to continue.

3D Plots

To plot a function $z(x, y)$, create an $x - y$ grid (rectangular), evaluate the function at each grid point, and plot.

Create the x grid: `x = 0:0.05:1`.
Create the y grid: `y = 1:0.2:3`.
Create the combined mesh: `[X,Y] = meshgrid(x,y)`.

Evaluate the function: $Z = fn(x, y)$ where fn is an expression or function; for $z = x^2 + y^2$:

```
Z = X.*X + Y.*Y
```

Plot the 3D plot: `mesh(X,Y,Z)`

Color in the surface: `surf(X,Y,Z)`
Create contours in 2D: `contour(X,Y,Z,20)`
Create contours in 3D: `contour3(X,Y,Z,20)`

More Complicated Plots

Possible symbols are

```
+ plus             ^ triangle (up)
o circle           v triangle (down)
* star             > triangle (right)
. point            < triangle (left)
x cross            p pentagram
s square           h hexagram
d diamond          none no marker (default)
```

The colors are in the default order for multiple plots, but white is not used.

```
b blue             Line types
g green            - solid
r red              -- dashed
c cyan             : dotted
m magenta          _. dashdot
y yellow
k black
w white
```

Use Greek Letters and Symbols in the Text

These are TEX commands.

```
α \ alpha      Γ \ Gamma      ∞ \ infty
β \ beta       Δ \ Delta      ≥ \ geq
γ \ gamma      Θ \ Theta      ≤ \ leq
. .            . .            ∂ \ partial
. . .          .              ± \ pm
ω \ omega      Ω \ Omega      ℜ \ Re
                              ℑ \ Im
```

Bold, Italics, and Subscripts

To make the title, $x$label, or $y$label appear in boldface type, with a font size of 12, use

```
ylabel('specific volume', 'fontsize',12,'fontweight','b')
```

To make the label appear in italics, use

```
ylabel('specific volume','fontangle','italic')
```

To add a subscript, use Q_1 to make Q_1. To add a superscript use Q^1 to make Q_1. To increase the size of the text, which may be required for publication, use

```
set(gca,'LineWidth',2,'FontSize',18)
```

For other options, of which there are hundreds, use the help menu. To increase the size of the legend use the regular legend command and

```
hx=legend; set(hx,'FontSize',18)
```

It is possible to create plots by issuing commands (as shown later) and by creating them interactively. The Videos "Using Basic Plotting Functions" and "Creating a Basic Plot Interactively" provide additional details (find in "help/MATLAB/Demos" section). To use the interactive feature, plot something and then use the menus in the figure to perform some of the same actions as done with commands.

OTHER APPLICATIONS

Examples solving two-point boundary value problems are shown in Chapter 9, and examples solving time-dependent problems in one spatial dimension are shown there, too. In these problems you want to know the solution at each x-position as a function of time. It is simple to plot the solution versus time, but it is more instructive to plot the solution for all positions at several different times. This is done as follows.

Plotting Results from Integration of Partial Differential Equations Using Method of Lines

```
tspan = [0 10], y0 = [...]
[t , y] = ode45(@rhs,tspan,y0)
plot(t,y)
```

This gives a plot of each variable $y(t,i)$ versus time. Here, $y(t,i)$ is the solution at point $x(i)$ and time t. To plot the solution versus $x(i)$, at various times, specify that you want the solution at a select number of times.

```
tspan = [0:1:10]   % 11 different times
y0 = [...]
[t , y] = ode45(@rhs,tspan,y0) % compute the solution
          % y(j,i) is a matrix: the column number is i,
          % and the row number j
          % identifies the time, t(j)
x = [...]  % specify the x-positions of the i-th variable
          % there are the same number of entries for x as for y0
hold on
for i=1:11
       plot(x(:),y(i,:))
end
hold off
```

IMPORT/EXPORT DATA

There are a variety of methods for importing and exporting data. Discussed here are methods for exchanging numbers between text files and variables in MATLAB as well as between Excel and MATLAB. An illustration is also given in which an Excel application is programmed with text, text blocks, and data, which can be read by Excel. Additional applications are described in the video "Importing Data from Files," available from the help menu, MATLAB, Demos.

Suppose you have a text file with two columns of numbers and five rows. This could have been written in Notepad on the PC or TextEdit on the Macintosh. The easiest way to

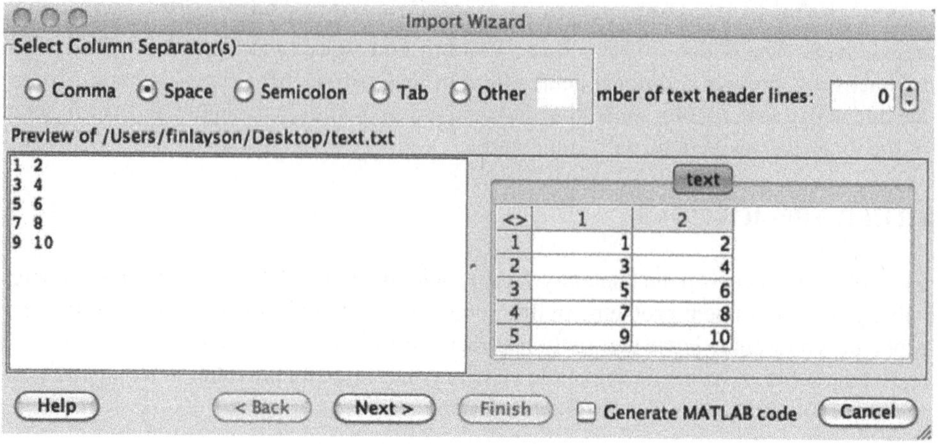

FIGURE B.5   Import wizard.

read the data is to use the Import Wizard. Select File > Import Data. Then select the text file you wish to import from. The Import Wizard will display the text file on the left-hand side and the data in MATLAB on the right-hand side. You can choose to import the data as an array, or as vectors in columns or rows. Choose next and the data is in MATLAB as an array. See Figure B.5 by choosing File/Import Data. In this case the text file had the numbers 1–10 in two columns, five rows, and that is how the array "test" appears in MATLAB, too. To export the data, use the command

```
save newfilename arrayname -ascii
```

Suppose you have an Excel file with only numbers. You can easily import it using the command:

```
[num] = xlsread('filename.xlsx').
```

The array num now contains the numbers from the Excel file filename.xlsx. To export an array "A" from MATLAB, use the command:

```
[num] = xlswrite('filename',A).
```

The data goes into the Excel file. It is also possible to specify where it goes in (sheet and range). See the help menu for functions, Data Import and Export, Spreadsheets.

It is more complicated to do the import and export when both text and numerical values are involved. Illustrated here is just one way of doing that, but it allows you to easily create what is equivalent to a GUI in Excel, import the information into MATLAB, and perform calculations. The problem posed is the same as shown in Figure 2.1, but the Excel spreadsheet will be used to enter the data, it will be transferred to MATLAB, and the program "specvol" will calculate the specific volume. The Excel spreadsheet is prepared as shown in Figure B.6a. The darkened blocks need to be filled in by the user. Then one saves the spreadsheet in a Tab delimited format, as pvt_data.txt, giving the result shown in

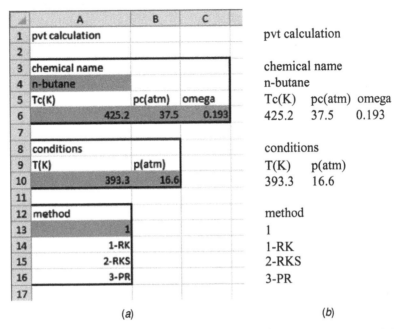

	A	B	C
1	pvt calculation		
2			
3	chemical name		
4	n-butane		
5	Tc(K)	pc(atm)	omega
6		425.2	37.5	0.193
7			
8	conditions		
9	T(K)	p(atm)	
10		393.3	16.6	
11			
12	method		
13	1		
14	1-RK		
15	2-RKS		
16	3-PR		
17			

(a)

pvt calculation

chemical name
n-butane
Tc(K)   pc(atm) omega
425.2   37.5   0.193

conditions
T(K)   p(atm)
393.3   16.6

method
1
1-RK
2-RKS
3-PR

(b)

FIGURE B.6 Excel spreadsheet to provide data to MATLAB: (*a*) spreadsheet; (*b*) tab delimited result.

Figure B.6*b*. The program to read the data and perform the calculations is shown later. The answers, of course, are the same as before.

```
% pvt calculation with input from Excel
global T p Tc pc R aRK bRK icode
inputfile = strcat('pvt_data.txt')
fid = fopen(inputfile,'r');
% read chemical data
for i=1:4;tline = fgets(fid);end % this line skips the first
   % two lines from Excel
chemicalname = sscanf(tline,'%s')
for i=1:1;tline = fgets(fid);end
Tc = fscanf(fid,'%g%')
pc = fscanf(fid,'%g%')
omega = fscanf(fid,'%g%')
% read conditions
for i=1:4;tline = fgets(fid);end
input1 = fscanf(fid,'%g',2);
T = input1(1)
p = input1(2)
% pick an equation of state
for i=1:3;tline = fgets(fid);end
icode = fscanf(fid,'%g',1)
% icode for EOS
% = 1 Redlich-Kwong
% = 2 Redlich-Kwong-Soave
% = 3 Peng-Robinson
```

```
fclose(fid); % close the file pvt_data.txt
% set up the problem
R=0.08206
if icode==1
    aRK=0.42748*(R*Tc)^2/pc
    bRK=0.08664*R*Tc/pc
    alpha = (Tc/T)^0.5;
elseif icode==2
    aRK=0.42748*(R*Tc)^2/pc;
    bRK=0.08664*R*Tc/pc;
    m=0.48+1.574*omega-0.176*omega*omega;
    alpha = (1+m*(1-(T/Tc)^0.5))^2;
else
    aRK=0.45724*(R*Tc)^2/pc;
    bRK=0.07780*R*Tc/pc;
    m=0.37464+1.54226*omega-0.26992*omega*omega;
    alpha = (1+m*(1-(T/Tc)^0.5))^2;
end
aRK=aRK*alpha
% solve the problem
vol=fzero('specvol',2)
Z=p*vol/(R*T)
```

Import/Export with Comsol Multiphysics

See Appendix D for ways to transfer data between Comsol Multiphysics and MATLAB. You need a license for LiveLink with MATLAB.

PROGRAMMING GRAPHICAL USER INTERFACES

You can also make GUIs in MATLAB. This may be convenient if you are preparing a program that can be used by someone not familiar with MATLAB. Data can be entered and the program told to run, with appropriate output. To learn about this, go to the help menu, Users Guide, Creating GUIs.

MATLAB HELP

When you are really stuck, issue the command

```
help commandname
```

This may lead you to the solution you are looking for. Issuing the command

```
help
```

gives you a table of contents; look at "help funfun," for example, to see all the methods and options for integrating differential equations. Alternatively, click on the arrow and look under MATLAB; there are several folders (see Figure B.7) that lead you to instructions to perform specific tasks.

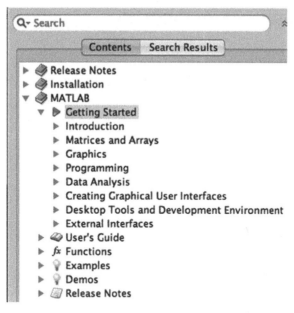

FIGURE B.7   Help/search window.

APPLICATIONS OF MATLAB

There are many chemical engineering examples in the book that use MATLAB; see the Table of Contents. Listed here are examples applying special techniques.

- Solving a single nonlinear equation, Chapter 2, p. 15, Chapter 3, p. 34, Chapter 4., p. 53, Chapter 8, p. 164
- Plotting, Chapter 2, p. 18
- Multiple equations, few unknowns, Chapter 4, p. 56
- Integrate ordinary differential equation, Chapter 8, pp. 141, 144, 146, 156, 166
- Fit straight line to data, book website on parameter estimation
- Polynomial regression, book website on parameter estimation
- Nonlinear regression, book website on parameter estimation

APPENDIX C

HINTS WHEN USING ASPEN PLUS®

This appendix gives a few hints and tips on using Aspen Plus.[1] It assumes that you are a beginner, so it shows the main steps you go through to use Aspen Plus, with screen shots to illustrate what you should see on the screen. You should work through this appendix using Aspen Plus to reproduce the examples shown here and review the examples again when you use Aspen Plus in other chapters. Keep in mind that this is a complex program with many options developed over decades by engineers working for Aspen Technology as well as the companies using the program. Some engineers' whole job is doing simulations to model chemical process, and it is unrealistic to expect a beginner to do the same. However, you can use Aspen Plus to do many powerful things, and it is important to know the full capabilities available to you.

INTRODUCTION

Figure C.1 shows the major functions. The three boxes as the lower left are most important (the Environment buttons). "Properties" is chosen when you want to specify chemicals or thermodynamic models, or to obtain thermodynamic data about them. "Simulations" is chosen when you want to prepare the flowsheet, provide parameters for all the units, and specify inlet conditions of flow rate, temperature, and pressure. "Energy Analysis" is used to summarize energy flows in your process. When you click one of these, the tabs and groups across the top change as described below. The Navigation Pane is shown on the left. This has things like Components, Methods (for Properties) or Flowsheet, Streams, Blocks, Utilities (for Simulation).

[1] Aspen Plus is a registered trademark of Aspen Technology, Inc. This book uses version 8.0.

Introduction to Chemical Engineering Computing, Updated Second Edition. Bruce A. Finlayson.
© 2014 John Wiley & Sons, Inc. Published 2014 by John Wiley & Sons, Inc.

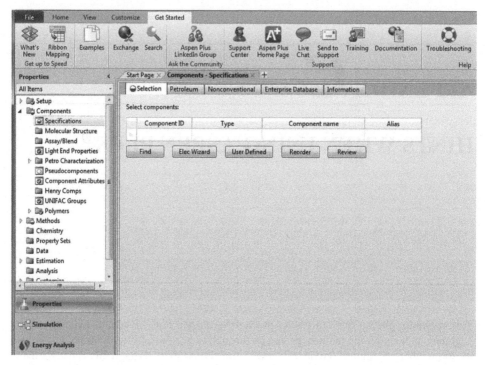

FIGURE C.1   Aspen Plus Screen.

Some of the tabs across the top (the ribbon) for the Properties Environment | Home are shown in Figure C.2a. You can hide the ribbon by choosing the up arrow on the upper, right hand side. Choose the down arrow to get it back. The options are grouped by function: Clipboard (not shown), Units, Navigate, etc. Use Components in the Navigate group to choose the components, and use Methods in the Navigate group to choose the thermodynamic model. Prepare thermodynamic plots by choosing Analysis and then Pure, Binary, etc. depending on your goal. NIST refers to the National Institute of Standards and Technology. This tab leads you to thermodynamic data available in the literature. The File tab (not shown) has the usual items: New, Open, Save, and so on. One menu item is Export, which allows you to export a report, a summary, the input, and any messages provided during the run. The report, in particular, is handy to have when done as a printed summary. Draw Structure allows you to represent chemicals not already in the database. The Methods Assistant provides an aid for choosing the method to calculate thermodynamic properties.

Some of the tabs across the top (the ribbon) for the Simulation Environment | Home are shown in Figure C.2b. Again the choices are grouped by function. In the Run panel, you can (from left to right, top to bottom): run the calculation, run it one block at a time, stop

FIGURE C.2a   Tabs for the Properties Environment.

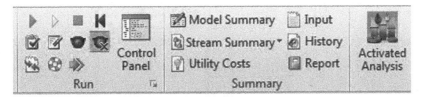

FIGURE C.2b   Tabs for the Simulation Environment.

the calculation, reinitialize (throw away any calculations that might have gone astray). The red and white life preserver is the Troubleshooting icon; choose it for help. The green arrow will take you to the next window that needs specifications. The Control Panel is where you find out what happened during the iteration. Information is given there about convergence, how the error changes from iteration to iteration, and so forth. In the Home tab, one can obtain a Model Summary, Stream Summary, Input Summary, History, and Report, which give information that is useful in your report.

The View tab allows you to decide which things you want displayed. The Model Palette is clicked to see the Model library as in Figure C.3. It is clicked again to make this go away. If the model has many units in it, Zoom enlarges the flowsheet, but an even better option is to use Flowsheet Sections (see below). The Modify tab allows you to specify what is shown on the flowsheet, such as temperature and pressure, along with the usual stream and block names.

On the upper, right part of the screen is a question mark, 🔵, which is the Help Button. This takes you to a window where you can search the Contents, Index, or Search. The troubleshooting tab, 🔵, does the same thing. The Get Started Tab, which is useful for a beginner, has sections for What's New, Ribbon Mapping (which shows how the program changes from version 7.3 to 8.0), Examples, Documentation, and Troubleshooting, among others. You can also look at the document "Getting Started Building and Running a Process Model"; get this from your system administrator.

FLOWSHEET

Model Library

The possible models are shown in the menu bar at the bottom (see Figure C.3.) If this does not appear, choose the View | Model | Palette menu. The models are organized by function, such as Mixers/Splitters, Separators, Heat Exchangers, Columns, or Reactors, and so on. Click on the tab to see the different possibilities: a new menu lists the different types within that class. Click on Columns, for example, and you see choices such as DSTWU, Distl, RadFrac, and so on. The Model Palette can be hidden and recovered by clicking the words Model Palette or using the F10 key.

FIGURE C.3   Model Palette.

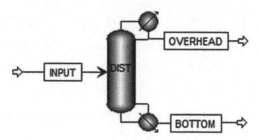

FIGURE C.4   Flowsheet of a distillation column.

Place Units on Flowsheet

To place a model or unit operation (now called a block), on your flowsheet click on the desired model and then click in the screen where you want the block to appear. Figure C.4 shows a DSTWU block. Continue in this fashion until you have the blocks you want. To delete a block, click on it and either choose Edit | Delete or right click and scroll down to delete. In the Navigation Pane, select Flowsheet | Section | GLOBAL | + and check that each unit has the correct streams in and out (see Figure 7.3). By using Modify | Global | Options or Modify | Display | Options you can cause the block (and stream) names to appear on the flowsheet (or not), and you can have them assigned numbers with a specified prefix (or not). You can also right-click on a block and delete it or rename it.

Connect the Units with Streams

Click on the "Material Streams" icon in the lower, left corner of the screen (see Figure C.1, where it appears in the lower left-hand corner, or Figure C.3 where it appears on the left). Then click in the flowsheet where you want the stream to originate and drag the cursor to where you want it to end. Continue in this way until you have placed all the streams in your flowsheet. When you are done, click the arrow to the upper left of the "Material Streams" box (Figures C.1 and C.3) to disconnect the drawing option. If you added streams you do not want, select them and choose Edit/Delete or right click and scroll down to delete. You can also rename units and streams by using right click and scrolling down to rename. If you select Tools/Options and the Flowsheet Tab, you can have the streams and blocks named automatically, or you can set it so that you have to supply a name.

Data

The Navigation Pane will normally be open. It is the list of items on the left. Close it by clicking the "<" and open it again by clickiing the ">". It provides a list of windows to be completed (see Figure C.5): Setup, Components, Properties, Flowsheet, Streams, Blocks, Utilities. These are discussed in turn.

Setup

You can specify a title for your problem, choose the units you want (these can be overridden in each stream and block), and create report options. To do this, click on the glasses (Data Browser) and double-click the folder Setup. In the Setup/Report Options window you put checks in the boxes for the properties you wish to include in your report, as illustrated in Figure C.5.

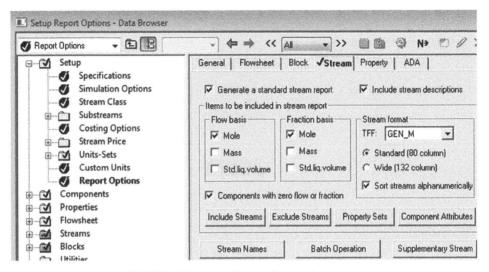

FIGURE C.5   Setup/Report Options for streams.

Data Entry

Aspen Plus has a menu that walks you through the procedure to specify a process. By choosing the green arrow one can step through the parts of the Navigation pane that are incomplete.

The menu items on the left that are in red are not complete, and must be completed. Double-click them in turn, starting from the top. Then click on the red submenus. A new window will appear, asking you to fill in the needed information. To close up the menu, click on a folder, and the subfolders beneath it will be hidden again. This takes you through each window one by one.

As you add information it might be convenient to go back and forth between the flowsheet and the window you are filling in. Below the Navigation Pane choose Simulation and then Main Flowsheet (at the top).

Specify Components

Choose Properties Environment and then Components; then Components appears in the Navigation Pane. Choosing Specifications gives Figure C.6. Type the name of the component in the left-hand column of the window. If you type in a name and the third column is not automatically filled in (as in Figure C.6 for C5H12), double-click on the blank box in the third column. Type the name into the window that appears, choose Find Now, and search for your chemical in the list provided. Close the window after specifying all the chemicals appearing in your process.

Specify Properties

You use Properties Environment I Navigate I Method to tell Aspen Plus what thermodynamic model you want to use. Initially all models are available. You can limit the choices by selecting an industry or application, and then only the pertinent choices are available. Chapter 6 discusses this choice in more detail, but this is one decision that must be

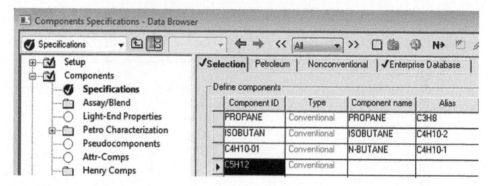

FIGURE C.6   Component identification.

validated by comparison with experimental data. In a design situation this is essential. You have the choice of a number of databases, but for introductory use, choose the latest and default database. When a program has been in use in industry for many years, each user wants their database to be retained, so that databases tend to get added on and little housekeeping takes place. The Home I Tools I Methods Assistant provides guidance for the choice of thermodynamic property calculation. The NIST Thermo Data Engine provides the parameters used in the thermodynamic models.

Specify Input Streams

With the Simulation Environment (bottom left) chosen, when you open up the stream folder (on the left), the streams that are input streams to the process will be shown in red. Click them in turn, filling in the properties of that stream. You must specify the flow rate of each species (although not necessarily those that are zero), the temperature, and the pressure, as shown in Figure C.7. You can specify mole fractions and the total flow rate if you prefer. You do not need to specify conditions in internal or output steams; the computer will calculate those. Sometimes it is helpful to specify an internal stream to speed convergence, but you need to know which one to use (see Chapter 7).

Specify Block Parameters

Click each block in turn and fill in block parameters (see Figure C.8). Sometimes the window has multiple options, which you need to complete if they are red. If you want some of the results (like temperature and pressure) to appear on the flowsheet, choose Main Flowsheet I Modify and click the boxes for the items to display.

RUN THE PROBLEM

To run the problem, choose the black arrow in the Run Panel or the F5 key. The open triangle allows you to calculate one unit at a time. (If some data is not yet specified, the program will tell you.) By choosing the N➔ button, you get the most information back from the computer, including information about convergence, as shown in Figure 7.7a.

If you run the problem, make a change in a parameter, and run the problem again, Aspen Plus will use the first solution as the starting guess for the second problem. If you do not

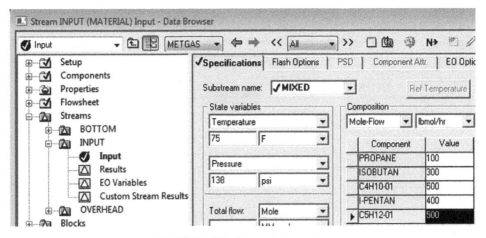

FIGURE C.7 Stream information.

want this to happen, choose Run/Reinitialize or the reinitialize button (see Figure C.2b), and the starting values are put back to the default values (flow rates of internal streams and output streams are usually zero). You can also do the iterative calculations one unit at a time by choosing the open triangle.

Scrutinize the Stream Table

In the left-hand window, click Results/Summary and Stream to get the mass and energy balances, the values of flow rate and enthalpy of each stream, as illustrated in Figure C.9. When the stream information is displayed on the screen, click Stream Table to get the information on the flowsheet. If you click on the upper left blank box of the stream table, the table will be copied to a format that can be pasted into a spreadsheet like Excel.

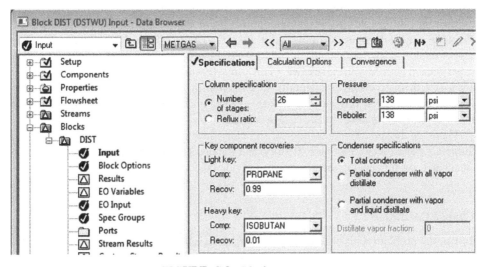

FIGURE C.8 Block parameters.

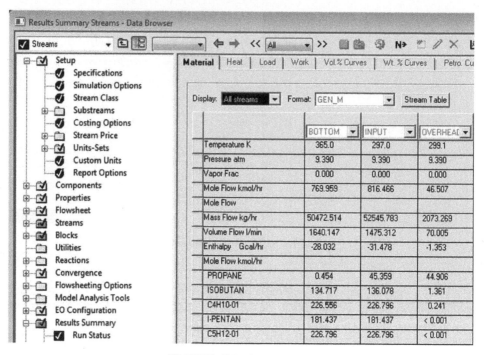

FIGURE C.9   Stream table.

Checking Your Results

Take the streams in the Stream Table and follow them through the process. Did the computer do what you intended? Are the results reasonable? By using the Navigation Pane I Blocks, you can look at the results for each block, or you can choose the ModelSummary icon and obtain the results for each collection of like models (i.e., all the heat exchangers).

You expect Aspen Plus to be correct, but there are two possible problems: lack of convergence and poor choices of thermodynamic correlations. By using the right arrow to run your problem you will get printed information about the convergence or lack of it. *Read the output!* You get this information from the Home I Control Panel menu, too. The proper choice of thermodynamic correlation can only be determined by comparison with experimental data or with experience. (This is one reason why chemical engineers are paid a lot—for their experience.) Naturally, at this point in your career, few of you have that experience. However, you can still look at your mass and energy balances and see if they make sense. *Every number needs to be examined.*

There are three important sources of summary information, all listed under the Home I Summary panel: Input Summary, History, and Report. You should look at these—you may detect an error by viewing the summaries provided there.

Change Conditions

To run the problem with different conditions, change the data and click the Run button. You can also set the Auto Run On button to have it run automatically whenever you change something. To rerun the problem you must change *something*. If you just want to run 30 more iterations, for example, add a space in the title and click **N⧐**.

REPORT

Transfer the Flowsheet and Mass and Energy Balance to a Word Processing Program

Select the flowsheet by dragging over the desired area and choosing Home | Copy, and pasting it into the application. You can also use the Print Screen key and then go to your word processing program and paste it. Use the Picture | Crop function to select the portion of the diagram that you want displayed. The best way to capture the Stream Table is to select the entire table and paste it into Excel. This is done from the Results Summary | Stream option.

If you would like the temperature and pressure to appear on the flowsheet, choose Main Flowsheet, then Modify and click the boxes for temperature and pressure. Figure C.10 shows the resulting change in the flowsheet.

Prepare Your Report

Detailed information can be obtained using the Home | Report menu. This gives details about individual blocks. The Home | History menu gives information about the convergence and what calculations have been done. The Home | Input Summary lists the parameters you have set. Every report should have a flow sheet with blocks and streams labeled, mass and energy balances referenced to the flow sheet, and a text description of the process. You should outline the problem, describe the choices you have made, and list all the ways you examined your results for validity. When you are *really* done, you can choose File and Export. In the "Save as type" window, scroll down to report Files (*.rep). Browse to the folder where you want to save it and click Save. A text version of the whole simulation is saved, giving all the detail you supplied and all the results. It is very useful to document your work (perhaps as an Appendix in a report), and the report gives the detail needed to write a descriptive report. It does use paper, though, if you print it. For example, the report for the four-column gas plant problem in Chapter 7 was 10 pages long. You can choose the

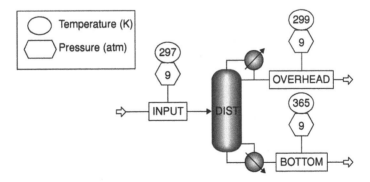

FIGURE C.10 Adding temperature and pressure to the flowsheet.

Model Summary icon and obtain the results for each collection of like models (i.e., all the heat exchangers).

Save Your Results

Save your program and results in the computer memory. A good strategy is to create a folder for each new problem and save your program and results in the folder. The reason is that when you choose File I Save As and Aspen Plus Document, three files are created. It is also useful to save your file periodically when working, just in case something goes wrong and you need to restart. The three files usually created are the .apw, .bkp, and .def. You cannot restart from the previous results using the backupfile (.bkp) alone and must have the .apw file.

Getting Help

Click Get Started and Troubleshooting. The Help menu provides detailed help in all areas, and you can search by contents, index, or key words.

ADVANCED FEATURES

Flowsheet Sections

If your process has already been divided into flowsheet sections, you can access each section by itself. In the Navigation Pane, choose Flowsheet I Flowsheet sections. In all cases the sections will be listed on the left and also in a window to the right as shown in Figure C.11. Do not click on any section on the left panel. Choose one section from the list on the right, and then click Make Current Section (lower right-hand corner). To see only that section, choose Main Flowsheet in the main window. Then under the Modify tab, unclick Show All and pick the section you wish to see (if it is different from your current section). Then the only flowsheet shown will be that section.

Another alternative is to zoom the flowsheet. Using CTRL up arrow or down arrow zooms in or out, respectively.

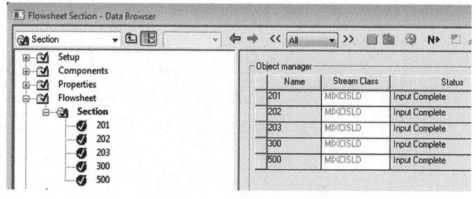

FIGURE C.11   Flowsheet sections.

When making your flowsheet, you can make selections as follows. Select the flowsheet sections as mentioned earlier. Global will be shown. Change its name to a section name of your choice. Then prepare the flow sheet for that section. Go back to flowsheet/sections and right click on Sections, choose New and provide a new name. Then make the new section current. All process units you enter after that will be in that new section. To connect the sections, choose Main Flowsheet in the main window. Then under the Modify Tab click Show All. Then the total flowsheet will be shown. Then you can choose a stream and change its source or destination so that one is in one section and the other is in the other section. If you wish to move a process unit from one section to another, select the unit and right click. Choose: change section.

Mass Balance Only Simulations and Inclusion of Solids

Sometimes it is desirable to simulate a process using mass balances only, without energy balances. To do that in Aspen Plus, click on the Browser, choose Simulation Environment | Setup | Calculation options, and uncheck the box "Perform Heat Balance Calculations." Then the parameters for energy balances will not be required. Units like FSplit, Sep, Sep2 will be usable; units like DSTWU, Flash2, RStoic, and so on, can be used if you do not use heat duties. A full list can be seen by choosing the Help button in Aspen and entering "mass balance only simulations" into the index.

In Aspen, the liquid/vapor stream is generally called "Mixed" or "Conventional." In versions 7.2 and earlier, solids suspended in the liquid (or even by themselves) are contained in a substream called "CISOLID." AspenTech supplies a pdf file "Getting Started Modeling Processes with Solids." Solids are generally given a molecular weight of 1 so that in chemical reactions the stoichiometric coefficient is an integer multiplying the molecular weight. Aspen Plus checks this and tells you if the mass is not balanced in a reactor. In version 7.3 simple solids can be contained in the "Mixed" stream; in this text, they have been left in "CISOLID."

Transfer between Excel and Aspen

It is possible to transfer information between Excel and Aspen. Going from Aspen to Excel is simple: the stream summary can be copied to the clipboard by clicking the blank square in the upper left corner; choose Copy or Select All; then open Excel and paste it. It is also possible to transfer from Excel to Aspen. One operation is to import *Txy* data set from Excel to Aspen Properties. Then you can estimate or regress the property parameters. To do this, see the instructions under Get Started | What's New, search on Excel in the index. It is also possible to have a block that is defined in Excel and linked to Aspen. Note also that Excel has an Aspen Tab, and you can open (enable) Aspen Simulation Workbook from Excel. Both of these operations are beyond the scope of this book though.

Block Summary

After doing the calculation, the results for each block should be examined. This is done most easily by choosing Results Summary in the Navigation Pane and look at Models, Equipment, Operating Costs, CO2 Emissions. All the blocks are listed there, with like blocks grouped together, and can be accessed with the tabs at the bottom.

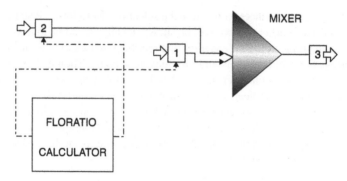

FIGURE C.12   Calculator block.

Calculator Blocks

Calculator blocks are blocks that the user defines to do a specific calculation that is not available in Aspen Plus. The programs can be from Excel, or they can be written using FORTRAN. Simple calculations can be done using the in-line FORTRAN that comes with Aspen. An example is given below. More complicated calculations can be done by writing a FORTRAN subroutine external to Aspen and compiling it. It can be linked dynamically to Aspen.

To demonstrate the calculator block, consider the flowsheet shown in Figure C.12. Stream 1 is defined as 100 lb/h of water. We would like stream 2 to be half the flow rate of stream 1. Obviously we could set it, but this block might be in the middle of a process for which we do not know the flow rate of stream 1 a priori. We choose Flowsheeting Options | Calculator in the Navigation Pane. The calculator block is named FLORATIO. Variables are named and set as follows: SECOND is the water flow rate in stream 2. FEED is the water flow rate in stream 1. The formula is

$$SECOND = 0.5 * FEED$$

To write this choose Flowsheet Options | Calculator | FLORATIO. In the Define tab, click on a left blank window and give the variable a name (FEED). Then click on the Definition tab choose Streams. In the Reference window, choose Mass Flow; then Stream

√Define	√Calculate	√Sequence	Tears	Stream Flash	

Calculator block execution sequence

Execute:		Block type:	Block name:
Before	▼	Unit operation ▼	MIXER ▼

List variables as import or export

Import variables:	FEED	

Export variables:	SECOND	

FIGURE C.13   Sequence for calculator block

TABLE C.1 Examples of Blocks in Aspen Examples

Example	Ssplit	RADFRAC	Extractor	RCSTR	Rplug	REquil	HeatX or MHeatX	Solids	Nonconventional	Filter	Decanter	Crusher	Screen
Ammonia					X								
Biodiesel		X	X							X			
Bioethanol from corn	X	X						X					
Biethanol from corn stover	X	X						X					
Cogeneration							X						
Entrained flow coal gasifier					X			X	X				
IGCC	X	X				X	X	X	X			X	X
Moving bed coal gasifier				X				X	X				
Oil shale retorting				X									
Physical solvents		X											
Pipeline gas	X					X							
Sulfuric acid		X		X			X					X	X
Urea		X						X					
Bauer								X					
3-Phase		X									X		

(and insert the stream number), then the Substream and Components. Finally in Info. Flow insert Import Variable, since this information will be input to the calculations. In the Calculate tab type in the equations. In the Sequence tab, choose Execute: Before, Block Type: Unit Operation, Block Name: Mixer; see Figure C.13. Now when the calculation sequence begins, the first thing that will be calculated will be FLORATIO; it will take FEED and calculate SECOND. Then when MIXER is calculated it will add them together. In this simple example, it was necessary to specify the temperature and pressure of stream 2, and a small amount of water, since it is an input stream (used 0.001).

Aspen Examples

Aspen comes with a number of examples. Access them from Get Started | Examples. Examples and some of the units that are in each example are listed in Table C.1. Simple units like Mixer, Heat Exchangers, RStoic, and so on are not listed there. If you are having trouble with one of the blocks listed in Table C.1 you should look at the appropriate example to see a working version of that block. In addition, Physical Solvents has an example of equilibrium absorption using RADFRAC, Amines has an example of rate-based absorption using RADFRAC, and Ammonia has an example of Flash2 used as an absorber. In addition, there are examples with polymers and scaling.

MOLECULE DRAW

Aspen Plus also has the capability to estimate physical properties from the structure. You can click on Property Environment | Home | Tools | Draw Structure (see Figure C.2a) and draw the molecule with different atoms, different bonds, and different configurations. To see how to do this, search on "molecular draw." The connectivity of molecules is used in estimation of physical properties by the Aspen Physical Property System or by the NIST Thermo Data Engine.

APPLICATIONS OF ASPEN PLUS

Many chemical engineering applications are described in the text and can be found using the Table of Contents. Specialized techniques are described in

- Test of Thermodynamics, Chapter 3, p. 39; Chapter 6, p. 97
- NIST Thermo Data Engine, Chapter 3, p. 41
- Sensitivity, Chapter 6, p. 102
- Design Specs, Chapter 6, p. 102
- Convergence, Chapter 7, p. 120
- Utility Cost, Chapter 7, p. 118
- Greenhouse Gas Emissions, Chapter 7, p. 120
- Optimization, Chapter 7, p. 122
- Chemical Reactor, Plug Flow, Chapter 8, p. 159

APPENDIX D

HINTS WHEN USING COMSOL MULTIPHYSICS®

This appendix gives an introduction to the use of Comsol Multiphysics. The program Comsol Multiphysics,[1] uses the finite element method to solve fluid flow, heat transfer, and mass transfer problems, as well as many other equations. In the past decade, computer software has become very powerful, and this allows chemical engineers to solve very complicated problems. This poses a problem though: you may be tempted to pose a problem that is way more complicated than you need. Instead, heed Einstein, "make it as simple as possible, and no simpler." Another problem is that many details of the computer program are hidden from you, or at least not readily accessible. By clicking buttons, or forgetting to make a particular choice, you may inadvertently solve the wrong problem.

On the other hand, engineers used to reduce all problems to ones that could be solved analytically, because then they could solve them. The assumptions used in the simplification might or might not be justified, but the analytical form of the answer aided their understanding of the problem. Modern computer programs can solve those simple problems, and the solution provides a test of the computer program and your use of it: Does the computer solution agree with the analytical solution when it should? But the programs can also solve realistic problems that are impossible to solve analytically. What you have to do is (a) pose a realistic form of the problem, and (b) verify you have solved it accurately enough.

The finite element method replaces a differential equation with a large set of algebraic equations. The details to make this switch are complicated, but fortunately, Comsol Multiphysics has done that for you. You still need to know how to use the program, because, after all, it is up to you to decide if you have solved the right problem, determined the accuracy, and derived useful properties from the solution.

The examples are made with the Chemical Reaction Engineering Module addition to Comsol Multiphysics, version 4.2a. However, most of them can be solved with the basic

---

[1]Comsol Multiphysics is a registered trademark of Comsol, Inc. This book uses Version 4.2a.

Introduction to Chemical Engineering Computing, Updated Second Edition. Bruce A. Finlayson.
© 2014 John Wiley & Sons, Inc. Published 2014 by John Wiley & Sons, Inc.

Comsol Multiphysics, although then the stoichiometry of a reaction will be up to you. If you would like to know what modules are included in your version, go to Options/Licenses or the book website. Appendix E describes the finite element method in one and two dimensions so you have some concept of the approximation: going from a single differential equation to a set of algebraic equations. This appendix presents an overview of many of the choices provided by Comsol Multiphysics. Illustrations of how Comsol Multiphysics is used to solve problems are given in Chapters 9–11. Thus, you may wish to skim this appendix on a first reading, and then come back to it as you use the program to solve the examples. More comprehensive accounts of Comsol Multiphysics are available for version 3.5a (Wilkes, 2006; Zimmerman, 2008; Datta and Rakesh, 2010).

BASIC COMSOL MULTIPHYSICS TECHNIQUES

The first step after you have installed Comsol Multiphysics is to open it, then choose the pull-down menu Options/Preferences. Click to show "Equation View" and Discretization, since these are used frequently to verify what equation you are solving and what approximation you are using for the dependent variables. If you do this, they will be available every time you open Comsol Multiphysics.

Opening Screens

Figure D.1 shows the top, left of the opening screen of Comsol Multiphysics. Look at the options across the top. The Model Builder is where you will define your problem; at the start it has the title "Untitled," but this will be replaced with the name you give the problem when you save it. To the right of the Model Builder is the Model Wizard. This is where you pick the dimensions of the problem, from no dimensions (ordinary differential equations), to 1D, 2D, or 3D, including axisymmetric problems. More information about these choices is given below and in the examples. The small, rectangular box and square box in the Model Wizard part of the screen act in the same way as the same icons in the Model Builder, namely to contract or expand the window.

To the right of the Model Wizard is the Graphics window. On its right, or perhaps below it, are three icons that represent Messages, Log, and Results (tables): see Figure D.2. The Log tab displays information about the calculations as they proceed; clicking on the Results tab shows tables that you have created (see later), and Messages tab gives error messages,

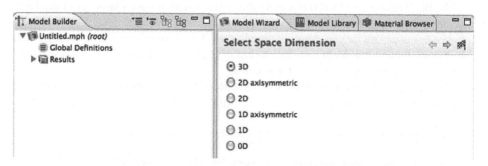

FIGURE D.1   Comsol Multiphysics opening screen, left.

FIGURE D.2   Comsol Multiphysics opening screen, tabs.

for example, when something is amiss. If you want to expand the Model Builder, Settings, or Graphics windows to cover the screen, double-click on the tabs. Double-click the tab again to reverse the process.

As you use the various tabs, the screen may end up in a form you do not want. One way to get back the original screen is to choose Options/Desktop Layout/ and then either "Widescreen Layout" (the three areas are across the screen from left to right), "Regular Screen Layout" (the first two areas are one above the other on the left whereas the Graphics window goes from top to bottom), or "Reset Desktop."

Equations

To solve a problem, one begins with the Model Wizard. Choose the dimensions: 0D, 1D, 1D Axisymmetric, 2D, 2D Axisymmetric, or 3D. Then click on the right arrow in the Model Wizard. The next screen lists the options for models (see Figure D.3*a*). The main ones are "Chemical Species Transport," "Fluid Flow," and "Heat Transfer." Each of these has additional options, some of which are shown in Figure D.3*b*. The available options depend on your license, but the ones shown in Figure D.3*b* are those that come with the Comsol Multiphysics and Chemical Reaction Engineering Module. The more limited choices with only the basic Comsol Multiphysics are displayed in Figure D.3*c*. Find the equation you want, click on the right arrow at the top of the Model Wizard, and choose Stationary or "Time Dependent." Then click on the finish flag: you have finished picking the differential equation!

The Model Library (see the tab in Figure D.1) contains models that have already been solved. You can open them and see both capabilities of Comsol Multiphysics and examine how the problem is set up. Examples setting up a problem are given in Chapters 8–11, but here they are summarized with a few more options displayed. Basically, we are going to work down the Model Builder window, opening options (called nodes) and acting on them: "Model 1," "Study 1," Results. A heat transfer problem in two dimensions is used as an example. After choosing in the Model Wizard: 2D, "Heat Transfer/Heat Transfer in Solids/Stationary" and the finish flag, one has the menu shown in Figure D.4 (although the Model 1 will be closed; click on the node arrow to open it.) Note the icons at the top, right of the Model Builder panel. The second one from the left is to "show" additional things, and we choose "Discretization" and "Equation View." In the root node we can choose a system of dimensions in the settings window; for the "Unit System" choose "none."

SPECIFY THE PROBLEM AND PARAMETERS

We step through "Model 1" in turn: Definitions, "Geometry 1," Materials, "Heat Transfer (*ht*)," "Mesh 1." If we click on heat_example, then in the settings window, near the bottom, is a place we can choose the units for data input. The default units are SI.

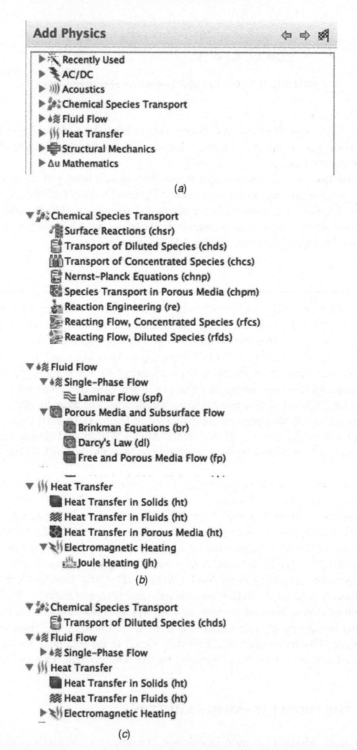

FIGURE D.3 Equation options for the Comsol Multiphysics with the Chemical Reaction Engineering Module: (*a*) subject areas, (*b*) chemical species, fluid flow, heat transfer, (*c*) options with only Comsol Multiphysics.

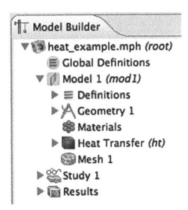

FIGURE D.4   Model choices for heat transfer example.

Physics

If you right click on Model 1, the following options are available: Add Physics, Mesh, Delete, Rename, Properties, and Dynamic Help. The first one allows you to add a second equation to the model (not done in this example); the second one allows you to add a second mesh so that you can solve the problem on two different meshes. The Properties option allows you to add comments that will be saved along with the model. The other options are self-explanatory.

Definitions

Right click on Definitions to see the options there; consider the Variables. Click on it and a table appears; you can define variables here, assign a number to them or define them in terms of an equation you write. A variety of functions are supplied, but you can write your own using the same format used in Excel and MATLAB: $+, -, *, /, \wedge$, and so on. For example, choose function step. In the settings window the default function step is a step change located at $x = 0$ going from 0 to 1 with a transition width of 0.1. These parameters can be changed. Then anywhere you want to use such a function, you can just use its name in the expression. The probes are useful for obtaining specific results, say the value of the solution at a point inside the domain or along an internal line (see later). A number of useful icons are identified in Figure D.5.

Geometry

Right click on Geometry. If you click on one of the shapes (rectangle, square, circle, etc.) a screen appears that allows you to define the shape and place it precisely in the geometry. If you have defined two shapes, another option appears: Boolean Operations. This option allows you to add or subtract the shapes to create a more complicated shape. Here a rectangle 0.2×0.1 m is constructed; then choose build. Next, define a circle inside with a radius of 0.01 m, and build. Now if you right click on Geometry new options appear. The Boolean Operation "Difference," with objects to add R1 and +, and objects to subtract C1 and +, and then build, results in a rectangle with a hole in the middle. In this example, this is not done because we want to have heat generation in the circle, C1; see Figure D.6.

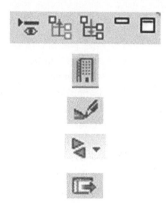

FIGURE D.5   Icons: Show, Collapse, Expand Minimize, Maximize; Build, Plot, Expressions, Export.

Materials

The Materials option allows you to import physical parameters of many compounds as well as define some yourself. Right click on Materials and open the Materials Browser. Listed there will be "Built-In" and "Liquids and Gases." Most of the Build-In materials are solids. If your simulation uses any of these materials, you can simply select it (e.g., "Built-In/Aluminum," "Liquids and Gases/Gases/Ethanol vapor," and "Liquids and Gases/Liquids/Ethanol"). Pick a material and right click it to "Add Material to Model." The Settings will list the properties that are included, along with typical units. You may have to add some properties, too. The properties can be functions of temperature (for liquids) or temperature and pressure (for gases). Be sure to use the units consistently throughout your model. If you are solving a flow problem, under "Laminar Flow" will be "Fluid Properties." The settings window indicates "from material"; leave that choice. You can pick the temperature at which the properties are evaluated for an isothermal problem. To see the value of a property, you can right click on Results/Derived Values and choose Point Evaluation; then pick a point in your domain and add it in the Settings window, use the Expressions icon in the settings window to find the property you want, and click =. The answer appears in the table. You can also change to "user-defined" and specify the value yourself. You specify units in brackets and the program converts them to a consistent set of units that agrees with

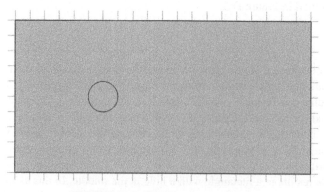

FIGURE D.6   Geometry for example.

the Unit System you have chosen in the root node. For example, if you are using the SI system but define a material density as 1[g/cm^3]; look at the density at a point and it will be listed as 1000 (kg/m^3).

Discretization

Click on "Heat Transfer in Solids." The geometry is shown along with numbers identifying different parts of it. Those numbers are usually in the box, but if they are not click on an object and choose the +. There are two other tabs that are very important. The Discretization tab allows you to select the finite element approximation: linear, quadratic, or cubic functions defined on the element. Usually we use quadratic functions. These are the default values for temperature. For concentration, though, the default values are linear, and best results are achieved in laminar flow with quadratic elements. For flow in two-dimensional cases, use "P2 + P1," which means the velocity is quadratic and the pressure is linear. For flow in three dimensions, keep "P1 + P1," the default. The Dependent Variables tab identifies the symbol used in the computer to represent quantities, in this case the value is T. You will not see the Discretization tab unless you have chosen it under the Show icon.

Boundary Conditions

Open the Heat Transfer option: it shows "Heat Transfer in Solids," "Thermal Insulation," and "Initial Values." The first one is the equation we chose originally; the boundary conditions are automatically set to thermal insulation until we change them. The "Initial Values" option is provided in case the problem cannot be defined if a variable is zero, for example. In a reaction rate, if the temperature appears as $1/T$, then T obviously cannot be zero. In that case one would put another value here to begin the iterations. Click on "Heat Transfer in Solids." Make sure the geometric objects we want are in the Selection box; if not add them. Click on the Equations tab. Then the equation will be displayed; you need this for checking the signs of things like heat sources, and so on. You must also specify the thermal conductivity, either by having picked a material in the Materials option or by choosing "user-defined" and inserting the value; here it is taken as 1 W/m K. Since the "Heat Transfer in Solids" equation has the options of a translational velocity, u_{trans}, it is necessary to choose "user-defined" for the density and heat capacity but these can be left as zero for this example.

Go back and right click on "Heat Transfer." Shown are a number of choices for boundary conditions. Here we choose insulated boundaries at the right-hand side and bottom and set the temperature to zero on the other two boundaries. Thus, choose "Temperature"; select the left-hand and top boundaries, and choose "+." Set the value to zero. If you go back and look at Thermal Insulation you will see that two boundaries have been marked "overridden"; those are the ones we just set. We want to have a heat source in the circle. Right click on "Heat Transfer," choose "Heat Source." Then select the circle, "+" to add it to the box. Set the value to 1000 W/m^3; the dimensions are set for an infinitely long cylinder. It is useful to click the Equation tab, too, to check the sign. Although Source is pretty descriptive for heat transfer, Reaction in the mass transfer option can be positive (for products) or negative (for reagents).

When specifying boundary conditions, there are some built-in functions that are useful. Under "Global Definitions" choose Functions to see a variety of them: step, ramp, rectangle, and triangle functions as well as MATLAB functions. The step function, for example, gives you options for defining a region over which the function will change (narrow or broad),

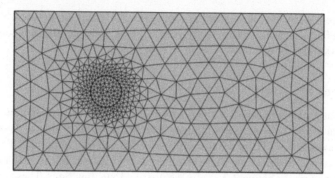

FIGURE D.7   Mesh for heat transfer example.

since the variables are represented by continuous functions. A simple alternative is to use an expression such as $1*(x < 0.5) + 0.5*(x > 0.5)$ to have the function be 1.0 over the region $x = 0$ to 0.5 and 0.5 for larger x. But the step function gives you more control over the specification.

Mesh

Click on "Mesh." The default mesh is always "normal." If you click on the "build" icon, a mesh is generated automatically. The mesh is shown in Figure D.7. Keep in mind that the mesh essentially determines the accuracy of the calculations. It is necessary for you to gain experience, and to do that it is useful to solve the problem on finer and finer meshes to see if the solution changes. It is possible to go back to "Model 1" and add a mesh. This is done and the "finer" mesh is chosen, followed by build.

SOLVE AND EXAMINE THE SOLUTION

Solve

The problem is solved next. Right click on "Study 1" and choose "= Compute." If for some reason there is an error, it will be listed under "Study 1" (several layers down). Open "Error" to see the cause of the error.

Plot

The solution is shown in Figure D.8 using contour lines. Here the temperature contours are all plotted in black, but they can be plotted in color with a scale showing, too. The temperature can also be plotted with different colors representing different temperatures; the temperature scale is shown at the right. This is called a surface plot. To make changes in the plot, right click on "2D Plot Group." You can go back and define a second mesh. Then under "Study/Stationary," select "Mesh 2" and right click "Study 1", =. This provides two solutions with different degrees of refinement.

Suppose we want to find the total heat flux out of the circle. Choose "Results/Line Integration," right click "Line Integration," and select the pieces of the circle: 5, 6, 7, and 8; click "+" to add them to the box. To find the heat flux, click on "Expression +" and choose heat transfer. There are about 30 variables listed, but some of them are not valid

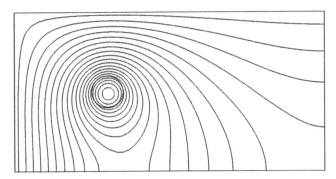

FIGURE D.8 Solution to heat transfer example.

for this problem; choose temperature gradient magnitude. In the computer program this is called ht.gradTmag. Click on the "=" and look at the results table. The value is 0.314574. Do this again, but choose lines 1 and 3 (the left and top). Now the value is 0.313233. For a perfect mass balance these should be the same, but the difference is 0.001341. If you go back and choose a "finer" mesh, the discrepancy is 0.000864, and with "extremely fine" it is 0.000374. Thus, as the mesh is refined the answers get better and better. There is always a trade-off between high accuracy and the solution time, but you need to be sure the mesh you use provides sufficient accuracy for your purposes. You can see other variables that can be plotted. Then click on "replace expressions" (the two green and orange arrows). A long list of variables can be found under the headings: definitions, heat transfer, geometry, and mesh.

If you want to plot the solution or an expression along a boundary, choose Results, right click "1D Plot Group," right click "Line Graph." Then select the boundary, choose "+" to add it to the box, define the expression and click the plot icon. The temperature along the bottom is shown in Figure D.9. When you have more than one solution, you need to identify in the Settings window which solution you want plotted. If you want to plot several variables on the same plot, right click "1D Plot Group," then right click "Line Graph" or "Point Graph" for each variable you want to plot. If you want to plot a variable along a line of a 2D domain, define the "Cut Line 2D" in Results/Data Set; then choose that Data Set in a Line Graph.

Publication Quality Figures

The plots can be saved in three formats: jpeg, bmp (bit-mapped), and png (portable network graphics). None of these are publication quality. Thus, the line plots in this book were made by exporting the data to a text file, which is imported into MATLAB. In MATLAB you have total control over the way the axes and legends are labeled (font size, bold or not, etc.)

Results

When you evaluate an integral, for example, the numerical result will appear in the Results section the Graphics window. Be careful, though, because the size of the column may be too small to include the exponent unless you expand it by dragging the border to the right. Unfortunately, many significant digits are displayed under the default "Full Precision"

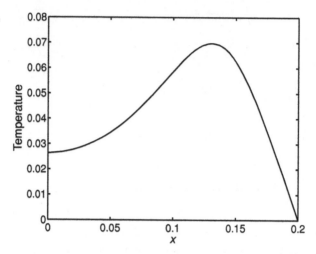

FIGURE D.9   Temperature along bottom boundary in heat transfer example.

option, many more than the accuracy of the finite element method justifies, and the exponent is more important than the eighth or ninth significant digit. To avoid this problem, click on the number icon when Results is showing and only four significant digits will be displayed.

Probes

If one wants to know the value of temperature or any expression at a specific point, right click on Definitions, choose Probes and Domain Point Probes. Then indicate the x and y locations you wish to use. After the solution is found, choose Results and right click on "Derived Values." Choose "Point Evaluation." Then in the Settings window, change the "Data Set" to "Domain Point Probe." Choose the "=" to get the value in the table. For $x = 0.04$ and $y = 0.02$, the value of T was 0.0758. Alternatively, define the point in "Cut Point 2D" in Results/Data Set and chooe that Cut Point 2D in "Derived Values/Point Expression."

Data Sets

If one wants to plot along a line that is not on the boundary, choose "Results/Data Set" and right click to choose "Cut Line 2D." In the settings window, define the x and y points of the beginning and end of the line. Then choose Results, right click to get "Plot 1D Group." Open it, and right click and choose "Line Graph." In the "Data Set" use "Cut Line 2D." Click the plot icon to get the plot. If you want to integrate something along a line, define "Cut Line 2D" the same way. Then use "Results/Derived Values" and right click "Line Integration." In the data set at the top of the settings window, choose "Cut Line 2D" and then "=." The answer is in the Results Table. Similar approaches work in 1D and 3D problems. In 3D you can define a data set to be a plane and plot an expression in that plane. To define the plane, you can use "Quick" to select a plane perpendicular to a coordinate axis or "General" to define a plane using three points with x, y, z values specified. In all these cases you need to select the solution you want to plot in the Settings window.

There are a variety of options for plotting. If the problem is in three dimensions, you can plot various expressions on surfaces, or in slices, and you can define 2D and 1D plots of part of the solution; usually this involves creating a Data Set first. The expressions arrows

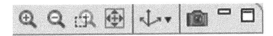

FIGURE D.10 Icons for graphics: zoom in, out, box, fill screen, view, take picture, minimize, maximize.

(green and orange arrows) allow you to choose various expressions that are predefined in Comsol Multiphysics. For example, in a heat transfer problem, the conductive heat flux, x-component, is called ht.dfluxx, ht.dfluxy, ht.dfluxz. The temperature gradient is ht.gradTx, ht.gradTy, ht.gradTz. There are many such entries that can be accessed once the expressions arrows are clicked.

There are a number of options to change how the plot is labeled. Once the plot appears, look at the top Graphics window. The icons there (shown in Figure D.10) allow you to make the plot bigger, smaller, use the cursor to drag across the part you would like to expand, and fill the Graphics Window. Click the camera, browse, select the folder you want, provide a name, save, and back in the camera window click OK. You can rotate a three-dimensional object, and to get back to the standard view, click the icon that looks like an axis. This icon has subicons, too, that view the object in xy, xz, yz views.

ADVANCED FEATURES

Mesh

The accuracy of the solution may depend on the mesh, so it is worthwhile to understand some options when preparing the mesh. Consider first two-dimensional problems. If one right clicks on the mesh node, several options are available: free triangular, free quadrilateral, mapped, boundary layer, and statistics. The first two cover the domain with either triangles or quadrilaterals. The mapped mesh causes the quadrilateral elements to follow the geometry more. Choose the method you want, click build, and the mesh is created. If you want to refine it, right click on refine; this will divide each and every element. If you want to refine the mesh locally, it is possible to do that by either refining the edges or creating a domain that overlaps the solution domain. In the first case, before creating a domain, right click mesh and choose refine edge. Pick an edge or edges and set the maximum element size to a small number. The edge will be divided, and that refinement will influence the domain around it when the whole domain is meshed. In the second case, create a domain on top of the geometry and refine just that domain. Still another option is to right click on Mesh and choose "Boundary Layers." Under "Boundary Layer Properties" (open the arrow) select the boundaries where you want boundary layer meshes to appear, add them to the box using the "+," and build.

As an illustration of one of these methods, consider the microfluidic device shown in Figure 11.4. If the Peclet number is large (here 2000 is used for illustration), only a small amount of material will diffuse sideways. If the mesh is regular (see Figure D.11a), the answer may look ragged and irregular, as seen in Figure D.11b. This is because the elements are not small enough. If one keeps $Pe\,\Delta x \leq 2$ the oscillations would disappear, but this is impractical for large Peclet number. However, in this problem, the sharp change in concentration is focused on a narrow band in the middle of the device. Thus, create a domain that is 0.1×2.5 along the centerline. Then right click Mesh and choose "Free triangular mesh" and build. Then choose "More operations/refine." Choose the narrow band, set the number of refinements to 2, and click build. The revised mesh is shown

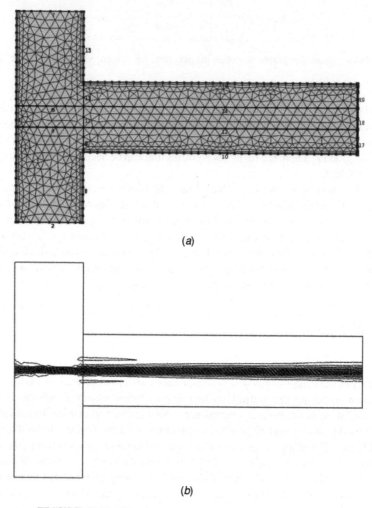

(a)

(b)

FIGURE D.11   Regular mesh (*a*) and solution (*b*), $Pe = 2000$.

in Figure D.12*a* and the concentration contours are shown in Figure D.12*b*. Clearly, the irregularities in Figure D.11*b* have gone away. The concentration profiles at the exit are shown in Figure D.13 for the two cases. They are close to each other at the exit, but the variances are slightly different (0.1979 for mesh 11a and 0.2043 for mesh 11b). Remember that the highest variance indicating no mixing whatever is 0.25. This two-dimensional case is relatively easy, since you can use lots of elements, but in three-dimensional problems that may not be practical.

Transfer to Excel

The solution and mesh can be written to a text file that can be entered into an Excel spreadsheet or a word processor. Under "Results/Export" right click to get "Data 1" and put in an expression. For the output, Browse and put in a file name and move to the desired folder, then choose OK. Back in the Data window, click the export icon (box with a right-facing arrow at the top) and a text file is made. This can be read in any program that reads

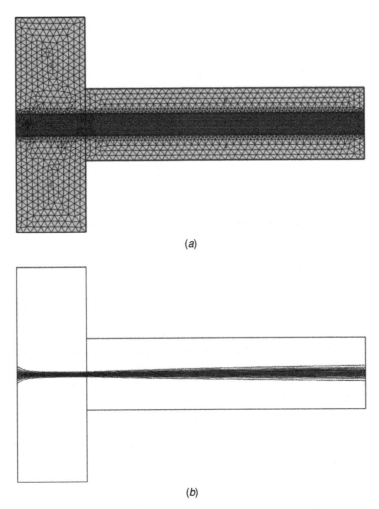

(a)

(b)

FIGURE D.12   Refined mesh (*a*) and solution (*b*), *Pe* = 2000.

text files, but it can also be opened, copied, and pasted into an Excel spreadsheet. If you want to do this with the mesh, right click on "Results/Export" and choose Mesh. Name the file in a similar manner, but this time you have a choice of a text file or a binary file (which can be read faster if you are reading it in another computational program). This file gives the locations of the mesh points and element information (which mesh points are in an element). Open Excel and choose Import. The files are delimited with spaces and columns in the text file will go into columns in Excel. You can also do this using MATLAB. For example, A=[then paste the file]; you can then create the columns as a = A(:,1), b = A(:,2). You can do the same thing with the solution by choosing Results/Data Set and choosing a Solution. Then "add to Export" and pick the file type (.txt, .csv, .dat); save it by Browsing to set the area and then use the export icon.

LiveLink with MATLAB

If you have a license for "Comsol LiveLink™ for MATLAB" (part of the Comsol offerings) and MATLAB, you can use MATLAB functions inside Comsol Multiphysics. To do so,

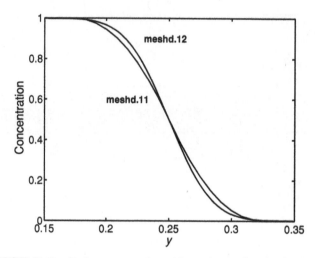

FIGURE D.13 Outlet concentration with regular mesh and refined mesh.

first open "Comsol with MATLAB." The first time you do this you will be asked to supply a username and password. You can create the m-file in MATLAB or create it in a text file and save it as filename.m. Create your Comsol model, or open a previous model. Right click "Global Definitions" and go to "Functions/MATLAB." One window asks you to define the function by giving it a name (e.g., testfn) and identifying the arguments (e.g., T). The arguments can be variables in the Comsol model, and the function name can then be used elsewhere in the model. Also, you can give the derivative by specifying the same function name, the same variable name, and the derivative of the function with respect to the variable. In this example, the MATLAB function is

```
%saved as testfn.m
function y=f(T)
y=T.*T
```

Then in the model you can call the function testfn(T) and the MATLAB problem will be executed. Note that T (in a one-dimensional problem) is a vector of values at nodes, so the notation T*T would not work; T.*T is needed. Before running the code, be sure the MATLAB m-file and the Comsol file are in the same directory; otherwise the program will respond that it cannot find the function.

In future calculations, you can open just Comsol Multiphysics; the link with MATLAB will be established when it sees that there is a MATLAB function defined. Information about LiveLink with MATLAB is available in "Comsol LiveLink™ for MATLAB" and External MATLAB functions. This connection is much more powerful than in this simple example. You can build models, create geometries and meshes, and do a variety of operations, including loops, if it is more convenient to do so in MATLAB.

Variables

You can define these in either "Global Definitions" or Definitions under a model. But, if you want to parameterize the solution, the parameters need to be identified in "Global Definitions." The x-derivative of the variable "vary" is d(vary,x).

Animation

To make a movie, choose "Results/Report" and right click to get Animation. You can save it as a file or open it in your browser. Usually this is done where the solution is plotted versus time, and you can select which time values you want to include in the movie. Near the top of the Settings window is Subject with possible choices, such as "2D Plot Group." Here you pick the one you want, which was set beforehand using Results and right clicking on one of the "Plot Groups." Thus, *what* you plot is determined in the "Plot Group," which you choose in the Animation settings. You have a choice for output of the movie: gif, flash, or avi. The avi format only can be chosen if you are using Windows. Then choose the export icon. The movie will appear in the graphics window, and a file will be saved, too. If you select "Open in Browser," your browser will be opened and the plot will occur there over an over. Both gif and flash format will run in your browser (on a Macintosh, too), and gif will run in QuickTime Player on a Macintosh.

Studies

When you right click on the name of your problem (root) one option is to "Add Study." Suppose you want to compare the solution obtained by using two different meshes (or even two different models). To compare the solution by using two different meshes, in the Model, right click and add a mesh; now there is a "Mesh 1" and "Mesh 2." In the first Study, when you open Stationary, for example, in Settings, "Mesh 1" is chosen by default. Now you could solve the problem on this mesh ("Study 1," right click =), get the figure or results, save them, then go back and change to "Mesh 2" and solve the problem again. This time the second mesh would be used. But if you want to quickly go back and forth between the solutions, and their plots, go to the "Study 2" and choose "Mesh 2" for it. Then solve the problem ("Study 2," right click =). Then in Results, create two "Plot Groups," define the plot, but in one of them choose the data set: "Solution 1" and in the other one choose the data set: "Solution 2" (or similar ones). Click on the plot icon to see the plot of each. If you click on nodes in the "Model Builder/Results," you can see first one and then the other. You can do the same thing with "Derived Results." For a transient problem, to plot the value of the solution at a point for all time, define the point as above and choose "Cut Point 2D" in 1D Plot Group. For a transient problem the time variable will already be chosen. To plot the solution along a line for several time values, define the line as above and choose "Cut Line 2D" in the 1D Plot Group. To evaluate an integral of an expression, choose Results/Derived Value/Average or Integration. Clicking the equal sign causes a table of values to be displayed under Results, and this Table can be plotted in a 1D Plot Group by choosing Table Graph. This can also be done by choosing Definitions/Domain Probe and choosing the integral of an expression, called dom1. Then in Results/1D Plot Group, choose Global and the expression dom1.

Help with Convergence

If a non-linear problem does not converge, try putting a good guess in "Initial conditions." Also click on the question mark, type "convergence," and look at "Achieving Convergence when Solving Nonlinear Equations."

Help with Time-Dependent Problems

Implicit methods are used to solve time-dependent problems. This generally involves solve a linear algebra problem, $Ax = b$, at each iteration. In a nonlinear problem, the problem is linearized about the current iterative solution, which requires evaluating the derivative of the equations with respect to each variable. The result is called the Jacobian. It is possible to speed things up by evaluating the Jacobian only every few iterations, since this takes considerable time, but when doing so the iterations may not converge. Thus, you have the option to control some of the time integration methods. To do so, click on "Study 1/Solver Configurations/Solver 1/Time-Dependent Solver 1." Then you can choose Direct, Advanced, Fully Coupled, and Direct. Click on one of them and the dynamic help icon (looks like a laptop) to see the options. Generally for a nonlinear problem, the time integration is more robust when choosing "Fully Coupled," "Jacobian Update" every iteration, allowing several iterations, and possibly increasing the tolerance factor. Of course, it is sometimes best to *decrease* the tolerance factor to make sure the solution is good, since once it gets off, it may behave badly. There are other options associated with solving the linear algebra problem, and these involve how the problem is solved, how it incorporates the sparsity of the matrix (zeros), and perhaps it will be solved iteratively even for a linear problem to keep the memory requirements small. Consequently, you may need to experiment with these tools to get the best performance for your problem.

Jump Discontinuity

To illustrate how to program a jump discontinuity in Comsol Multiphysics, consider the following problem:

$$\frac{d^2c}{dx^2} = 0, \quad 0 < x < 1; \quad \frac{d^2n}{dx^2} = -10n, \quad 1 < x < 2$$

$$c(0) = 1, \quad \left.\frac{dc}{dx}\right|_{x=1} = \left.\frac{dn}{dx}\right|_{x=1}, \quad n(1) = \frac{c(1)}{1 + 10 * c(1)}, \quad \left.\frac{dn}{dx}\right|_{x=2} = 0 \tag{D.1}$$

There is diffusion in the left region and diffusion and reaction in the right region. Set the diffusivity to 1.0 for both problems. Be sure to unclick convection if it is clicked at the start. At the interface the flux is continuous but the value of n at $x = 1$ is a nonlinear function of the value of c at the same point. Open Comsol Multiphysics and create the two problems using Transport module as Model 1 and Model 2. To connect the models, in Model 2, open Linear Extrusion and set the left point in Model 2 to be identified with the right point in Model 1 (identified by the Geometry 1 and 2). Then in Model 1, create a variable flux = mod2.linext1(mod2.c2x). In Model 1, for the right-hand boundary, set the value of the flux to this variable flux. In Model 1, open Linear Extrusion and set the right point in Model 1 to be identified with the left point in Model 2 (again identified by the Geometry 1 and 2). In Model 2 define the variable cb=mod1.linext2(mod1.c). Also define nofc=cb/(1+10*cb). Then in Model 2, set the concentration on the left point to nofc. Eliminate Study 2 and only use Study 1 solving Model 1 and 2. The solutions are shown in Figure D.14. The values of c and n obey the equilibrium constraint at the inner boundary and the fluxes are continuous across that boundary. The fluxes can be plotted, too, to get their values.

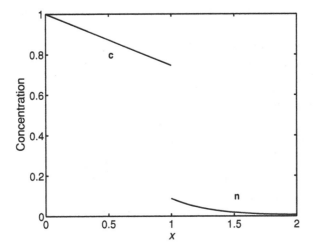

FIGURE D.14   Solution with a jump discontinuity.

TABLE D.1   Selected Model Library Problems and Techniques in Comsol Multiphysics

Chemical Engineering

    Surface diffusion and surface reaction
    Tubular reactor; weak form and boundary interface

Fluid Dynamics

    Time-dependent solution of a falling sand particle
    Micromixer
    Sloshing_tank; moving mesh, free surface flow

Multiphysics

    Free_convection
    Marangoni_convection; heated from side; how to include surface tension in BC

Help

When you are stuck figuring out a command, or node, click on it and then click the question mark. A window will appear with information about that specific command. The information may be too general, telling you what it does, but not how to use it, but that at least is a start.

Another source of help is the collection of problems in the Model Library. This is accessed by changing from Settings to Model Library in the middle window. Table D.1 lists some key problems and techniques.

APPLICATIONS OF COMSOL MULTIPHYSICS

This Comsol Multiphysics options described in this Appendix merely touch the surface of possibilities. Additional examples and applications are provided throughout the book, and special techniques are listed here:

- Solving ordinary differential equations that are initial value problems, Chapter 8, p. 149
- Solving ordinary differential equations that are boundary value problems, Chapter 9.
- Icons in Comsol Multiphysics, Chapter 9, p. 183–184
- Parametric solutions, Chapter 9, p. 188
- Use of defined variables, Chapter 9, p. 193, 203, Chapter 10, p. 228
- Effect of P1 and P2 elements, Figure 9.13
- Problems with strong convection, Chapter 9, p. 201, Chapter 11, p. 267, 278, 280
- Integration in cylindrical geometry, Chapter 9, p. 192
- Viscous dissipation, Chapter 11, p. 271
- Wall reactions, Chapter 11, p. 274
- Model couplings, Chapter 10, p. 248

APPENDIX E

MATHEMATICAL METHODS

The software packages Excel®, MATLAB®, Aspen Plus®, and Comsol Multiphysics® use mathematical methods that are preprogrammed. Engineers can solve chemical engineering problems without knowing the details of those mathematical methods, but some of you may be interested in the methods themselves. A brief overview is given here. There are numerous books that describe the methods in greater detail, including Finlayson (1980, 1992), Constantinides and Mostoufi (1999), and Cutlip and Shacham (1999), as well as the Mathematics Section of Perry's Chemical Engineers' Handbook (Finlayson and Biegler, 2008) and Ullmann's Encyclopedia of Industrial Chemistry (Finlayson, 2006). The subjects include

- Nonlinear algebraic equations.
- Ordinary differential equations as initial value problems.
- Ordinary differential equations as boundary value problems.
- Partial differential equations in time and one space dimension.
- Partial differential equations in two or more spatial dimensions (and possibly time, too).

In the case of differential equations, we are always replacing the differential equation with a finite approximation, with a finite number of grid points or elements, and it is always necessary to assess the accuracy by solving the problem with more than one mesh, or at least comparing integral balances, like flow in versus flow out. These considerations are emphasized by the discussion in this Appendix as well as several problems in Chapters 8–11.

Introduction to Chemical Engineering Computing, Updated Second Edition. Bruce A. Finlayson.
© 2014 John Wiley & Sons, Inc. Published 2014 by John Wiley & Sons, Inc.

ALGEBRAIC EQUATIONS

A single nonlinear algebraic equation can be written as

$$f(x) = 0 \qquad \text{(E.1)}$$

Iterative methods are used to solve such equations, and the successive substitution and Newton–Raphson methods are described here. Let x^k be the value of x at the kth iteration. The goal is to make x^k satisfy Eq. (E.1) as k increases.

Successive Substitution

In this method, the equation is multiplied by a constant, β, and an iterative equation is used:

$$x^{k+1} = x^k + \beta f(x^k) \qquad \text{(E.2)}$$

Under certain conditions, it can be proved that this method converges (Finlayson, 1980). The conditions require a bounded derivative df/dx and also a particular sign of β. The smaller the value of β the slower is the convergence to the solution, but also then it is more likely that a solution can be found. It is sometimes hard to predict a good value of β.

Newton–Raphson

In this method, a Taylor series is written for Eq. (E.1), evaluated at the xk value:

$$f(x^{k+1}) = f(x^k) + \left.\frac{df}{dx}\right|_{x^k} (x^{k+1} - x^k) + \left.\frac{d^2 f}{dx^2}\right|_{x^k} \frac{(x^{k+1} - x^k)^2}{2!} + \cdots \qquad \text{(E.3)}$$

The terms higher than linear in $x^{k+1} - x^k$ are discarded, and the left-hand side is set to zero. You want the left-hand side to be zero, of course, and when you are close to the solution the higher order terms are small. The iterations then proceed as follows:

$$x^{k+1} = x^k - \frac{f(x^k)}{df/dx|_{x^k}} \qquad \text{(E.4)}$$

Convergence proofs are available under certain conditions (Finlayson, 1980), and once the iterate value gets close to the solution, the convergence is very rapid. This method is generally better than the successive substitution method, except for special cases, but sometimes a good initial guess is required.

The Newton–Raphson method can be extended to sets of nonlinear equations. When the equations are

$$f_i(\{x_j\}) = 0, \, i, j = 1, \ldots, n \qquad \text{(E.5)}$$

the expansion is

$$f_i(\{\mathbf{x}^{k+1}\}) = f_i(\{\mathbf{x}^k\}) + \sum_{j=1}^{n} \left.\frac{\partial f_i}{\partial x_j}\right|_{\mathbf{x}^k} (x_j^{k+1} - x_j^k) + \cdots \qquad \text{(E.6)}$$

You set the left-hand side to zero, neglect the second order and higher terms, and rearrange the equations as

$$\sum_{j=1}^{n} \frac{\partial f_i}{\partial x_j}\bigg|_{\mathbf{x}^k} (x_j^{k+1} - x_j^k) = -f_i(\{\mathbf{x}^k\}) \tag{E.7}$$

This is a linear problem, which can be written in the form

$$\sum_{j=1}^{n} A_{ij}^k (x_j^{k+1} - x_j^k) = -f_i(\{\mathbf{x}^k\}), \ A_{ij}^k = \frac{\partial f_i}{\partial x_j}\bigg|_{\mathbf{x}^k} \tag{E.8}$$

This problem is now a set of linear equations, which is easily solved. The matrix A is called the Jacobian. Linear algebra is used to solve this set of equations (Finlayson, 1980, 2003):

$$\mathbf{A}^k(\mathbf{x}^{k+1} - \mathbf{x}^k) = -\mathbf{f}^k, \mathbf{x}^{k+1} = \mathbf{x}^k - (\mathbf{A}^k)^{-1}\mathbf{f}^k \tag{E.9}$$

After each step, the Jacobian and function f are reevaluated and the procedure is repeated. The Newton–Raphson method for sets of equations also converges quickly once one is close to the solution, but iterations can easily diverge unless special precautions are taken. The programs used in this book take those precautions, which makes them more reliable than programs you might write yourself. Problems that ask you to program the Newton–Raphson method include 2.10, 2.11, 3.16, 3.17, 4.15, 4.16, and 8.21.

The Newton–Raphson method requires that you differentiate the function with respect to all the variables. The secant method avoids that mathematical step and uses a numerical difference to calculate the derivative:

$$\frac{\partial f_i}{\partial x_j}\bigg|_{\mathbf{x}^k} = \frac{f_i(x_j^k) - f_i(x_j^{k-1})}{x_j^k - x_j^{k-1}} \tag{E.10}$$

The value of the functions is kept after each iteration in order to make this calculation. The Wegstein method is essentially a secant method, with some constraints on the parameters as described in Chapter 7. It is also possible to use a numerical derivative:

$$\frac{\partial f_i}{\partial x_j}\bigg|_{\mathbf{x}^k} = \frac{f_i(x_j^k + \varepsilon) - f_i(x_j^k)}{\varepsilon} \tag{E.11}$$

One of the advantages of Comsol Multiphysics is that it differentiates the equations symbolically, including any terms that you add to the problem, and this enhances convergence. The parametric solver in Comsol Multiphysics is possible because the Newton–Raphson method is being used. Consider Eq. (E.1) when it depends upon a parameter:

$$f(x(\alpha), \alpha) = 0 \tag{E.12}$$

Differentiate Eq. (E.12) with respect to α:

$$\frac{\partial f}{\partial x}\bigg|_{\alpha} \frac{dx}{d\alpha} + \frac{\partial f}{\partial \alpha}\bigg|_{x} = 0 \tag{E.13}$$

Once the solution to Eq. (E.12) is known, it is easy to find the derivative of the solution with respect to the parameter. Because the matrix inverse is already known, this calculation is fast. [Actually, an LU decomposition would be done rather than an inverse, as it is twice as fast. See Finlayson (1980) and the book website.] When changing the parameter, a linear extrapolation about the current solution is then possible:

$$x(\alpha + \Delta\alpha) = x(\alpha) + \frac{dx}{d\alpha}\Delta\alpha \qquad (E.14)$$

This often gives an excellent guess of the solution for the next value of α. Sometimes even these methods do not solve the very hardest problems. In that case, a *homotopy method* may be necessary (Finlayson, et al. 2006).

ORDINARY DIFFERENTIAL EQUATIONS AS INITIAL VALUE PROBLEMS

There are excellent methods for solving sets of first-order ordinary differential equations that are initial value problems (ODE-IVP). An example of an ODE-IVP is

$$\frac{dy}{dt} = f(t, y), \, y(0) = y_0 \qquad (E.15)$$

Note that all the conditions are known at one time, $t = 0$. Thus it is possible to calculate the function on the right-hand side at $t = 0$ to get the derivative there. This makes the set of equations initial value problems. The equations are ordinary differential equations because there is only one independent variable. Any higher-order ordinary differential equation can be turned into a set of first-order ordinary differential equations; they are initial value problems if all the conditions are known at the same value of the independent variable (Finlayson, 1980, 2008, pp. 3 and 48–51; Finlayson, et al. 2006, vol. B1, pp. 47–59]. The methods for initial value problems are explained here for a single equation; extension to multiple equations is straightforward. These methods are used when solving plug-flow reactors (Chapter 8) as well as time-dependent transport problems (Chapters 9–11).

Euler's Method

In numerical solutions of ODE-IVP, the solution y at the point t^n is represented by y^n. The simplest method is Euler's method, which is obtained by writing a difference expression for the derivative:

$$\frac{y^{n+1} - y^n}{\Delta t} = f(t^n, y^n) \qquad (E.16)$$

The first-order approximation of the derivative is not very accurate, so that this method is not very accurate either. It is easy to program, though. Rearrange Eq. (E.16) as

$$y^{n+1} = y^n + \Delta t f(t^n, y^n) \qquad (E.17)$$

This equation can be applied repeatedly to go from $t = 0$ to any time, t. This can be done easily in Microsoft Excel. To solve

$$\frac{dy}{dt} = -y, \, y(0) = 1 \qquad (E.18)$$

put the time step in cell E2; let column A be the time, column B the y, and column C the first derivative, dy/dt. The cells representing dy/dt are calculated using $-y$ in the same row and put in column C: C2 = –B2. Then the value of y in the next row is taken as

$$B3 := B2 + \$E\$2 * C2. \qquad (E.19)$$

which is the equivalent of Eq. (E.17).

The stability limit for the Euler method applied to

$$\frac{dy}{dt} = -\lambda y \qquad (E.20)$$

is $\lambda \Delta t \leq 2$ (Finlayson, 1980). For other problems, $dy/dt = f(y,t)$, it is

$$\left| \lambda \frac{df}{dy} \right| \leq 2 \qquad (E.21)$$

For systems the λ is the absolute value of the largest eigenvalue of the Jacobian, $\partial f_i / \partial y_j$, assuming all eigenvalues are negative.

There are three major problems for the Euler method. First, the accuracy is poor, since the method is based upon Eq. (E.16) in which only a first-order difference expression is used. The errors in the method are proportional to Δt. Second, stability is difficult to achieve for many problems. The only way to have a stable Euler method is to use a small enough time step-size, but you may not know what value is sufficient. Furthermore, a value that is sufficient at the beginning may not be sufficient later on, and it may take an excessively long time to finish the computation. Third, to validate the results it is necessary to solve the problem at least twice, with different time-steps. The method can, however, be programmed in Excel, and problems 8.19 and 8.20 use this method.

Runge–Kutta Methods

The Euler method can be improved by doing the calculation in two steps:

$$\bar{y} = y^n + \frac{\Delta t}{2} f(t^n, y^n)$$
$$y^{n+1} = y^n + \Delta t f\left(t^n + \frac{\Delta t}{2}, \bar{y}\right) \qquad (E.22)$$

This is a second-order Runge–Kutta method (Finlayson, 1980), sometimes called the mid-point rule. The first step is an approximation of the solution halfway between the beginning and ending time, and the second step evaluates the right-hand side at that mid-point. The error goes as $(\Delta t)^2$, which is much smaller than achieved with the Euler method. The second-order Runge–Kutta methods (there are several) also have a stability limitation. Problem 8.22 requires programming the Runge–Kutta method, and the results can be compared with those of 8.19 which use the Euler method. The stability limit for this method is also $\lambda \Delta t \leq 2$.

MATLAB and ode45 and ode15s

The method used in MATLAB is a fourth-order Runge–Kutta method with a variable step-size. The Δt is changed to guarantee the accuracy of the calculation. The accuracy is estimated each step, and the step-size is reduced to meet the specified accuracy. The method still has a stability limitation. The variable step size overcomes two of the problems—knowing what step size to use and using a small enough step-size to guarantee a specified accuracy. This makes the method very robust, which is why it is the workhorse in MATLAB. The method is called ode45 because the fourth-order method is chosen in such a way that it achieves fifth-order accuracy. It is based on an older program called RKF45 that was written by Boeing Computer Services many decades ago. More details can be found in Finlayson et al. (2008, pp. 3–49) and Finlayson et al. (2006, vol. B1, pp. 1–49, 60–61) and the book website. The stability limit of this method is $\lambda \Delta t \leq 2.8$.

In some problems, certain parameters vary quickly and others vary more slowly. For example, in packed-bed chemical reactors, the concentration can change quickly in time, but the temperature will not change rapidly because of the large heat capacity of the solid. Such problems are called "stiff." The mathematical definition is that the eigenvalues of the Jacobian are widely separated, certainly by factors of thousands, perhaps by factors of millions. It turns out that the stable step size of an explicit method is limited by the largest eigenvalue (fastest responding variable), but the time for something to happen is controlled by the smallest eigenvalue (slowest responding variable). Thus, for stiff problems, many time steps would need to be taken with a small time step, and the calculations would be very slow.

Implicit methods have been developed to overcome this problem. The Backward Euler method is

$$\frac{y^{n+1} - y^n}{\Delta t} = f(t^{n+1}, y^{n+1}), \quad \text{or} \quad y^{n+1} - \Delta t f(t^{n+1}, y^{n+1}) = y^n \tag{E.23}$$

This method is still first order, but it is stable for any step size. The only problem is that Eq. (E.23) is a nonlinear algebraic equation for y^{n+1}, and if there are several differential equations, there will be a set of nonlinear algebraic equations to be solved. This is not an easy task. Some higher order methods are stable, too; the best are Gear's backward difference formulas (Gear, 1971; Finlayson et al., 2008, pp. 3–49) and the book website. Methods such as these are also included in MATLAB, as ode15s and ode23s. These programs adjust the step-size to maintain a specified accuracy. They also use sophisticated methods to solve the set of nonlinear equations. If a solution cannot be found, the step-size is reduced; then a solution is easier to find.

It is also possible to have differential–algebraic equations. There will be both differential and algebraic equations, with all the variables possibly occurring in all the equations. This essentially means that the variables determined from the algebraic equations are acting infinitely fast; this is merely a system of stiff equations taken to an extreme. Special methods exist for these problems, too. Because some of the variables are supposed to change immediately, they cannot be included in the error tests used to select the time step. Information is available by using the help menu, MATLAB/User's Guide/Mathematics/Calculus/Ordinary Differential Equations/Initial Value Problems/Differential–Algebraic Equations.

ORDINARY DIFFERENTIAL EQUATIONS AS BOUNDARY VALUE PROBLEMS

Transport problems involve second-order differential equations. Although a single second-order differential equation can be reduced to two first-order differential equations, the problem is not an initial value problem because there are boundary conditions at two different values of the independent variable. This section describes methods for solving such problems and illustrates them with the finite difference method as well as the finite element method.

The prototype problem is heat conduction in a slab or reaction and diffusion in a flat layer, as described in Chapter 9. Here, the methods are demonstrated on a problem with reaction and diffusion in a flat layer with a first-order chemical reaction:

$$D\frac{d^2c}{dx^2} = kc, \text{ with boundary conditions } \frac{dc}{dx}(0) = 0, c(R) = c_0 \qquad (E.24)$$

Finite Difference Method

This method is explained first because it is the oldest and the simplest to explain, even though it is not the method embedded in Comsol Multiphysics. The first step is to divide the spatial region into smaller regions by placing grid points on the line (see Figure E.1). Here, the grid points are evenly spaced; although this is not necessary, it does make the presentation easier to grasp. The unknown function $y(x)$ is expanded in a Taylor series about the (unknown) value at a grid point:

$$y(x) = y(x_i) + \frac{dy}{dx}\bigg|_i (x - x_i) + \frac{d^2y}{dx^2}\bigg|_i \frac{(x - x_i)^2}{2!} + \cdots \qquad (E.25)$$

This equation is evaluated for points on either side:

$$y(x_{i+1}) = y(x_i) + \frac{dy}{dx}\bigg|_i (x_{i+1} - x_i) + \frac{d^2y}{dx^2}\bigg|_i \frac{(x_{i+1} - x_i)^2}{2!} + \cdots$$
$$y(x_{i-1}) = y(x_i) + \frac{dy}{dx}\bigg|_i (x_{i-1} - x_i) + \frac{d^2y}{dx^2}\bigg|_i \frac{(x_{i-1} - x_i)^2}{2!} + \cdots$$
$$(E.26)$$

If the last two equations are subtracted and rearranged; you can obtain the following approximation for the first derivative when the grid spacing is uniform:

$$\frac{dy}{dx}\bigg|_i = \frac{y(x_{i+1}) - y(x_{i-1})}{2\Delta x} \text{ when } \Delta x = (x_{i+1} - x_i) = -(x_{i-1} - x_i) \qquad (E.27)$$

This equation is second order, meaning that the errors decrease proportional to Δx^2 once Δx is small enough. The two equations can also be rearranged by neglecting the second-order terms, to obtain two other approximations for the first derivative:

$$\frac{dy}{dx}\bigg|_i = \frac{y(x_{i+1}) - y(x_i)}{\Delta x}$$
$$\frac{dy}{dx}\bigg|_i = \frac{y(x_i) - y(x_{i-1})}{\Delta x}$$
$$(E.28)$$

FIGURE E.1   Finite difference grid.

These approximations are only first order, and errors are proportional to Δx when it is small enough. Another rearrangement gives an approximation for the second derivative. Take the sum of Eq. (E.26) and rearrange it to give

$$\left.\frac{d^2y}{dx^2}\right|_i = \frac{y(x_{i+1}) - 2y(x_i) + y(x_{i-1})}{\Delta x^2} \tag{E.29}$$

This approximation can be shown to be second order in Δx.

The idea in the finite difference method is that the differential equation, valid for all positions x, is replaced by a set of equations representing the equation only at the grid points, using the equations derived above. Thus, for problem Eq. (E.24), the equation at grid point i is

$$D\frac{c_{i+1} - 2c_i + c_{i-1}}{\Delta x^2} = kc_i, \text{ where } c_i = c(x_i) \tag{E.30}$$

Such an equation is written for each interior grid point shown in Figure E.1. At the boundaries, the finite difference method uses the boundary conditions, again with the finite difference representation of derivatives:

$$\left.\frac{dc}{dx}\right|_1 = \frac{c(x_2) - c(x_1)}{\Delta x}, c_{n+1} = 1 \tag{E.31}$$

It is possible to use a more accurate (second order) derivative, at the cost of extra programming. Equations (E.30) and (E.31) represent a set of $n + 1$ algebraic equations in $n + 1$ unknowns. The most accurate way to apply the boundary condition at $x = 0$ in Eq. (E.24) is to introduce a false boundary point at $x = -\Delta x$; call the value there c_0. The differential equation at $x = 0$, point 1, is then

$$D\frac{c_2 - 2c_1 + c_0}{\Delta x^2} = kc_1 \tag{E.32}$$

But by the boundary condition at $x = 0$,

$$\frac{c_2 - c_0}{2\Delta x} = 0, \quad \text{or} \quad c_2 = c_0 \tag{E.33}$$

Thus the differential equation becomes

$$D\frac{2(c_2 - c_1)}{\Delta x^2} = kc_1 \tag{E.34}$$

This must be programmed separately from Eq. (E.30).

If you are using MATLAB, you can use linear algebra to solve the set of equations. For this simple, one-dimensional problem, a special method is used for a tridiagonal matrix. See Finlayson (1980) for complete details.

Finite Difference Method in Excel

It is possible to program the finite difference method in Excel and use the "Calculation" feature to handle the circular reference. Turn off the iteration, prepare the spreadsheet, and then turn the calculation back on. Whether this converges depends on the initial guess.

TABLE E.1   Finite Difference Method in Excel for Boundary Value Problems

D1	E1	F1
c_{i-1}	c_i	c_{i+1}

To see how this is done, rearrange Eq. (E.30) to the following form:

$$c_i = \frac{c_{i+1} + c_{i-1}}{2 + k\Delta x^2 / D} \tag{E.35}$$

Identify the cells in the spreadsheet with the value of concentration at a node, i, as shown in Table E.1. Put the value of $k\Delta x^2 / D$ in cell A2. The equation for cell E1 would be different:

$$= (D1 + F1)/(2 + \$A\$2) \tag{E.36}$$

Next, copy this equation over a series of cells that correspond to the number of grid points. For the first and last cell (grid point) use a different equation, appropriate to the boundary condition. Then turn on the iteration feature to get the solution.

The next step is to resolve the problem using more grid points, and a smaller Δx. This means you have to program the spreadsheet again, but it does give you an indication if the answer changes much as the mesh is refined, which is what assures you that you have solved the problem accurately enough. Problems 9.30 (linear and nonlinear problems) and 9.31, employ this method.

After the linear algebra problem is solved, the result is an approximation for $y(x_i)$ at each ith grid point. This can be plotted, and the points can be joined by straight lines, giving an approximation to the solution of the differential equation. More elaborate methods of interpolation are also possible. The solution to the finite difference equations, though, is an approximation to the solution to the differential equation. To assess its accuracy, you must refine the grid and solve another problem. By doubling the number of points once, and then twice, it is possible to see if the solution changes appreciably. If not, you have found an adequate approximation.

Finite Element Method in One Space Dimension

The finite element method proceeds in a slightly different way. The unknown solution $y(x)$ is expanded in a series of functions, called trial functions:

$$c(x) = \sum_{i=1}^{n+1} c_i N_i(x) \tag{E.37}$$

The functions in the finite element method are constant, linear, or quadratic functions of position in a small region (called a finite element). The idea is explained for interpolation of a known function, Eq. (E.38), which is plotted in Figure E.2:

$$z = \frac{11x^4 + x}{12} \tag{E.38}$$

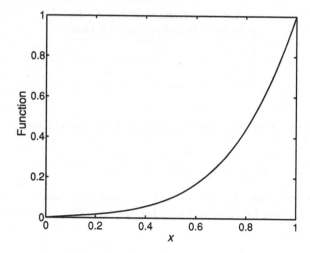

FIGURE E.2   Function $z = (11 \times 4 + x)/12$

The region $0 \le x \le 1$ is divided into finite elements using the same grid points shown in Figure E.1. If the approximations are taken as constant in each element, two interpolations are shown in Figure E.3. Even though only 8 or 16 elements are used, the general shape of the function is clear.

Next, the same mesh is used, but with linear trial functions. Interpolation with 4, 8, and 16 elements is shown in Figure E.4. This representation is much better than that in Figure E.4. Indeed, with 16 elements, the curve—although composed of straight-line segments—looks like a smooth curve. Whatever the trial function, the approximation is better when more elements are used, and mesh refinement is the easiest way to ensure that the approximation is accurate enough.

It is beyond the scope of this book to describe the method used to obtain the coefficients in Eq. (E.30), and how the boundary conditions are included, but complete details are available (Finlayson, 1972, 1980). There are a variety of books available about the finite element method. A book by Gresho and Sani (1998) focuses on flow and convection/diffusion is. In the finite element method, the representation of the second derivative is the same as given by the finite difference method, but the representation of the function is different. The finite element method applied to Eq. (E.24) with linear trial functions is

$$D\frac{c_{i+1} - 2c_i + c_{i-1}}{\Delta x^2} = \frac{k}{6}(c_{i+1} + 4c_i + c_{i-1}), \text{ where } c_i = c(x_i) \qquad \text{(E.39)}$$

If quadratic functions are used, there are more differences; in fact, the only time the representations are the same is for first and second derivatives that are second order and when using linear trial functions in the finite element method.

The boundary conditions are applied in the finite element method in a different way than in the finite difference method, and then the linear algebra problem is solved to give the approximation of the solution. The solution is known at the grid points, which are the points between elements, and a form of the solution is known in between, either linear or quadratic in position as described here. (Comsol Multiphysics has available even higher order approximations.) The result is still an approximation to the solution of the differential equation, and the mesh must be refined and the procedure repeated until no further changes

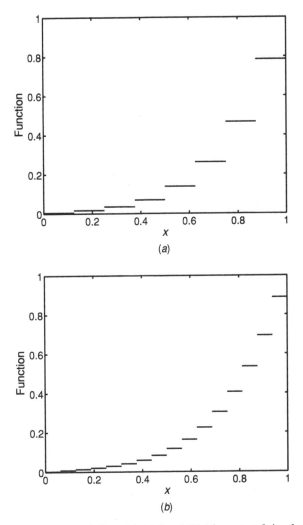

FIGURE E.3   Interpolation with (*a*) 8 and (*b*) 16 constant finite elements.

are noted in the approximation. The finite element can be programmed in Excel, too, but the details are more complicated and best left to real finite element programs.

Initial Value Methods

Another option for solving boundary value problems is to treat them like initial value problems. Since a second-order equation can be reduced to two first-order equations, two initial conditions are necessary. One condition will be known at a boundary. Simply assume a value for the other dependent variable at that same boundary, integrate to the other side, and check if the required boundary condition is satisfied. If not, change the initial value and repeat the integration. The success of this method depends upon the skill with which you program the iterations from one trial to the next and successive substitution or Newton–Raphson methods can be used to control the iterations. In the case of the Newton–Raphson methods,

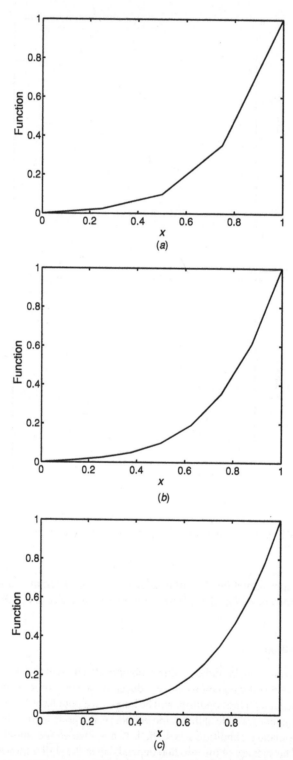

FIGURE E.4   Interpolation of $z = (11 \times 4 + x)/12$ with (a) 4, (b) 8, and (c) 16 linear finite elements.

one must differentiate the solution with respect to the unsatisfied boundary condition and solve a problem with twice as many unknowns (Finlayson, 1980, pp. 139–140).

PARTIAL DIFFERENTIAL EQUATIONS IN TIME AND ONE SPACE DIMENSION

Time-dependent transport problems lead to partial differential equations, but these are easily solved by combining the methods for initial value problems and boundary value problems. Take the heat conduction problem in Eq. (9.28):

$$\frac{\partial T}{\partial t} = \alpha \frac{\partial^2 T}{\partial x^2} \tag{E.40}$$

Write the temperature as a function of position, x, and time, t, but consider only the positions where there is a grid point, x_i. The function $T(x_i,t)$ is a function of only one independent variable, t, and it can be written as $T_i(t)$. It can be differentiated in time as follows:

$$\frac{\partial}{\partial t}[T(x_i, t)] = \frac{dT_i(t)}{dt} \tag{E.41}$$

Thus, the differential equation (E.40) defined at the x_i grid point using the finite difference method is

$$\frac{dT_i}{dt} = \alpha \frac{T_{i+1} - 2T_i + T_{i-1}}{\Delta x^2} \tag{E.42}$$

This equation is written for every grid point, and you have a set of initial value problems for the set of $\{T_i(t)\}$:

$$\frac{dT_2}{dt} = \alpha \frac{T_3 - 2T_2 + T_1}{\Delta x^2}, \frac{dT_3}{dt} = \alpha \frac{T_4 - 2T_3 + T_2}{\Delta x^2}$$

$$\frac{dT_i}{dt} = \alpha \frac{T_{i+1} - 2T_i + T_{i-1}}{\Delta x^2}, \frac{dT_n}{dt} = \alpha \frac{T_{n+1} - 2T_n + T_{n-1}}{\Delta x^2}$$

$$\text{or } \frac{dT_i}{dt} = \sum_{j=1}^{n+1} A_{ij} T_j, i = 2, \ldots, n \tag{E.43}$$

The boundary conditions also need to be included, leading to a set of differential–algebraic equations as an initial value problem. It is easier, though, if the algebraic equations be separated out. A boundary condition involving the first derivative can be combined with the differential equation as is done above for boundary value problems. A boundary condition involving the function can just be applied in whatever equation requires that specific value. Then the remaining equations are all ordinary differential equations.

Although the standard methods can be applied, the set of Eqs. (E.43) is stiff, and becomes stiffer as the grid resolution becomes smaller. One eigenvalue of the matrix on the right-hand side of Eq. (E.43) approximates the slowest time response of the differential equation, but the largest eigenvalue is some constant divided by the grid-spacing squared. Thus, as you refine the mesh, the points become closer together, and the eigenvalues get further apart, making the problem stiffer. Thus, when solving partial differential equations, stiff methods are used; in MATLAB, for example, ode45 is replaced by ode15s or ode23s.

If you want to solve this problem with Excel using the Euler method, you write the equation as

$$T_i^{n+1} = T_i^n + \frac{\alpha \Delta t}{\Delta x^2}(T_{i+1}^n - 2T_i^n + T_{i-1}^n) \tag{E.44}$$

and put T^n in one row and $T^n + 1$ in the row below. Insert the equations for interior points and the boundary conditions and copy down as many rows as you want time steps. You will find that the method is stable only if

$$\frac{\alpha \Delta t}{\Delta x^2} \leq \frac{1}{2} \tag{E.45}$$

This method is used in Problem 9.32.

Problems with Strong Convection

Another complication can be important: in a problem with convection, the rate of convection may be much faster than the rate of diffusion. The solution may be like a wave moving along in space. Such problems lead to difficulties when solved on a fixed mesh, and these difficulties apply to both finite element and finite difference methods. For example, the convective diffusion equation representing convection with velocity v down a packed bed is

$$\frac{\partial c}{\partial t} + u\frac{\partial c}{\partial x} = D\frac{\partial^2 c}{\partial x^2} \tag{E.46}$$

The boundary and initial conditions are

$$-D\frac{\partial c}{\partial x} + uc\bigg|_{x=0} = uc_0, \quad \frac{\partial c}{\partial x}\bigg|_{x=L} = 0, \, c(x,0) = c_{in}(x) \tag{E.47}$$

If the inlet condition changes in time, then a wave moves downstream. This equation can be made nondimensional, and this is instructive here:

$$\frac{\partial c'}{\partial t'} + Pe\frac{\partial c'}{\partial x'} = \frac{\partial^2 c'}{\partial x'^2} \tag{E.48}$$

The Peclet number is the rate of convection divided by the rate of diffusion:

$$Pe = \frac{uL}{D} \tag{E.49}$$

The finite element method applied to this equation gives the following equation at each grid point:

$$\frac{1}{6}\left[\frac{dc_{i+1}}{dt} + 4\frac{dc_i}{dt} + \frac{dc_{i-1}}{dt}\right] + Pe\frac{c_{i+1} - c_{i-1}}{2\Delta x} = \frac{c_{i+1} - 2c_i + c_{i-1}}{\Delta x^2} \tag{E.50}$$

It is possible to show that the finite element solution will oscillate wildly above and below the exact solution whenever there is a sharp change in the concentration and the mesh Peclet number is too large (Finlayson, 1992, p. 29 and the book website). Thus, you need

$$Pe\Delta x \leq 2 \tag{E.51}$$

One way to fix this problem is to refine the mesh until Eq. (E.51) is satisfied, but for large problems that may require too many points (especially in 2D and 3D). Another way is to add some "fake" diffusion to the problem. In this case, a Petrov–Galerkin method is

$$\frac{1}{6}\left[\frac{dc_{i+1}}{dt} + 4\frac{dc_i}{dt} + \frac{dc_{i-1}}{dt}\right] + Pe\frac{c_{i+1} - c_{i-1}}{2\Delta x} = \left[1 + \frac{Pe\Delta x}{2}\right]\frac{c_{i+1} - 2c_i + c_{i-1}}{\Delta x^2}$$

$$(E.52)$$

and this makes the oscillations smaller or eliminates them. The trick is to add enough "fake" diffusion to dampen the oscillations but not so much as to change the solution. In Comsol Multiphysics this is called streamline diffusion or stabilization. There are a number of other techniques for treating this difficulty as summarized by Finlayson (1992). The MacCormack flux-corrected method is a good one, and the program to prepare Figure 9.18b is available from the book website. The finite volume method is what is used with finite difference methods, and good discussions of those are in LeVeque (1992, 2002).

PARTIAL DIFFERENTIAL EQUATIONS IN TWO SPACE DIMENSIONS

You can solve problems in two space dimensions using the same methods as applied in one dimension, except that the details become more complicated. The problems are stiffer, and implicit methods are usually required. The problem of unwanted oscillations is more severe because added diffusion may smooth the solution in the flow direction, as desired, but it may also diffuse it in the direction perpendicular to flow, which is not desired. Careful treatment, though, allows for streamline diffusion, that is, the extra diffusion is only added in the flow direction. These details have been handled by Comsol Multiphysics, which frees you to concentrate on the physics and the problem you want to solve; see stabilization in the settings window for the physics equation. Simple problems in rectangular geometries can be solved in Excel.

Finite-Difference Method for Elliptic Equations in Excel

The finite-difference method is easy to program in Excel, provided the equation is not too complicated. Solving the problem in Excel does require solving it more than once, though, to assess the accuracy. As an example, take the heat conduction equation with a heat generation term, Eq. (E.53):

$$k\left(\frac{\partial^2 T}{\partial x^2} + \frac{\partial^2 T}{\partial y^2}\right) = Q \qquad (E.53)$$

The finite difference form of this equation is

$$\frac{T_{i+1,j} - 2T_{i,j} + T_{i-1,j}}{\Delta x^2} + \frac{T_{i,j+1} - 2T_{i,j} + T_{i,j-1}}{\Delta y^2} = \frac{Q}{k} \qquad (E.54)$$

where $T_{i,j}$ is the temperature at the ith location in the x-direction and the jth location in the y-direction. Take the case when $\Delta x = \Delta y$ and rearrange as shown in Eq. (E.55):

$$T_{i,j} = \frac{T_{i+1,j} + T_{i-1,j} + T_{i,j+1} + T_{i,j-1}}{4} - \Delta x^2\frac{Q}{k} \qquad (E.55)$$

TABLE E.2   Finite Difference Method in Excel for Elliptic Boundary Value Problem

$D5\text{-}T_{i-1,j}$	$E4\text{-}T_{i,j-1}$ $E5\text{-}T_{i,j}$ $E6\text{-}T_{i,j+1}$	$F5\text{-}T_{i+1,j}$

The spreadsheet is arranged as shown in Table E.2. When the value of $\Delta x^2 Q/k$ is placed in cell A2 the equation for cell E5 is

$$= (D5 + F5 + E4 + E6)/4 - \$A\$2 \qquad (E.56)$$

You copy this equation for every internal grid point, set the boundary equations, and turn on the iteration feature to get the solution. Then you have to do it again with a finer mesh to assess the accuracy. If the heat generation term depends on temperature, it is easy to include that complication just by inserting the formula in place of $\$A\$2$.

Finite Element Method for Two-Dimensional Problems

In the finite element method that is used by Comsol Multiphysics, the trial functions are defined on an element. A square element in two dimensions with bilinear trial functions has four trial functions per element. The element covers $0 \le u \le 1, 0 \le v \le 1$, and the trial functions are

$$
\begin{aligned}
N_1 &= (1-u)(1-v), & N_2 &= u(1-v) \\
N_3 &= uv, & N_4 &= (1-u)v
\end{aligned}
\qquad (E.57)
$$

The domain is covered with elements to represent a function on the entire domain. Consider the function

$$z = x^2 \exp(y - 0.5) \qquad (E.58)$$

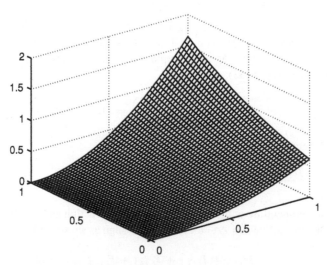

FIGURE E.5   Function $z = x^2 \exp(y - 0.5)$.

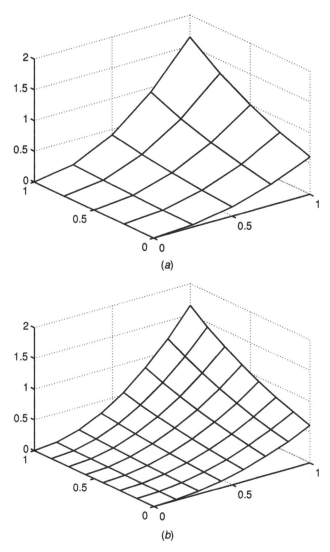

FIGURE E.6   Interpolation of $z = x^2 \exp(y - 0.5)$ with 16 (a) and 64 (b) bilinear finite elements.

which is plotted in Figure E.5. Plots of the finite element interpolation on $4 \times 4 = 16$ bilinear elements or $8 \times 8 = 64$ bilinear elements are shown in Figure E.6. Sixty-four elements do a very good job of representing the function. If quadratic functions are used, fewer elements are needed to approximate a smooth function. Either variational principles or the Galerkin method are used to determine these trial functions to represent the solution to differential equations. The method is beyond the scope of this book and is covered by Finlayson (1972, 1980, 2003).

The recommendation in this book is to use quadratic elements for concentration or temperature in 2D and 3D cases, use quadratic elements for flow in 2D cases and linear elements for flow in 3D cases. For flow the options will be displayed as P1 + P1 and P2 + P1, where the first order of polynomial refers to the velocity components and the second to the pressure. Other options are also available in Comsol Multiphysics and the interested reader is invited to solve a problem using a variety of choices of elements. A real question is "is it better to increase the number of elements or the degree of the polynomial?" In both cases the degree of freedom increases, but accuracy, computer memory used, and time to solution will be affected. The trade-off may depend on the problem you are solving, but generally practitioners make a choice of polynomial and then refine the mesh, always comparing results from different meshes to assess the accuracy.

SUMMARY

This appendix gives a brief overview of the methods that are preprogrammed in the software packages Excel, MATLAB, Aspen Plus, and Comsol Multiphysics. The interested reader may pursue the references, including the book website, for more specific details.

PROBLEMS

$E.1_1$ Verify the stability condition given by Eq. (E.45). To do so, write the equation in the form

$$T_i^{n+1} = AT_{i-1}^n + BT_i^n + CT_{i+1}^n \qquad (E.59)$$

Consider this equation when A, B, and C are positive. Take absolute values of both sides and make the right-hand side bigger by replacing each term by its absolute value ($|a+b| \leq |a|+|b|$). On the right-hand side, replace each T term by the maximum over all values of i. Then if $A + B + C \leq 1$, the maximum value at the $n + 1$ time level is no bigger than the time level at the n time level. Thus, the solution cannot become unbounded. The condition that B be positive gives the stability condition (this is a sufficient condition). You can see what happens when Δt is too big by solving one of the transient heat transfer problems when $B < 0$.

$E.2_1$ Expand on Problem E.1 by writing the differential equation for transient heat transfer in a cylinder and a sphere. Using the condition that A, B, and C are positive, how does the stability condition change from that for planar geometry? Be sure to consider the differential equation at $r = 0$.

$E.3_1$ Expand on Problem E.1 by writing the differential equation for transient heat transfer in two and three dimensions. Applying the condition that all the coefficients have to be positive, what is the stability limitation as you go from 1 to 2 to 3 dimensions?

$E.4_1$ To determine the truncation error in Eqs. (E.27) and (E.29), keep terms up to $(x-x_i)4$ in Eq. (E.25). Perform the same steps used to generate Eqs. (E.25)–(E.29) to get the truncation error.

$E.5_2$ Solve the T-sensor problem in Chapter 11 with $Pe = 500$. Investigate the accuracy of the finite element method by calculating the variance at the outlet. To do this, start with linear discretization and solve the problem on increasingly dense meshes. Keep a record of the variance and the number of degrees of freedom. Then do this again with quadratic discretization, then with cubic discretization. Keep the flow simulation fixed as you improve the discretization for concentration. Plot the variance versus number of degrees of freedom. Plot the absolute value of the residual versus the number of degrees of freedom. For this problem the residual is

$$\text{Residual} = u * cx + v * cy - (cxx + cyy)/Pe \tag{E.60}$$

When you use linear discretization the second derivatives will not be defined. Integrate the absolute value of the residual over the entire domain; the residual will be oscillating around zero so the absolute value is necessary.

REFERENCES

Abramowitz, M. A., Stegan, I. A. (eds.), *Handbook of Mathematical Functions*. National Bureau of Standards: Gaithersburg, MD, 1964.

Aden, A., Ruth, R., Ibsen, K., Jechura, J., Neeves, K., Sheehan, J., Wallace, B. (NREL); Montague, L., Slayton, A., Luka, J. (Harris Group, Seattle), *Lignocelluosic Biomass to Ethanol Process Design and Economics Utilizing Co-Current Dilute Acid Prehydrolysis and Enzymatic Hydrolysis for Corn Stover*, NREL/TP-510–32438, National Renewable Energy Laboratory, 1617 Cole Boulevard, Golden, CO 80401–3393, Contract No. DE-AC36–99-GO10337, June, 2002. The description of the full process is at: http://www.nrel.gov/docs/fy02osti/32438.pdf.

Aguiló, A., Penrod, J. D., Acetaldehyde. In *Encyclopedia of Chemical Processing and Design*, Vol. 1, McKetta, J. J. (ed.). Marcel Dekker: New York, 1976.

Anonymous, The post-alcohol world. *The Economist* Oct. 30, 2010; 84–86.

Aris, R., On the dispersion of a solute in a fluid flowing through a tube. *Proc. Roy. Soc. Lond.* 1956; *A235*, 67–77.

Aspen, The Aspen Model. 2011; http://devafdc.nrel.gov/biogeneral/Aspen_Models./ Download the BKP file (the Aspen backup file), the input file and the spreadsheet.

Barron, R. F., *Cryogenic Systems*. Oxford University Press: New York, 1985.

Beltrami, E. J., *Mathematical Models for Society and Biology*. Academic Press: San Diego, 2002.

Berg, J. C., *An Introduction to Interfaces and Colloids: The Bridge to Nanoscience*. World Scientific: Hackensack, NJ, 2010.

Bird, R. B., Armstrong, R. C., Hassager, O., *Dynamics of Polymeric Liquids*. Wiley-Interscience: New York, 1987.

Bird, R. B., Stewart, W. E., Lightfoot, E. N., *Transport Phenomena*, 2nd ed. Wiley: New York, 2002.

Broyden, C. G., A class of methods for solving nonlinear simultaneous equations. *Math. Comput.* 1965; *19*, 577–593.

Broyden, C. G., The convergence of a class of double-rank minimization algorithms. *J. Inst. Math. Appl.* 1970; *6*, 76–90.

Cavett, R. H., Application of numerical methods to the convergence of simulated processes involving recycle loops. *Am. Petrol. Inst.* 1963; *43*, 57–76.

Chang, M. W., Finlayson, B. A., On the proper boundary condition for the thermal entry problem. *Int. J. Num. Methods Eng.* 1980; *15*, 935–942.

Choi, Y., Stenger, H. G., Water-gas shift reaction kinetics and reactor modeling for fuel cell grade hydrogen. *J. Power Sources* 2003; *124*, 432–439.

Chow, A. Lab-on-a-chip: opportunities for chemical engineering. *AIChE J.* 2002; *48*, 1590–1595.

Chu, S., Carbon capture and sequestration. *Science* 2009; *325*, 1599.

Chung, B. G., Flanagan, L. A., Rhee, S. W., Schwartz, P. H., Lee, A. P., Monuki, E. S., Jeon, N. L., Human neural stem cell growth and differentiation in a gradient-generating microfluidic device. *Lab. Chip* 2005; *5*, 401–406.

Constantinides, A., Mostoufi, N., *Numerical Methods for Chemical Engineers with MATLAB Applications.* Prentice-Hall: Englewood Cliffs, NJ, 1999.

Cooper, H. W., Producing electricity and chemicals simultaneously. *Chem. Eng. Prog.* 2010; *106*, 24–32.

Cowfer, J. A., Gorensek, M. B., Vinyl chloride. In *Encyclopedia of Chemical Technology*, Vol. 24, 4th ed. Kroschwitz, J. I. (Exec. ed.), Mary Howe-Grant (ed.). Wiley: New York, 1997.

Cutlip, M. B., Shacham, M., *Problem Solving in Chemical and Biochemical Engineering with Numerical Methods.* Prentice-Hall: Englewood Cliffs, NJ, 2007.

Dagan, Z., Weinbaum, S., Pfeffer, R., An infinite-serries solution for the creeping motion through an orifice of finite length. *J. Fluid Mech.* 1982; *115*, 505–523.

Datta, A., Rakesh, V., *An Introduction to Modeling of Transport Processes.* Cambridge University Press: Cambridge, UK, 2010.

Deen, W. M., *Analysis of Transport Phenomena.* Oxford University Press: Oxford, UK, 1998.

deGroot, S. R., Mazur, P., *Non-Equilibrium Thermodynamics.* Dover: New York, 1954.

De Leeuw, M. L., Z. *Phys. Chem.* 1911; *77*, 2841; as reported by Hirata et al. (1975).

Denbigh, K., *The Principles of Chemical Equilibrium,* 3rd ed. Cambridge University Press: Cambridge, UK, 1971.

Ellis, S. R. M., Garbett, R. O, *Ind. Eng. Chem.* 1960; *52*, 385; as reported by Hirata et al. (1975).

Finlayson, B. A., *The Method of Weighted Residuals and Variational Principles.* Academic Press: London, 1972.

Finlayson, B. A., *Nonlinear Analysis in Chemical Engineering.* McGraw-Hill: New York, 1980; [reprinted, Ravenna Park: Seattle, WA, 2003.]

Finlayson, B. A., *Numerical Methods for Problems with Moving Fronts.* Ravenna Park: Seattle, WA, 1992.

Finlayson, B. A., Biegler, L. T., Grossmann, I. E. Mathematics in chemical engineering. In *Ullmann's Encyclopedia of Industrial Chemistry*, Electronic Release, Vol. B1, Wiley-VCH, Weinheim, 2006.

Finlayson, B. A., Biegler, L. T., Mathematics. Section 3. In *Perry's Chemical Engineers' Handbook*, 8th ed. Perry, R. H., Green, D. W. (eds.). McGraw-Hill: New York, 2008.

Finlayson, B. A., Shaw, R. A., Modeling microfluidic separations using Comsol Multiphysics. *Comsol Conference,* Boston, 2010; http://www.comsol/papers/8117/.

Finlayson, B. A., Young, L. C., Mathematical models of the monolith catalytic converter: Part III. Hysteresis in carbon monoxide reactor. *AIChE J.* 1979; *25*, 192–196.

Finlayson, B. A., Drapala, P. W., Gebhardt, M., Harrison, M. D., Johnson, B., Lukman, M., Kunaridtipol, S., Plaisted, T., Tyree, Z., VanBuren, J., Witsara, A., Microcomponent flow characterization, Chapter 8, Koch, M. V. et al. (eds.), Wiley-VCH, Weinheim 2007.

Fogler, H. S., *Essentials of Chemical Reaction Engineering*, 4th ed. Prentice Hall: Englewood Cliffs, NJ, 2010.

Frank-Kamenetskii, D. A., *Diffusion and Heat Transfer in Chemical Kinetics*. Plenum Press: New York, 1969.

Freemantle, M., Finding the best conditions rapidly. *C&E News*, February 2005; *83*, 11.

Gear, C. W., *Numerical Initial Value Problems in Ordinary Differential Equations*. Prentice-Hall: Englewood Cliffs, NJ, 1971.

Gilat, A., *MATLAB: An Introduction with Applications*, 4th ed. Wiley: New York, 2010.

Gokhale, S. V., Jayaraman, V. K., Tayal, R. K., Kulkarni, B. D., Microchannel reactors: applications and use in process development. *Int. J. Chem. Reactor Engr.* 2005; *3*, Review R2.

Gomez, F. A., *Biological Applications of Microfluidics*. Wiley-Interscience: Hoboken, NJ, 2008.

Gray, J., Biomolecular modeling in a process dynamics and control course. *Chem. Eng. Ed.* 2006; *40*, 297–306.

Gresho, P. M., Sani, R. L., *Incompressible Flow and the Finite Element Method*. Wiley: New York, 1998.

Gresho, P. M., Sani, R. L., Engleman, M. S., *Incompressible Flow and the Finite Element Method. Vol. 2: Isothermal Laminar Flow*. Wiley: New York, 1998.

Groenier, W. S., Thodos, G., Pressure-volume-temperature behavior of ammonia in the gaseous and liquid states. *J. Chem. Eng. Data* 1960; *5*, 285–288.

Hatch, A., Kamholz, A. E., Hawkins, K. R., Munson, M. S., Schilling, E. A., Weigl, B. H., Yager, P., A rapid diffusion immunoassay in a T-sensor. *Nat. Biotechnol.* 2001; *19*, 461–465.

Hellinckx, L., Grootjans, J., Van den Boxch, B., Stability analysis of catalyst particle through orthogonal collocation. *Chem. Eng. Sci.* 1972; *27*, 644–647.

Henry, C. M., Go with the flow. *C&E News*, March 82, 2005; *83*, 57–59.

Hess, G., Incentives boost coal gasification. *C&E News*, January 16, 2006; *84*, 22–24.

Hirata, M., Ohe, S., Nagahama, K., *Computer Aided Data Book of Vapor-Liquid Equilibria*. Kodansha Ltd. and Elsevier: Tokyo and New York, 1975.

Hlaváček, V., Marek, M., Kubíček, M., Modelling of chemical reactors, X: Multiple solutions of enthalpy and mass balances for a catalytic reaction within a porous catalyst particle. *Chem. Eng. Sci.* 1968; *23*, 1083–1097.

Hughes, T. J. R., Brooks, A. N., A multi-dimensional upwind scheme with no crosswind diffusion. In *Finite Element Methods for Convection Dominated Flow*, AMD Vol. 34, Hughes, T. J. R. (ed.). ASME: New York, 1979.

Jackson, N. R., Finlayson, B. A., Calculation of hole pressure: I. Newtonian fluids. *J. Non-Newtonian Fluid Mech.* 1982a; *10*, 55–69.

Jackson, N. R., Finlayson, B. A., Calculation of hole pressure: II. Viscoelastic fluids. *J. Non-Newtonian Fluid Mech.* 1982b; *10*, 71–84.

Jensen, K. F., Microchemical systems: status, challenges, and opportunities. *AIChE J.* 1999; *45*, 2051–2054.

Jones, C. A., Schoenbom, E. M., Colburn, A. P., Equilibrium still for miscible liquids data of ethylene dichloride-tolune and ethanol-water. *Ind. Eng. Chem.* 1943; *35*, 666–672.

Kirby, B. J., *Micro- and Nanoscale Fluid Mechanics*. Cambridge University Press: Cambridge, UK, 2010.

Koch, M. V., VandenBussche, K. M, Chrisman, R. W., *Micro Instrumentation*. Wiley-VCH: Weinheim, 2007.

Kolavennu, P. K., Telotte, J. C., Palanki, S., Design of a fuel processor system for generating hydrogen for automotive applications. *Chem. Eng. Ed.* 2006; *40*, 239–244.

Koretsky, M. D., *Engineering and Chemical Thermodynamics.* Wiley: New York, 2004.

Kusmanto, F., Jacobsen, E. L., Finlayson, B. A., Applicability of continuum mechanics to pressure drop in small orifices. *Phys. Fluids* 2004; *16*, 4129–4134.

Ladisch, M. R., Mosier, N. S., Kim, Y. M., Ximenes, E., Hogsett, D., Converting cellulose to biofuels. *Chem. Eng. Prog.* 2010; *106*, 56–63.

LeVeque, R. J., *Numerical Methods for Conservation Laws.* Birkhäuser Verlag: Boston, 1992.

LeVeque, R. J., *Finite Volume Methods for Hyperbolic Problems.* Cambridge University Press: Cambridge, UK, 2002.

Lin, Y., Enszer, J. A., Stadtherr, M. A., Enclosing all solutions of two-point boundary value problems for ODEs. *Comput. Chem. Eng.* 2008; *32*, 1714–1725.

Liu, K., Song, C., Subramani, V., *Hydrogen and Syngas Production and Purification Technologies.* Wiley: Hoboken, NJ, 2010.

Marsh, H. D., Severin, M., Schuster, H. G., Steam reforming. In *Encyclopedia of Chemical Processing and Design*, Vol. 47, McKetta, J. J. (ed.). Marcel Dekker: New York, 1994.

McKetta, J. J. (ed.), *Encyclopedia of Chemical Processing and Design.* Marcel Dekker: New York, 1976–1994.

Middleman, S., *Fundamentals of Polymer Processing.* McGraw-Hill: New York, 1977.

Murti, P. S., Van Winkle, M., *Chem. Eng. Data Series* 1959; *3*, 72, [as reported by Hirata ;et al. (1975)].

National Academy of Engineering, What you need to know about energy. 2008; http://sites .nationalacademies.org/Energy/Energy_043338.

National Energy Technology Laboratory (www.netl.doe.gov), DOE/NETL-2007/1281, *Cost and Performance Baseline for Fossil Energy Plants*, Aug. 2007.

Neils, C., Tyree, Z., Finlayson, B., Folch, A., Combinatorial mixing of microfluidic streams. *Lab-on-a-Chip*, 2004; *4*, 342–350.

Orazem, M., Newman, J., Primary current distribution and resistance of a slotted-electrode cell. *J. Electrochem. Soc.* 1984; *137*, 2857–2861.

Otani, S., Matsuoka, S., Sato, M., Tatoray "Tatoray" process—a new transalkylation process of aromatics developed by Toray. *Jap. Chem. Q.* 1968; *4*(4), 16–18.

Othmer, D. F., Morley, F. R., Composition of vapors from boiling binary solutions apparatus for determinations under pressure. *Ind. Eng. Chem.* 1946; *38*, 751–757.

Peng, D.-Y., Robinson, D. B., A new two-constant equation of state. *Ind. Engr. Chem. Fund.* 1976; *15*, 59–64.

Perry, R. H., Green, D. W. (ed.), *Perry's Chemical Engineers' Handbook*, 8th ed. McGraw-Hill: New York, 2008.

Peyret, R., Taylor, T. D., *Computational Methods for Fluid Flow.* Springer Verlag: Dusseldorf, 1983.

Pignet, T., Schmidt, L. D., Kinetics of NH_3 oxidation of Pt, Rh, and Pd. *J. Cat.* 1975; *40*, 212–225.

Platon, A., Wang, Y., Water-gas shift technologies. Chapter 6. In *Hydrogen and Syngas Production and Purification Technologies.* Lin, et al. (eds.) Wiley: Hoboken, NJ, 2010.

Prasad, S., Gautam, A., Integrated coal gasification combined cycle route. *Chem. Engr. World.* 2004; *39*(9), 60–61.

Redlich, O., Kwong, J. N. S., On the thermodynamics of solutions. 5. An equation of state—fugacities of gaseous solutions. *Chem. Rev.* 1949; *44*, 233–244.

Regalbuto, J. R., Cellulosic biofuels—got gasoline? *Science* 2009; *325*, 822–824.

Reid, R. C., Prausnitz, J. M., Sherwood, T. K., *The Properties of Gases and Liquids,* 3rd ed. McGraw-Hill: New York, 1977.

Rhee, H. K., Aris, R., Amundson, N. R., *First-Order Partial Differential Equations, Vol. I, Theory and Application of Single Equations*. Prentice-Hall: Englewood Cliffs, NJ, 1986.

Rosen, E. M., Pauls, A. C., Computer aided chemical process design: the FLOWTRAN system. *Comp. Chem. Engr.* 1977; *1*, 11–21.

Rosensweig, R. E., *Ferrohydrodynamics*. Cambridge University Press: Cambridge, MA, 1985; reprinted by Dover Publications: New York, 1997.

Ruthven, D. M., Farooq, S., Knaebel, K. S., *Pressure Swing Adsorption*. VCH: New York, 1994.

Sage, B. H., Webster, D. C., Lacey, W. N., Phase equilibria in hydrocarbon systems: XIX. Thermodynamic properties of *n*-butane. *Ind. Eng. Chem.* 1937; *29*, 1188–1194.

Sandler, S. I., *Chemical, Biochemical, and Engineering Thermodynamics*, 4th ed. Wiley: New York, 2006.

Sandler, S. I., *An Introduction to Applied Statistical Thermodynamics*. Wiley: New York, 2011.

Schattka, B., Alexander, M., Ying, S. L., Man, A., Shaw, R. A., Metabolic fingerprinting of biofluids by infrared spectroscopy: modeling and optimization of flow rates for laminar fluid diffusion interface sample preconditioning. *Analytical Chemistry*, 2011; *83*, 555–562.

Schmidt, L. D., *The Engineering of Chemical Reactions*. Oxford University Press: New York, 2005.

Service, R. F., Another biofuels drawback: the demand for irrigation. *Science* 2009; *326*, 516–517.

Shuler, M. L., Immobilized biocatalysts and reactors. In *Chemical Engineering Education in a Changing Environment*. Sandler, S. I., Finlayson, B. A. (eds.). AIChE: New York, 1988, pp. 37–56.

Sircar, S., Golden, T. C., Pressure swing adsorption technology for hydrogen production. In *Hydrogen and Syngas Production and Purification Technologies*, Chapter 10, pp. 414–450, Liu et al. (ed.). Wiley, New York, 2010.

Soave, G., Equilibrium constants from a modified Redlich-Kwong equation of state. *Chem. Eng. Sci.* 1972; *27*, 1197–1203.

Stephenson, R. M., *Introduction to the Chemical Process Industries*. Reinhold: New York, 1966.

Stirling, R. G., Ethyl chloride. In *Encyclopedia of Chemical Processing and Design*, Vol. 20, McKetta, J. J. (ed.). Marcel Dekker: New York, 1984.

Stone, H. A., Kim, S., Microfluidics: basic issues, applications, and challenges. *AIChE J.* 2001; *47*, 1250–1254.

Straty, G. C., Palavra, A. M. F., Bruno, T. J., PVT properties of methanol at temperatures to 300°C. *Int. J. Thermophys.* 1986; *7*, 1077–1089.

Taylor, G., Dispersion of soluble matter in solvent flowing slowly through a tube. *Proc. Roy. Soc. Lond.*, 1953; *A219*, 186–203.

Toray Industries, Inc., UOP Process Division: benzene and xylenes (TATORAY). *Hydro. Proc.* 1975; *54*, 115.

Weekman, V. W. Jr., Gazing into an energy crystal ball. *Chem. Eng. Prog.* June, 2010; *106*, 23–27.

Wegstein, J. H., Accelerating convergence of iterative processes. *Commun. Assoc. Comput. Mach.*, 1958; *1*(6), 9–13.

Weigl, B. H., Yager, P., Microfluidic diffusion-based separation and detection. *Science* 1999; *283*, 346–347.

Wiebe, R., Tremeame, T. H., Solubility of nitrogen in liquid ammonia at 25° from 25 to 1000 atmospheres. *J. Am. Chem. Soc.* 1933; *55*, 975–978.

Wiebe, R., Tremeame, T. H., The solubility of hydrogen in liquid ammonia at 25, 50, 75 and 100 C and at pressures to 1000 atmospheres. *J. Am. Chem. Soc.* 1934; *56*, 2357–2360.

Wilkes, J., *Fluid Mechanics for Chemical Engineers with Microfluidics and CFD*, 2nd ed. Prentice Hall: Englewood Cliffs, NJ, 2006.

Wong, D. S. H., Sandler, S. I., A theoretically correct mixing rule for cubic equations of state. *AIChE J.* 1992; *38*, 671–680.

Young, L. C., Finlayson, B. A., Axial dispersion in non-isothermal packed bed chemical reactors. *IEC Fund.* 1973; *12*, 412–422.

Young, L. C., Finlayson, B. A., Mathematical models of the monolith catalytic converter: Part I. Development of model and application of orthogonal collocation. *AIChE J.* 1976a; *22*, 331–343.

Young, L. C., Finlayson, B. A., Mathematical models of the monolith catalytic converter: Part II. Application to automobile exhaust. *AIChE J.* 1976b; *22*, 343–353.

Zimmerman, W. B. J., *Multiphysics Modeling with the Finite Element Method.* World Scientific: Singapore, 2008.

INDEX

Introduction to Chemical Engineering Computing, Updated Second Edition. Bruce A. Finlayson.
© 2014 John Wiley & Sons, Inc. Published 2014 by John Wiley & Sons, Inc.